Renewable Energy in East Asia

H0234704

Energy is crucial to the functioning of any human society, and it is central to understanding East Asia's 'economic miracle.' The region's rapid development over the last few decades has been inherently energy-intensive, and the impact on global energy security, climate change, and the 21st-century global system generally is now very significant and will become more so over foreseeable years and decades to come. The region is already the world's largest energy consumer and greenhouse gas emitter, so establishing cleaner energy systems in East Asia is both a regional and global challenge, and renewable energy has a critically important part to play in meeting this challenge.

This book presents a comprehensive study of renewable energy development in East Asia. It begins by examining renewable energy development in global and historic contexts and situates East Asia's position in the recent worldwide expansion of renewables. This same approach is applied on sector-specific chapter studies on wind, solar, hydropower, geothermal, ocean (wave and tidal), and bioenergy and to general trends in renewable energy policy. Governments play a critical role in promoting renewables and their contribution in tackling climate change and other environmental challenges. Christopher M. Dent contends that this is particularly relevant to East Asia, where state capacity practice has been increasingly allied to ecological modernisation thinking to form what he calls 'new developmentalism,' the principal foundation on which renewables have developed in the region as well as how East Asia's low carbon development is being generally promoted.

Renewable Energy in East Asia will be of huge interest to students and scholars of Asian studies, economics, political economy, energy studies, business, development, international relations, and environmental studies. It will also appeal to researchers working on the subject matter in government, business, international organisations, think tanks, and civil society organisations.

Christopher M. Dent is Professor of East Asia's International Political Economy at the University of Leeds, UK.

Routledge contemporary Asia series

Renewable Energy in East Asia

Towards a new developmentalism

Christopher M. Dent

Routledge
Taylor & Francis Group

LONDON AND NEW YORK

First published 2014
by Routledge
2 Park Square, Milton Park, Abingdon, Oxon OX14 4RN

and by Routledge
711 Third Avenue, New York, NY 10017

First issued in paperback 2017

Routledge is an imprint of the Taylor & Francis Group, an informa business

© 2014 Christopher M. Dent

British Library Cataloguing in Publication Data
A catalogue record for this book is available from the
British Library

Library of Congress Cataloging-in-Publication Data
Dent, Christopher M.
Renewable energy in East Asia : towards a new
developmentalism / Christopher M. Dent.
pages cm. – (Routledge contemporary Asia series ; 50)
Includes bibliographical references and index.
1. Energy policy–East Asia. 2. Energy industries–East Asia.
3. Renewable energy sources–East Asia. I. Title.
HD9502.E182D46 2015
333.79'4095–dc23
2014022478

ISBN 13: 978-1-138-09511-3 (pbk)
ISBN 13: 978-1-138-80719-8 (hbk)

Typeset in Baskerville
by Cenveo Publisher Services

To the memory of

Joanne Lesley Francis
(1945–2012)

A life of loving … mother, nana, sister, friend

Contents

Illustrations

Tables

Figures

Boxes

Preface

When I was 20 years old I read E.F. Schumacher's seminal work *Small is Beautiful*. It made me look at economics in a completely new way, as well as the world around me and the future of humanity. Schumacher's message was simple: either we find sustainable paths of economic development or there will be no future. A few years later, when I studied for my master's in economics at the University of Leeds, I decided to write my dissertation on the prospects for renewable energy in Britain. I have now returned to the subject of renewables—but with East Asia as the focus.

Energy is crucial to the functioning of any human society, and it is central to understanding East Asia's 'economic miracle.' The region's rapid development over the last few decades has been inherently energy-intensive. Moreover, East Asia's impact on global energy security, climate change, and the 21st-century global system generally is now very significant and will become more so over foreseeable years and decades to come. The region is already the world's largest energy consumer and greenhouse gas emitter. Establishing cleaner energy systems in East Asia is vital not just for the region but also for the rest of humanity. It is both a regional and global challenge, and renewable energy has a critically important part to play in meeting this challenge. Although themselves not without controversy and challenges, renewables offer vital pathways to sustainable energy development. This book examines how those pathways have thus far evolved in East Asia and could evolve in the future. I argue that the region's governments, companies, and societies all have crucial roles in advancing renewable energy development in the frame of 'new developmental' partnership.

The four main aims and approach of this book are as follows:

1. *Comprehensive and holistic*: To study each part of East Asia (China, Japan, South Korea, Taiwan, and Southeast Asia) in both global and historic contexts, covering each mainstream sector (wind, solar, hydropower, geothermal, ocean, and bioenergy) from broad political economy-related perspectives and making important empirical and theoretical/conceptual contributions to the subject matter.

2. *Accessible*: While seeking to make robust scholarly arguments covering a range of inter-related technical subjects, the book is written to appeal to a broad readership.
3. *Empirical*: To make a wide-ranging survey on East Asia's renewable energy development using a large number and wide variety of different source materials, as well as to provide the reader with useful data and data trends.
4. *Theoretical and conceptual*: Presenting a new concept and analytical framework of 'new developmentalism,' which was founded on a synthesis of state capacity and ecological modernisation theories, to explain the rise of renewable energy in East Asia as part of broader efforts to decarbonise economic activity in the region, as well as to critically evaluate this approach in light of studied empirical evidence.

This book took more than three years to research and write, and I owe great debts of gratitude to many people who have provided valuable research information and assistance–namely Po-Yao Kuo, Kaare Sandholt, Chen Chung-Tsien, Christopher Len, Jørgen Delman, Lee Jae-Seung, Elsie Minshull, Ken Koyama, Ole Odgaard, Yukari Yamashita, Sarah Leitner, Pan Jiahua, Elspeth Thomson, Liao Mei, Simon Smith, and Thidarat Sawai. Many thanks also to Hannah Mack and Stephanie Rogers at Routledge for their cheerful help and encouragement throughout the book project's development. I also would like to thank my three anonymous reviewers for their helpful comments on improving the book. In addition, a special acknowledgement of gratitude to my ever-reliable espresso machine for consistently delivering fabulous cappuccinos that provided caffeine-inspired moments of lucidity during my morning book-writing sessions. Finally, I would also like to thank my family and friends for their love, support, and patience. This book is dedicated to the memory of my mother, Joanne Lesley Francis.

Christopher M. Dent
Leeds, Yorkshire
June 2014

1 Renewable energy
An introduction

1.1. Renewable energy and East Asia

1.1.1. Introducing renewable energy

Renewable energy (RE) is derived from replenishable natural processes, sources, or phenomena such as wind, solar, hydropower, geothermal, waves, tides, and biological matter. This energy is renewable in actual or theoretical terms because it can provide an inexhaustible supply of energy to meet the needs of humankind. In contrast, fossil fuels (oil, natural gas, coal) and nuclear energy (based on uranium) are non-renewable because they are derived from depletable mineral resources. Today, we often refer to renewables in terms of their technological applications, such as wind turbines, solar modules or panels, tidal barrages, and biofuels. These applications can vary enormously in terms of scale, from small solar photovoltaic cells generating a few watts of electricity to power electronic devices such as calculators to huge hydropower dams that provide power to millions of households through national grid systems. Renewable energy technologies can hence be deployed on a *micro-scale*, where individuals and local communities own installations and are both the producer and consumer, and on a *macro-scale*, in which RE plants and products serve the energy needs of large parts of the country or even the whole country.

In addition to the environmental virtues of renewable energy, its micro-scale applications are also making an important contribution to reducing global energy poverty (IPCC 2011). Roughly 1.5 billion people currently do not have access to electricity, and many more only have access to unstable supplies of it (REN21 2013). Wind, solar, bioenergy, and small hydropower in particular have scope for deployment in remote areas located far from national grid infrastructures, thus helping to improve livelihoods by using sustainable methods to provide power for schools, hospitals, small enterprises, communications, and other utility services. Renewables have played an important part in the rural electrification programmes of many East Asian countries.

Renewable energy is arguably the main element of a broader green-energy cluster of technologies that additionally includes energy efficiency and saving, electric vehicles, fuel cells, and other eco-industry sectors. The ability of renewables to produce cleaner forms of energy makes them crucial to decarbonising economic activity, achieving sustainable development, and tackling climate change. This is especially relevant to East Asia given that no other part of the world is making, and will continue to make, as big an impact on climate change and global energy security. The region is also the world's most important production hub and is experiencing rapid urbanisation; thus, it is where much of the planet's most carbon-intensive activity is concentrated. East Asia's demand for energy across all fields—electricity power generation, transportation, and thermal heat—has grown almost three times faster than the global average since the 1970s, currently accounts for around a third of the world's total, and is set to rise further still. Renewable energy in East Asia is thus critical not only to the region's future sustainable development but also to the world's.

1.1.2. East Asia: an economic and energy profile

East Asia comprises two sub-regions: Northeast Asia (China, Japan, South Korea, North Korea, Taiwan, and Mongolia) and Southeast Asia (Brunei, Cambodia, East Timor, Indonesia, Laos, Malaysia, Myanmar, Philippines, Singapore, Thailand, and Vietnam). Over the last five or six decades, East Asia has been the world's most dynamic and fastest growing economic region. Annual double-digit percent increases in economic growth have not been uncommon for many East Asia countries during this period. Table 1.1 reveals how the region consists of a diverse set of economies that together accounted for 25.7 percent of the global economy in 2012—a higher share than both the European Union (23.3 percent) and the United States (21.9 percent). It is also the world's second most integrated regional economy after the European Union (Dent 2008, 2010). Current projections suggest that East Asia's global economic importance will continue to grow into the 21st century if the growth trajectories of China and the region's other 'tiger economies' are maintained.

Significant increases in energy use in most parts of the region have been caused by rapid industrial development, rising incomes, and expanding populations. Manufacturing has been core to the development strategies of East Asia's largest economies, resulting in carbon-intensive economic growth. Overall, East Asia's energy use over doubled from 1990 to 2010, from 1,739 million tonnes of oil equivalent (mtoe) to 3,888 mtoe, during a time when world energy use only rose by half in comparison (Table 1.1). China accounted for much of these increases regionally and globally, reaching 2,417 mtoe in 2010 and thus surpassing the United States as the world's largest national energy consumer. Even more dramatic increases have occurred in East Asia's electricity grid capacity, quadrupling from 391.7 GW to 1,578.9 GW during the same 20-year period, doubling the region's global share from 14.7 percent to 31.2 percent. However, while China's and other East Asian countries' energy use per capita level has risen notably over recent decades, they remain well below those of the United States and European Union. Looking at the GDP per unit of energy use figures in Table 1.1 as a general proxy for energy efficiency, we can see that countries such as China, Mongolia, Indonesia, the Philippines, and Vietnam have made significant improvements, but they still have some distance to go before attaining the same standards as developed nations. Together with renewable energy, energy efficiency is the other key field of East Asia's green energy industry, and it is also a core element of most climate change strategies.

Approximately three-quarters of aggregated greenhouse gases result from energy use (IPCC 2011). The combination of burgeoning energy demand and still relatively low energy efficiency levels has steadily increased East Asia's impact on global carbon emissions and climate change. From 1990 to 2010, the region saw its carbon emission levels more than double, from 4,638 million tonnes (20.8 percent of the world's total) to 11,622 (34.6 percent of the world's total). Carbon emission per capita rates have correspondingly risen among the region's fastest growing developing economies, although these remain below Western levels (e.g. China is still approximately three times lower than the United States).

Let us use an illustrative example to help explain the relevance of renewable energy in combating historic increases in carbon emissions. Based on average power ratings, wind energy typically generates around 2.5 GWh per 1 MW installed capacity over the course of a year; this is based on a 28–30 percent average operational efficiency rating, as explained in Box 1.1. Also using a generalised world average, each gigawatt hour (GWh) of clean energy produced yields an estimated 600 tonnes reduction in carbon dioxide (CO_2) emissions (GWEC 2009). Thus, electricity generated by a 100 MW wind farm would annually result in a nominal average savings of 150,000 tonnes CO_2, or the equivalent emissions produced on a yearly average basis in 2010 by 24,194 individuals collectively in China, 20,000 in the European Union, or 8,523 people in the United States (see Table 1.1).

1.1.3. Main drivers of renewable energy development in East Asia

East Asia's economic and energy profile provides the context for the three main inter-related drivers behind renewable energy development in the region, the first being *environmental and sustainable development imperatives.* High-carbon economic activity is causing acute pollution and other environmental problems in East Asia and correspondingly adverse social

Table 1.1 East Asia's economic and energy profile

	GDP (US$ billion)	GDP per capita (US$)	Population (millions)	Energy use[1] (million tonnes of oil equivalent)		Energy use per capita (kilogrammes of oil equivalent)		Net energy imports[2] (% of energy use)	
	2012	**2012**	**2012**	**1990**	**2010**	**1990**	**2010**	**1990**	**2010**
China	8,227	6,091	1,359	872	2,417	768	1,807	−2	9
Japan	5,960	46,270	127	439	497	3,556	3,898	83	81
South Korea	1,130	22,590	50	93	250	2,171	5,060	76	82
Taiwan	474	20,328	23	55	109	2,946	4,712	95	98
Hong Kong SAR	263	36,796	7	9	14	1,518	1,963	99	100
Macau SAR	44	78,275	0.6	2	3	3,936	5,651	–	–
North Korea	12	506	25	33	19	1,645	756	13	−12
Mongolia	10	3,673	3	3	3	1,564	1,208	20	−357
Northeast Asia	16,120	10,107	1,595	1,506	3,312	–	–	–	–
Brunei	17	41,127	0.4	2	3	6,860	8,274	−788	−460
Cambodia	14	946	15	0.6	5	52	350	–	28
Indonesia	878	3,557	238	99	208	552	864	−71	−84
Laos	9	1,399	7	0.8	4	164	467		–
Malaysia	304	10,381	30	22	73	1,183	2,569	−120	−18
Myanmar	53	955	53	11	14	237	274	0	−61
Philippines	250	2,588	98	29	40	462	433	40	42
Singapore	274	22,590	5	12	33	3,779	6,456	99	99
Thailand	366	5,474	66	42	117	741	1,768	37	40
Vietnam	142	1,596	89	18	59	271	681	−2	−11
Southeast Asia	2,307	3,839	601	236	556	–	–	–	–
East Asia	**18,427**	**8,391**	**2,196**	**1,739**	**3,888**	–	–	–	–
% world total	25.7	–	30.9	20.3	31.5	–	–	–	–
United States	15,685	49,636	316	1,915	2,216	7,672	7,165	14	22
% world total	21.9	–	4.4	22.3	18.0	–	–	–	–
European Union	16,690	32,854	508	1,275	1,408	3,532	3,597	55	60
% world total	23.3	–	7.1	14.9	11.4	–	–	–	–
India	1,842	1,494	1,233	317	693	365	575	8	25
% world total	2.6	–	17.4	3.7	5.6	–	–	–	–
World	**71,666**	**10,172**	**7,106**	**8,575**	**12,328**	**1,665**	**1,852**	–	–

Sources: World Bank database (data.worldbank.org), US Energy Information Administration (www.eia.gov), IRENA (2014a), REN21 (2014).

Notes:
1. Use of primary energy before transformation to other end-use fuels, which is equal to indigenous production plus imports and stock changes, minus exports and fuels supplied to ships and aircraft engaged in international transport;
2. Minus figure indicates net exporting position;
3. To quote the data source: 'The ratio of GDP to energy use indicates energy efficiency. To produce comparable and consistent estimates of real GDP across economies relative to physical inputs to GDP – that is, units of energy use – GDP is converted to 2005 international dollars using purchasing power parity (PPP) rates' (data.worldbank.org). GDP, gross domestic product.

GDP per unit of energy use[3] (2005 PPP US$ kgoe)		Electrcity grid capacity (gigawatts)		Renewable energy installed electricity capacity (gigawatts)		Carbon dioxide (CO_2) emissions (million tonnes)		CO_2 emissions per capita (tonnes)	
1990	2010	1990	2010	2010	% total	1990	2010	1990	2010
1.4	3.8	126.6	987.9	250.2	25.3	2,461	8,287	2.2	6.2
7.5	8.0	169.2	287.0	32.1	11.2	1,095	1,171	8.9	9.2
5.2	5.3	20.0	84.7	2.9	3.4	247	568	5.8	11.5
6.1	6.6	17.9	48.9	3.1	6.3	129	259	6.2	10.3
15.8	21.8	7.5	12.6	0.0	0.0	28	36	4.8	5.2
–	–	0.2	0.5	0.0	0.0	1	1	2.9	1.9
–	–	9.5	9.5	5.0	52.6	244	72	12.2	2.9
1.6	3.0	0.9	0.8	0.0	0.0	10	12	4.6	4.2
–	–	351.8	1,431.9	293.3	20.5	4,214	10,406	–	–
7.2	5.5	0.4	0.8	0.0	0.0	6	9	25.0	22.9
–	5.5	0.0	0.4	0.02	4.9	0.5	4	0.1	0.3
3.8	4.5	12.7	34.1	6.1	17.9	150	434	0.8	1.8
–	–	0.2	1.9	1.8	97.3	0.2	2	0.1	0.3
5.8	5.4	5.0	25.4	2.1	8.3	57	217	3.1	7.7
–	–	1.1	1.7	0.8	46.7	4	9	0.1	0.2
5.5	8.2	6.8	16.3	5.4	33.1	42	82	0.7	0.9
6.7	8.1	3.4	10.3	0.02	0.2	47	14	15.4	8.6
5.4	4.5	8.3	48.2	4.3	8.9	96	295	1.7	4.4
3.3	4.2	2.0	15.2	5.5	36.2	21	150	0.3	1.7
–	–	39.9	154.2	26.0	16.9	424	1,216	–	–
–	–	**391.7**	**1,578.9**	**319.3**	**20.2**	**4,638**	**11,622**	–	–
–	–	*14.7*	*31.2*	*26.0*	–	*20.8*	*34.6*	–	–
4.2	5.9	714.7	1,039.1	144.0	14.0	4,879	5,433	19.5	17.6
–	–	*26.9*	*20.5*	*11.7*	–	*21.9*	*16.2*	–	–
6.9	8.5	669.3	981.9	246.0	26.1	3,236	3,689	8.6	7.5
–	–	*25.2*	*19.4*	*20.0*	–	*14.5*	*11.0*	–	–
3.3	5.4	71.8	208.1	52.8	27.9	691	2,008	0.8	1.7
–	–	*2.7*	*4.1*	*4.3*	–	*3.1*	*6.0*	–	–
4.2	**5.5**	**2,658.3**	**5,066.8**	**1,230.0**	**25.4**	**22,274**	**33,615**	**4.2**	**4.9**

Box 1.1 Energy units of measurement and statistics

Many readers may not be familiar with energy units of measurement and their use when studying renewable energy. Starting with a simple example, a typical household in a developed country would need a 2 to 4 kilowatt (kW) solar photovoltaic (PV) rooftop based system to cover all of its daytime electricity demands; this figure is notably lower in developing countries. Small-scale renewable energy devices used at this level tend to be rated in kilowatts. Larger installations or equipment operated normally by companies at the power plant level, such as wind farms and solar parks, are invariably rated in megawatt (MW) terms, with 1 MW equalling 1,000 kW. In certain cases, very large RE plants, such as hydropower dams, may have gigawatt (GW) scale generation ratings, but most often this unit of power is used when examining industry, national, or international capacity levels. A single gigawatt is equal to 1,000 MW, and 1 terawatt (TW) equals 1,000 GW.

The actual power generated by any energy system or plant is expressed in terms of the number of energy unit hours over a particular period of time, normally a year (e.g. GWh per annum). Watt-based power ratings assigned to energy devices or systems relate to their peak or maximum power output levels when working at full potential capacity. Actual output levels achieved will depend on the operational efficiency (or capacity factor) of the equipment used. Due to the intermittency of wind, even high-performing wind turbines will normally have around an average 30 percent capacity factor rating. Thus, a 5 MW turbine would produce the following power output: 5 MW $\times$ 0.3 $\times$ 365 days $\times$ 24 hours = 13,140 MWh or 13.14 GWh. In contrast, more constant and dependable energy streams, such as hydropower, have capacity factors of 90 percent or greater. Watt-thermal units (e.g. MW_{th}) are used to explain the energy produced in heat generation. Meanwhile, tonnes of oil equivalent (toe) is often used to compare energy consumption or production levels across different energy sources.

welfare effects on its societies. When compared to other parts of the world, East Asia's energy-intensive development structures appear to be especially unsustainable. East Asian nations are also highly vulnerable to the climate change risks of extreme weather and rising sea levels. In addition to local and national pressures to establish cleaner energy systems, many governments in the region are under increasing international pressure on this front due to East Asia's growing impact on global climate change.

The second main driver is *energy security*, which can be understood in simple conceptual terms as addressing supply risk (securing reliable sources of energy fuels), price risk (maintaining predictable or at least stable energy prices), and environmental risk (mitigating the adverse environmental impacts of energy use), as discussed under the first main driver. The promotion of renewable energy in East Asia can too be explained in relation to the first two risk types. Sustained high growth rates in energy demand have led to a rapid depletion of the region's own fossil fuel deposits, and many of its countries have moved to or are moving toward a net energy-importing position. Table 1.1 also shows that East Asia's most advanced economies (Japan, South Korea, Taiwan, and Singapore, which are also among the largest) have chronically high energy-import dependency ratios, making them especially susceptible to the vicissitudes of foreign supply sources (Dent 2013a). This is potentially further compounded by volatility in international prices for oil, gas, and coal. Renewables have the advantage of being inherently indigenous energy sources, hence providing a long-term solution to foreign energy supply risk. Although some RE sectors are prone to occasional commodity price spikes (e.g. polysilicon and solar PV cell manufacturing), they have been historically less exposed to price risk than fossil fuel sectors.

The third main driver relates to the promotion of renewables as *emerging strategic industries*. Much of East Asia's contemporary economic success has been based on fostering strategic industries through exercises of state capacity aimed at forging developmental relationships between government and business. Chapter 2 discusses this subject in considerable detail. Strategic industries can have any or all of the following general attributes:

- New or emerging sectors having substantial growth potential, which are therefore important for generating future prosperity, income, and welfare
- Sectors performing essential functions in the economy, such energy and material supply
- Industries providing a material or other kind of foundation on which other sectors are based, such as steel and chemicals for various manufactured products.

Most RE sectors possess the first general attribute, with wind, solar, and bioenergy being perhaps the most relevant in this respect. Although the core technological foundations of many modern renewables can be traced back some decades, their rates of techno-innovatory advance have substantially quickened since the mid-2000s, generally corresponding with their rapid sector growth; hence, they may still be considered as 'emerging.' Renewables are also being increasingly viewed in the region as possessing the second attribute, particularly as a long-term prospect to resolve the aforementioned energy security and environmental predicaments of over-reliance on fossil fuels. Renewables may not yet be seen as strategic industries regarding the third attribute but could possibly be in the future.

1.1.4 The approach and structure of this book

This book presents a comprehensive study of East Asian renewable energy sectors from the perspectives of government economic policies and strategies, business and industry development, techno-innovation and production, socio-technical issues, international political economy, multi-level governance, and institutionalism. Its main argument is that the expansion of renewable energy in East Asia forms an integral part of the region's new developmentalism, which can be defined as revitalised and refocused forms of state capacity aimed at realising the transformative economic objectives associated with low carbon development. Chapter 2 presents an analytical framework of new developmentalism that fuses the theories of state capacity and ecological modernisation to explain the broader context in which the region's renewable energy development has occurred. In the meantime, this introductory chapter gives an overview of recent trends in renewable energy, looking at key matters such as the following:

- The evolution of renewables in the world's energy systems
- A brief introductions to RE sectors
- Investment and finance
- Challenges, constraints, and contentious issues
- Techno-innovation and production
- International political economy issues, such global competition, trade barriers and disputes, and rare earths and other resource dependencies
- Governance and institutionalism of renewable energy development at the national, global-multilateral, and regional levels.

Chapter 2 outlines the theoretical approach of the book to the subject matter based on the new developmentalism concept. Chapter 3 presents an overview of recent trends and key issues in RE policy from global and East Asia standpoints. Each renewable energy sector is examined in turn from Chapters 4 to 9, beginning with a general summary and thereafter studying each part of East Asia. Chapter 10 offers a conclusion of the main findings and arguments overall.

1.2. Renewables in the world's energy systems

1.2.1. Historic and contemporary perspectives

Although renewable energy use has expanded considerably in recent years, it continues to represent a relatively small fraction of global energy production and consumption. Today's energy structures remain strongly dominated by the fossil fuel industries of oil, natural gas, and coal. Significant structural and technical changes in the world's energy systems can take many decades to occur, which is relevant to the decarbonising of global economic activity generally. Although renewables are often seen as the most important *destination* energy sources, most parts of the world will remain highly dependent on oil, coal, and gas as *transition* fuels for many decades to come.

Renewable energy technologies have existed for centuries and, in some cases, millennia. Human societies around the world have devised ingenious and different ways to harness natural phenomena as sources of energy. Mechanical hydropower (water mills) played a vital part in the birth of the Industrial Revolution; later, hydroelectric technology contributed significantly to powering the earliest national electricity grids, which were established in the late 19th century. However, modern industrial development has been largely dependent on fossil fuel consumption, and the inexorable rise of carbon-based economies became one of the 20th century's defining features. Experiments with certain non-hydro renewables (e.g. wind, solar) were being undertaken during this time, but it was not until the early 1970s that the range of RE technologies we know today began to emerge. Governments then turned to renewables in response to both growing concerns over the long-term adverse environmental impacts of industrialism and the energy security challenges arising from the 1973–1974 oil crisis. Despite the crisis helping to initiate many modern RE policies worldwide, non-hydro renewable energy made limited inroads into national energy systems up to the 1990s. Governments invested far more substantially in the nuclear power industry during this period.

It was not until the early or mid-2000s that significant growth and development in renewables began, as new RE technologies were scaled up and approached commercialisation. Wind, solar, and bioenergy deserve special mention here, and electricity generation is arguably where renewables have made the biggest impact on global energy systems. Worldwide renewable power generation capacity increased from 889 GW in 2005 to 1,561 GW in 2013 (Table 1.2); according to REN21 (2014), this represents 26.4 percent of the global total (up from 21.6 percent in 2005), leading in addition to a rising share of global electricity generation consumption or output from 15.6 percent to 22.1 percent. As Table 1.2 indicates, hydropower accounted for 16.9 percent of the global electricity generation capacity in 2013, which was approximately three-quarters of the total renewables contribution. In terms of global primary energy demands, renewables accounted for 19.0 percent of the total in 2012, with traditional biomass accounting for a large (9.0 percent of the total primary energy demand) but gradually diminishing share, which was mainly related to cooking and heating in developing countries.

East Asia has been at the forefront of the recent global expansion of renewable energy. As Table 1.1 shows, for example, by 2010 the region was responsible for 319.3 GW (26.0 percent) of total world electricity generated

Table 1.2 Renewable energy development, sector overview

	Electricity generation (GW installed capacity, world)			Main East Asian producers, 2013 (installed capacity level, world ranking)
	2005	**2013**	**Added**	
Hydropower	770.5	1,000.0	229.5	China 260.2 GW (1)
Wind	59.0	318.1	259.1	China 91.4 GW (1) Japan 2.7 GW (13) Taiwan 0.6 GW (28)
Solar photovoltaic	5.4	138.9	133.5	China 18.8 GW (2) Japan 13.6 GW (4) South Korea 1.5 GW (14)
Biomass	44.0	88.0	44.0	China 8.5 GW (4) Japan 3.3 GW (5)
Geothermal	9.3	12.0	2.7	Philippines 1.8 GW (2) Indonesia 1.3 GW (3) Japan 0.5 GW (8)
Concentrating solar power	0.4	3.4	3.0	China 10 MW (9) Thailand 5 MW (10)
Ocean	0.3	0.5	0.2	South Korea 0.25 GW (1)
Renewables sector total	889	1,561	672	China 378.9 GW (1)
Nuclear sector	368	370	2	
Fossil fuel sector	2,858	3,978*	1,120	
All sectors	**4,115**	**5,909***	**1,794**	
Bioethanol	33bn litres	87bn litres		China 2.0bn (3); Thailand 1.0bn (5)
Biodiesel	4bn litres	26bn litres		Indonesia 2.0bn (6); Thailand 1.1bn (8)

Sources: EIA (2014), EPIA (2011, 2014), GWEC (2011, 2014), IEA (2006, 2014), REN21 (2006, 2014), Schneider and Froggatt (2013).

*Based on REN21 (2014) estimates.

by renewables, more than the European Union (246 GW, 20.0 percent) and almost double the US level (144 GW, 11.7 percent). East Asia's installed RE capacity level continues to rise faster than any other global region in absolute terms. By 2013, East Asia's total renewables power generation capacity was 457 GW, and its global share had risen to 29.2 percent.[1] Furthermore, China's installed RE power capacity (379 GW by 2013) is greater than any other nation by far,[2] and it has added approximately 40 percent of the new additions to global renewable energy capacity since the late 2000s (REN21 2010, 2014). In a similar vein, China is now the largest manufacturer of renewable energy products, helping make them more affordable worldwide. However, as we later discuss, the impressive growth of renewables in East Asia has not been without notable controversy, such as the environmental and socio-economic problems arising from hydropower development. In addition, renewable energy shares of total electricity generation are not that high in East Asia's most developed economies (Table 1.1). Moreover, the relatively high share figures for many of the region's developing countries often depend heavily on singular sectors (e.g. Laos, Myanmar, and North Korea on hydropower; Indonesia and Philippines on geothermal). Some smaller economies (e.g. Brunei, Hong Kong, Macau) have little or no renewables power generation capacity.

Fossil fuels are likely to remain dominant in the worldwide energy mix for some time to come. Table 1.2 shows that fossil fuels accounted for 67.3 percent of global electricity generation capacity in 2013, with a net increase of 1,120 GW over the 2005–2013 period compared to 672 GW for renewables. Many East Asian countries plan to continue expanding their coal and gas-fired power capacity simultaneously to pursuing ambitious renewable energy strategies. Nuclear power, meanwhile, has been a relatively static energy industry over the last decade or so, even generally declining globally in recent years. In 2002, total world nuclear power capacity stood at 362 GW from 444 reactors, increasing by just 13 GW in 2010 at a total capacity of 375 GW that year. However, this level fell slightly to 373 GW and 435 reactors worldwide by June 2014 (International Atomic Energy Agency 2014). Most of this drop is attributable to Japan's post-Fukushima energy policy, but many of the world's 31 nuclear power–producing countries have also been decommissioning old reactors and plants without replacing them (Schneider and Froggatt 2013). Only four East Asian economies have nuclear power capacity: Japan (38.0 GW in 2013), South Korea (20.7 GW), China (13.8 GW), and Taiwan (5.0 GW). Although there are plans in East Asia to construct new reactors in the future (28 in China and up to possibly 17 in South Korea), renewables have nevertheless gained ground over nuclear as the most viable long-term alternative energy source to fossil fuels. By the mid-2010s, the combined installed generation capacity of solar and wind energy alone was greater than that of nuclear power[3] (Table 1.2). Momentum for renewable power generation is growing: RE sectors together accounted for more than 50 percent of all investment in new generation capacity worldwide from 2008 to 2013 (REN21 2014), up from an estimated 20–25 percent share enjoyed by renewables in the early 2000s (REN21 2006).

1.2.2. Renewable energy sectors

Renewable energy comprises a broad spectrum of notably different technologies. The growth patterns and dynamics of each have in turn varied very significantly over recent years, as detailed in each of the following sector-specific sections.

- *Hydropower:* Hydropower is the oldest and most established mainstream renewables sector and by far the most dominant in terms of power generation. It is a mature energy technology with a relatively slow rate of annual growth (just 3.7 percent over 2008–2013), but nevertheless it has still made the second largest contribution to additional RE capacity in recent years. Total hydropower capacity had reached 1,000 GW worldwide by 2013 (REN21 2014). China's ambitious dam construction programme has been a central factor in the sector's development globally, being responsible for roughly half of the recent new additional hydropower capacity worldwide. The country's total capacity increased from 117.4 GW in 2005 to 260.2 GW by 2013, which is 26.0 percent of the world's total capacity. Hydropower has also figured highly in Southeast Asia's RE plans. However, serious environmental and socio-economic costs arising from large-hydropower projects in East Asia and globally have highlighted the need for shifting future sector development more toward small-hydropower applications.
- *Wind energy:* Wind energy is the largest non-hydro RE sector and one of the fastest growing, recording an average annual capacity expansion rate of 21 percent during 2008–2013 and adding more installed new capacity for the renewables than any other sector in recent years (Table 1.2). In 2000, total global wind energy capacity was just 18.0 GW, rising to 74.1 GW in 2006 and 318.1 GW by 2013 (GWEC 2014). China has made by far the most substantial contribution globally, overtaking the United States in 2010 as the world's leading nation on installed capacity terms and dwarfing the contributions made by other East Asian states. East Asia is also the world's largest manufacturing region of wind turbines–an industry where Chinese firms are again having a profound impact.
- *Solar PV:* Solar PV has been the fastest expanding mainstream renewable energy sector for the last decade or so, with an annual average growth rate of 55 percent over 2008–2013. Installed capacity was just 1.5 GW in 2000, rising to 5.4 GW by 2005 and then to 138.9 GW by 2013 (EPIA 2011a, 2014). Although Europe still dominates with well over half of total global capacity, the rapid ascendance of China's solar PV industry and a recent Japanese resurgence on installation has helped turned around East Asia's recent falling share. Japanese and Korean firms are also among the sector's most important technology leaders. Like wind energy, the impressive expansion of solar PV is primarily based on dynamic techno-innovation that has significantly improved cost efficiency and technical performance.
- *Bioenergy:* Bioenergy concerns unlocking the latent chemical energy found in biological matter. Biomass power generation capacity reached an estimated 88.0 GW globally by 2013, almost double the 2005 level of 44 GW (Table 1.2). The United States and the European Union (especially Germany, Sweden, and Britain) remain the major players, although

China and Japan are ranked fourth and fifth, respectfully. Biofuels have also experienced significant recent growth, accounting for 2–3 percent of total road transportation fuels in the early 2010s. Bioethanol production increased from 33 billion litres in 2005 to 87 billion litres by 2013. The United States and Brazil duopolised the sub-sector, with global shares of 58 percent and 29 percent, respectively. China was positioned third globally, producing 2.0 billion litres in 2013, and Thailand ranked fifth with 1.0 billion litres. The much smaller biodiesel fuel sub-sector has grown with similarly fast rates, increasing from 4 billion litres of production in 2005 to 26 billion litres by 2013. Indonesia is the highest globally ranked East Asian producer, with 2.0 billion litres (sixth); Thailand ranked eighth with 1.1 billion litres. As noted above, traditional uses of biomass for heat and cooking represents a large share of renewables contribution to global primary energy demand.

- *Geothermal*: Geothermal has remained a relatively small and static renewable energy sector, growing at only 3.2 percent per annum on average over 2008–2013 and with just 12.0 GW installed electricity capacity by 2013. Rates of techno-innovation and investment growth have been slow. There were, however, at least 78 countries using geothermal energy for heat generation of some kind, and 24 countries with installed geothermal power plants for electricity generation. The Philippines (1.8 GW) and Indonesia (1.3 GW) were ranked second and third, respectively, behind the United States (3.4 GW), whereas Japan was ranked eighth (0.5 GW). Direct use of geothermal energy for water heating (i.e. non-electricity generation) has risen by an annual average of approximately 9 percent from 2000, reaching a worldwide total energy output of 91.0 TWh by 2013 (IGA 2014; REN21 2011, 2013, 2014).

- *Concentrating solar power (CSP)*: CSP is currently a micro-scale sector, but it has notable growth potential. Over 2008–2013, it has experienced an annual average growth rate of 48 percent, with global installed capacity rising from 354 MW in 2005 to 3,425 MW by 2013. The increasing number of current large-scale investments in CSP plants across the world (mostly in Spain and, to a lesser extent, the United States) should lead to continued rapid growth, although its presence in East Asia remains low (REN21 2014). By 2013, China had installed 10 MW of a 50 MW CSP plant, giving it a global ranking of ninth for capacity; tenth-ranked Thailand had been operating a singular 5 MW plant since 2011.

- *Solar heating and cooling*: Solar heating and cooling, the third solar energy sub-sector, primarily concerns solar water heaters (SWH). Global solar heating (sometimes referred to as solar thermal) capacity increased to $283GW_{th}$ by 2012; of this, SWH accounted for $258GW_{th}$. China has been responsible for more than 80 percent of new additional capacity in recent years and currently hosts approximately two-thirds of global installed SWH capacity. China also manufactures around three-quarters of global SWH production. Japan is the only other notable East Asian country in this field, with 1.1 percent of world SWH capacity in 2012 and a global ranking of ninth (REN21 2014). There has been fast growth in solar space heating and cooling applications worldwide, but these remain small sectors of activity.

- *Ocean energy*: Ocean energy may be considered a micro-sector based primarily on experimental technologies. Installed global capacity was just 535.5 MW by 2012, of which 519.6 MW came from the small number of tidal range barrages currently in operation; another 8.8 MW came from test-bed tidal current turbines and 7.1 MW from a handful of wave-energy devices installed across a few countries (Ernst & Young 2013, Ocean Energy Systems 2013). The most significant recent development in the sector by far was South Korea's 254 MW Sihwa Lake tidal power station that became operational in August 2011, surpassing France's 240 MW Rance tidal barrage as the world's largest tidal power plant. South Korea has plans to complete other large plants by 2017, which could boost its national ocean energy capacity level to around 3 GW. China has hitherto installed much smaller tidal and wave-energy devices, whereas Japan, Taiwan, and some Southeast Asian nations are also testing prototype technologies in this field.

To summarise, the patterns and stages of renewable energy development vary considerably across the different sectors. Some are well established and relatively unchanged in their current technical form (hydropower and traditional bioenergy). Some have experienced strong techno-innovatory development, leading to scaled-up commercialisation and impressive growth (solar PV, wind energy, and biomass power generation). Others have conversely remained relatively static (geothermal and tidal energy) or have achieved comparatively modest advances (biofuels). Finally, some are still micro-sectors (concentrated solar power and ocean energy) or at the experimental stages of development (wave energy and solar cooling technologies).

1.2.3. Investment and finance

The rapid growth of renewable energy's more dynamic sectors, such as wind and solar, can be largely explained by increasing levels of public and private sector investment, which have helped improve their commercial viability. Annual global investments in renewables (excluding the large-hydro and SWH sectors) steadily increased from the mid-1990s to the mid-2000s, from around US$7 billion in 1995 to US$14 billion in 2000 and then to US$64.8 billion in 2005. This rapid growth pattern continued thereafter, with annual investment levels peaking in 2011 at US$279.6 billion and then falling back somewhat to US$214.4 in the 2013 year. If large-hydro (estimated at

US$35.5 billion in 2013) and SWHs (estimated at US$8.0 billion in 2012) were included, this level would be even higher.

The very recent decline in renewables investments can be mainly explained by lack of policy certainty in Europe and redirected investments into unconventional or shale gas. In addition, the significant advance of cost efficiencies and market competitiveness in solar PV and wind energy have reduced the need for state subsidy investments in these sectors. Furthermore, this was the fifth consecutive year that, with regard to net additions to power generation capacity, global investments in renewables exceeded those of fossil fuels.[4] Solar energy emerged as the dominant RE sector for new investments in the early 2010s, with US$113.7 billion (96 percent of which was solar PV); wind energy was second with US$80.1 billion that year. Together, these two sectors accounted for 90.4 percent of that year's total. This combined share has increased steadily over time, from just over half the same total back in the mid-2000s (REN21 2005, 2006, 2012, 2014; UNEP 2011).

Concerning investment type, a majority share (62.2 percent, US$133.4 billion in 2013) is attributable to *asset finance* of utility-scale infrastructural projects, such as wind farms and solar parks. Much of this investment is sourced from national and international development banks, such as the China Development Bank and the World Bank Group. According to Pew Charitable Trusts (2011), asset financing has been especially relevant in East Asian countries. Most notably, those such as China and South Korea have embarked on ambitious strategic plans on RE sector development as part of their new green stimulus packages in the wake of the 2008–2009 global financial crisis (Chapter 2). Meanwhile, investments in small-scale distributed capacity (e.g. residential solar PV) was 27.9 percent (US$59.9 billion) of the 2013 total, with solar PV being the most important beneficiary. Investments in small-scale renewables generally has been particularly robust in recent years, indicative of growing societal support. Other types of RE investments include public market investments, as well as research and development; venture capital has been a particularly important investment method in the United States.

Regarding national and regional trends, Europe has historically been the greatest investor in renewables, accounting for at least 40 percent of the global total during the 2000s, but this dropped to just 22.6 percent in 2013. US investment levels have been mostly around a third to a half of those of Europe over the last decade, and they have oscillated somewhat. Meanwhile, China has emerged as the world's largest investor in renewable energy, with its annual investment levels significantly increasing from just US$2.4 billion in 2004 to US$56.3 billion in 2013; China even overtook Europe as a whole, accounting for 26.3 percent of the global total in 2013 (Table 1.3). If China's large-hydro and SWH sectors are included, its global contribution increases notably. The country's burgeoning investments in renewables has been the major factor driving Asia's investments overall. Japan is currently

Table 1.3 Investment trends in renewable energy, 2004–2013

	US$ billion									
	2004	2005	2006	2007	2008	2009	2010	2011	2012	2013
China	2.4	5.8	10.1	15.8	24.9	37.1	36.7	51.9	59.6	56.3
India	2.5	2.9	4.4	6.3	5.4	4.2	8.7	12.6	7.2	6.1
Other Asia and Oceania*	6.8	8.2	9.0	10.9	11.4	12.9	20.7	25.3	29.5	43.3
Europe	19.7	29.4	39.1	61.8	73.4	75.3	102.4	114.8	86.4	48.4
United States	5.5	11.7	28.2	33.6	35.9	23.5	34.7	53.4	39.7	35.8
Brazil	0.6	2.6	4.6	11.0	12.2	7.8	7.7	9.7	6.8	3.1
Other Americas	1.4	3.3	3.2	4.9	5.8	6.1	11.5	8.7	9.9	12.4
Africa and Middle East	5.0	0.5	0.9	1.6	2.4	1.4	4.3	3.2	10.4	9.0
World	**43.9**	**64.4**	**99.5**	**145.9**	**171.4**	**168.3**	**226.7**	**279.6**	**249.5**	**214.4**
Solar	12.1	16.3	21.7	38.7	59.5	62.9	100.3	157.8	142.9	113.7
Wind	14.5	25.1	32.1	56.6	69.3	73.0	94.8	85.9	80.9	80.1
Biomass power	6.2	8.0	10.6	13.2	14.1	13.6	14.2	15.5	11.1	8.0
Small hydro (<50 MW)	1.7	4.9	5.4	5.5	7.2	5.4	4.8	6.8	6.0	5.1
Biofuels	3.7	9.2	27.6	29.3	19.2	10.4	8.9	9.4	6.6	4.9
Geothermal	1.3	1.0	1.4	1.9	1.8	2.7	3.5	3.7	1.8	2.5
Ocean	0.0	0.1	0.9	0.7	0.2	0.3	0.2	0.3	0.2	0.1

Source: REN21 (2014).

Note: Figures exclude the investments in the large-hydro (>50 MW) and solar water heater sectors.

*Mostly represented by East Asian countries.

ranked third globally in national terms, with its investment levels dramatically rising in 2012 (US$16.1 billion) and 2013 (US$28.6 billion), spurred by new policy support measures introduced after the March 2011 Fukushima nuclear disaster and mainly aimed at solar PV (REN21 2014). It is significant that East Asia's investment in renewables remained robust and even increased in 2012 and 2013, during a time of sharply declining investments globally and when investments in both Europe and the United States notably faltered.

1.3. Key challenges, constraints, and contentious issues

Notwithstanding the impressive growth figures surveyed above, significant challenges are ahead for substantially developing renewable energy in East Asia and globally into the 21st century. Generally speaking, this development will entail considerable structural changes to incumbent systems of energy production, distribution, and consumption. Because energy underpins many—if not most—forms of economic activity, the greater use of renewables would also shape key aspects of the economy, as well as relationships between economic agents. For example, increasingly affordable RE equipment (e.g. solar PV, bioenergy) has made individuals and local communities more energy independent and self-sufficient, creating 'prosumers'–those who both produce and consume their own electricity and thermal power, thus largely negating the need for energy supply companies. However, transformative changes of this kind require society to trust renewables to securely and predictably meet their energy needs, as well as make the necessary adjustments to economic habits and practices. There are hence important socio-technical issues to consider when scaling up the use of renewables. In this context, the following key constraints and contentious issues should be considered.

- *Intermittent power problems*: The rated power of an energy installation, such as a 5 MW wind turbine, is its technical or theoretical maximum power output. However, on-shore wind turbines normally operate at 20–40 percent of their rated power capacity due to wind frequency patterns. Similarly, solar energy only generates power during daytime hours, and ocean power is dependent on wind speeds, currents, and tidal movements. Hence, estimated power generation levels expressed by time period (e.g. GWh per annum) are essential for gauging the actual (not just the potential) power produced by a renewable energy installation. A well-located 2 MW wind turbine can produce more energy overall than a much larger one placed in a site with low wind speeds. The intermittent power problem is essentially one of unreliable supply, and it relates closely to the next constraint.
- *Non-dispatchability and capacity factor ratings*: Dispatchable sources of electricity generation are those able to immediately respond to fluctuations in energy demand from power grids. This is not possible for renewables such as wind, solar, and ocean energy given their generation intermittencies noted previously. However, hydropower, geothermal, and bioenergy are not faced with this problem; they offer constant energy streams like fossil fuels and nuclear. Due to intermittency and non-dispatchable issues, wind, solar, and ocean energy are likely to remain complementary to dispatchable forms of power generation. Furthermore, nuclear plants work at similarly high operating efficiency levels as hydropower dams, normally having a 90 percent or greater capacity factor rating. Thus, a 100 MW nuclear power station can produce similar or even greater amounts of generated power as a 300 MW rated wind farm (see Box 1.1).
- *Infrastructural issues*: Scaling up most types of renewables presents important challenges to existing energy infrastructures. The future expansion of a prosumer market in, for example, solar PV will require grids to switch from being traditionally one-way energy transmission systems

to essentially two-way systems. Moreover, renewables bring a much greater diversity of sources to energy infrastructures, which have to be accommodated and integrated into existing systems, such as wind farms located in remote areas far from the nearest grid network point. Problems often arise regarding who pays for this new infrastructure development.

- *Environmental and sustainability predicaments*: All energy systems, including renewables, are in some way materially based and require material processes to produce installation equipment. Consequently, no RE technology is completely carbon neutral or, in an absolutely pure sense, completely renewable because most materials on which they depend are depletable. For example, wind turbines are made from steel and other metals, requiring substantial concrete foundations and transportation fuels to move their components from the point of manufacture to site location. According to the China Council for International Cooperation on Environment and Development (CCICED 2011a), the approximately 20 GW of newly added wind capacity in China required the equivalent of about 2 percent of Chinese steel production that same year. Solar PV module production involves the use of various chemicals and mineral resources. Many of these resources are highly sought after, such as lithium, polysilicon, silver, and gallium for solar PV cells, as well as rare earth minerals, such as neodymium, for wind turbine magnets. Some RE installations cause their own localised environmental problems, such as ecological damage arising hydropower dam construction and operation, emissions from bioenergy combustion, and visibility and noise nuisances related to wind farms. They can additionally have land-intensive requirements, such as wind farms and solar parks, presenting spatial constraint challenges.

Policy-makers have to take these factors into account when formulating renewable energy strategies, and their decisions will also depend on prevailing governance structures and processes in the country or territory in question. For example, East Asia's more authoritarian governments have often exhibited little regard for local community consultation or involvement in the siting of RE plants. In the region's more open democratic societies, local and environmental impact assessments and consultations have become an integral part of RE policy. Moreover, the public's willingness to embrace the socio-technical changes involved with adopting RE technologies and practices in society generally depends a great deal on assessments of their local environmental impacts and how they affect lifestyles, income, and welfare (Walker et al. 2010). Cultural factors also play a significant part in a society's attitudes toward renewables (West et al. 2010). For example, Scandinavia's very internationalist-minded societies take a more positive view toward renewable energy compared to most other nations because its people generally believe it is consistent with acting in a globally responsible manner.

Whether East Asian societies will adopt a similar mindset in the future is vitally important to the region's renewable energy development. Although ambitious top-down approaches by governments for expanding renewables can make a notable difference, the long-run future of renewables ultimately relies on society's willingness to accept and use them.

1.4. Techno-innovation and production

Renewable energy sectors differ considerably, not only in terms of their scale and development but also regarding techno-innovation. Some may be considered relatively mature technologies (e.g. hydropower, geothermal), whereas others are still largely experimental and as yet unproven (e.g. wave energy). Wind energy and solar PV are viewed as mainstream energy sources with high rates of techno-innovation and investment in research and development (R&D). The principal technological improvements in wind energy centre on the increasing size of turbines, their corresponding higher power ratings, and improved technical and production efficiencies (Chapter 4). Arguably more impressive advances have been achieved in the solar PV industry, which is now the most commercially competitive renewables sector in many parts of the world. From 2008 to 2013, PV system costs were reduced by well over half while the cell efficiency ratings have steadily risen. Advances in second- and third-generation PVs are providing wider choices for installing solar energy applications, such as integrating them into building fabrics (Chapter 5).

Both the wind and solar energy industries have also benefitted techno-logically from their respective close links with other high-tech industry clusters. In wind energy's case, this includes nanotechnology (composite materials), aerospace (aerodynamics and advanced materials), energy storage electronics (advanced batteries), and meteorology software. Solar energy has strong connections with advanced materials and chemicals, satellite guidance technology, electronics, lasers, and high-grade concentrators used in space technology applications. Such links have allowed firms with techno-logical competences in related fields to develop competitive advantages in wind and solar energy. This being especially relevant to East Asia's highly diversified companies, such as Daewoo, Hitachi, LG, Mitsubishi, Samsung, Sharp, Sumitomo, Sony, Toshiba, and Matsushita Electric.

Although competition in wind and solar energy has intensified world-wide, both sectors are also characterised by dense patterns of strategic alliance networks, where inter-firm collaboration on new technology development is common (Zhou et al. 2012). This has led to a growth in international cross-licensing agreements between renewable energy firms, in which the intellectual property rights of one signatory party is extended either wholly or partly to another, consequently diffusing new technology and innovations. East Asian companies have benefitted from this techno-logy transfer process, and especially from developing economies such as

China. However, intensifying international competition has also led to companies more ardently protecting their intellectual property by registering patents. Such enforcement of intellectual property rights remains a contentious issue. On the one hand, patent regulation provides both confidence and incentives for firms to invest in new innovation. On the other hand, developing countries in particular question both the equity and efficacy of allowing firms (invariably from developed countries) to enjoy the patent-based monopolistic advantage of being the sole exploiter of new technology and innovation assets, 'especially when the knowledge is essential for promoting public policy goals such as climate change mitigation and adaptation' (Lee et al. 2009: 11). The expansion of compulsory licensing policies of many governments worldwide has helped to reinforce the position of patent holders in RE industries by making it mandatory for firms to adopt the latest technologies when constructing new or adapting existing installations.

There has been a notable increase in the number of new patent registrations for wind and solar PV, with both reaching approximately 1,400 per year globally by the late 2000s. A similar trend is evident in the bioenergy and concentrating solar power sectors (BNEF 2011, Lee et al. 2009). By the early 2010s, estimates indicated that there were more than 200,000 patents for RE technology, and Japanese firms have been among the world's most active patent registers (IRENA 2013a). Chapters that follow discuss how Japan and East Asia's other developed economies of Singapore, South Korea, and Taiwan have generally adopted a clear *technology-oriented* approach to renewable energy development over the *installation capacity-oriented* approach preferred by China and many Southeast Asian countries. However, this distinction has recently become more blurred. Governments and companies in the region's developing economies are now raising their ambitions on indigenous innovation in renewable energy, as revealed in recent macro-development and energy strategies, including China's 12th Five-Year Plan, Malaysia's Green Technology Strategy, and Thailand's Alternative Energy Development Plan 2012–2021. According to an IRENA (2013b) report, Chinese firms have been particularly active in patenting new innovations in the solar PV sector. Meanwhile, Japan, South Korea, and Taiwan have most recently embarked on bolder plans to expand their installed renewables capacity, especially in wind and solar energy.

Increased use of renewables has also been made possible from general techno-innovatory advances in energy systems. For example, the development of smart grids combines information and communication technology to electricity generation infrastructure in ways to make grids more operationally flexible and efficient, as well as better adapted to accommodate multiple sources of power generation, such as rooftop solar PV (Liserre et al. 2010, IEA 2011). In addition, advances in micro-grid technology offer new options for shared power generation at the local community level, which have also helped to encourage the deployment of small- to medium-scale renewable energy installations in various parts of East Asia, especially more remote locations. Meanwhile, new techno-innovation in the construction industry has provided more opportunities to integrate RE technology into new and existing buildings (Carbon Trust 2011).

Most renewable energy sectors are now perceived in East Asia as being relatively high-tech strategic industries; thus, the region's governments and companies have invested in expanding their production capacities and value chains. Chapter 2 discusses how this could be understood as a natural progression of East Asia's progressive techno-industrial development, as well as a functional outcome of the region being the world's largest production hub based on nurturing various sectoral competitive advantages. Over recent years, renewable energy sector manufacturing has accordingly shifted more from Europe and North America to Asia, especially to East Asia and in particular to China. In 2000, Chinese firms produced a mere 2 percent of solar PV equipment; by 2012, this figure was an estimated 63 percent. In the same year, European production was just 11 percent; the United States was much lower, at 3 percent of the global total. In 2013, nine of the world's top-ten PV module manufacturers were East Asian: six from China, two from Japan, and one from South Korea (see Table 5.1, p. 170) in 2012. Also, four Chinese companies were top-ten ranked globally in wind turbine production, and the country was estimated to be making approximately half of the world's wind energy equipment (Pew Charitable Trusts 2011).

Dynamic entrepreneurism in China's wind and solar PV industries was a major contributing factor to over-capacity problems, both nationally and globally in the early 2010s. The mass entry of new Chinese firms into both sectors brought intense competition to the national industry and knock-on effects in the international market. Many high-profile firms in Europe, Japan, and the United States have gone bankrupt, drastically cut back production, or been taken over by other firms (e.g. Hanwha Solar One's purchase of Germany's Q-Cells, once a global market leader). Despite the fact that in 2012 alone around 50 solar PV manufacturers filed for bankruptcy in China—five times the number of their European rivals—the country continues to expand PV production capacity as surviving companies take a positive view on long-term market growth and strong government support for the sector. State-owned banks have hitherto played a key role here, lending the nation's top-ten PV firms almost US$20 billion in 2012 (REN21 2013). Similar issues have arisen in the wind energy industry, where intensifying competition along with aforementioned advances in techno-innovation have exerted downward pressure on market prices.[5] Although this has diminished commercial profit margins, cost competitiveness will become increasingly vital for both wind and solar PV to compete against cheap shale gas in key markets. At present, this particularly applies to the United States, but Europe and East Asia could experience their own shale gas revolution in the not-so-distant future (IEA 2012).

The production structures of RE sectors vary significantly. Hydropower dams and tidal barrages are essentially large-scale construction projects involving huge amounts of basic materials, such as concrete. Biomass, tidal, and geothermal power plants are also still generally utility-scale constructions based on quite simple production value-chain configurations and mostly local sourcing. Value chains in the more dynamic sectors of wind and solar PV are far more complex. Table 1.4 provides an overview analysis of value chains for both industries across their respective upstream, manufacturing, and downstream sections. Companies become more vertically integrated when they venture into other section activities. As Table 1.4 shows, firms may outsource certain value-chain activities to other firms if it is more cost competitive to do so or if they lack the capacity or expertise themselves. Wind turbine and solar PV equipment are multi-component goods that have considerable scope for production network development among a large number of firms, but this depends on the value-chain segment. This, in turn, will determine the extent to which production is localised or internationalised.

1.5. The international political economy of renewable energy

1.5.1. Global competition in renewables

Fast-growing, dynamic, techno-innovative sectors such as wind, solar, and bioenergy are seen as integral to a broader 'green economy' development paradigm on which future world development could be based, as Chapter 2 will further discuss. Competition between nations and firms to secure stronger market positions in RE industries has accordingly intensified over time, especially between those from the world's three main economic regions of East Asia, Europe, and North America. Recent speeches made by key political leaders reveal how important this has now become:

> *We're in a competition all around the world, and other countries— Germany, China, South Korea—they know that clean energy technology is what is going to help spur job creation and economic growth for years to come. And that's why we've got to make sure that we win that competition. I don't want the new breakthrough technologies and the new manufacturing taking place in China and India. I want all those new jobs right here ... in the United States of America, with American workers, American know-how, American ingenuity.*

> – US President Barack Obama, speech given at Allison
> Transmission Headquarters, Indianapolis, IN,
> May 6, 2011, cited in Morris et al. (2012)

Table 1.4 Value chains in the solar PV and wind energy sectors

	Solar PV		Wind energy	
	Value chain segment	Key characteristics	Value chain segment	Key characteristics
Upstream	Research and development (R&D)	Mostly in house, some specialist R&D outsourcing.	Research and development (turbines and components)	Mostly in house, some specialist R&D outsourcing.
	Polysilicon mining and silicon ingot production	Global supply concentrated amongst a few small OECD firms but developing country suppliers rising.	Cast iron	Approx 15–20 tonnes p/MW capacity. Used for several components. Mostly outsourced production.
	Other materials	Glass, aluminium, etc for module production. Outsourced production and supply.	Forgings	Approx 10 tonnes p/MW capacity. Used for several components. Mostly outsourced production.
			Reinforcement fibres	Used in blades. Glass and carbon fibres. Woven and stiched fabrics. Outsourced production.
Manufacturing	Wafer manufacturing	Over 200 producers worldwide by 2010, most also involved in silicon ingot production.	Towers	Up to 200 metres high, up to over 300 tonnes. Mostly outsourced production.
	Cell manufacturing	Almost 250 producers worldwide by 2010.	Blades and hubs	Blades up to 80–90m in length. Blade and hub combined weight up to 45 tonnes. In house production.
	Module manufacturing	Typically range from 50–300 watts each module. Between 7–15m^2 surface area per kW capacity.	Gearboxes	Hitherto produced by a small number of large established OECD-based external suppliers. Increasing competition from Asian firms.

(continued)

Table 1.4 Value chains in the solar PV and wind energy sectors (*continued*)

	Solar PV		Wind energy	
	Value chain segment	*Key characteristics*	*Value chain segment*	*Key characteristics*
	Balance of system (BOS) components manufacturing	Inverters, mounting structures, batteries and other electronic equipment.	Generators	Most produced in house or by large generic electrical machinery firms. Growing number of Chinese firms entering the market.
			Large bearings	Used in gearbox, main shaft, movement controls and generator. Produced by large-scale external suppliers.
			Power convertors	Converts wind power into power grid frequency and voltage.
Downstream	System design	For rooftop and field installations. Vertically integrated solar PV producers, system integrators and external service firms.	Wind park site assessment	Outsourced to consultant firms.
	System integration	Integration of certain BOS components at installation site. Vertically integrated solar PV producers, system integrator firms.	Transport/logistics	Outsourced to logistics firms.
	Project development	For larger-scale projects. Vertically integrated solar PV producers and system integrator firms.	Wind park construction	Outsourced to construction services and installation services firms.
	Financing	Vertically integrated solar PV producers, project developers, banks and other financial service providers.	Financing	Outsourced to banks and other financial service providers.
	System installation and construction work	Outsourced to construction services and installation services firms.	Repowering and grid connections and wind power sales	Independent power producers and external utility companies.
	Operating and maintenance	Vertically integrated solar PV producers, system integrators and external service firms.	Operating and maintenance	Independent power producers and external utility companies.

Sources: Kierkegaard et al. (2009, 2010), EPIA (2011b).

We need to . . . seize the commanding point of having the world's best environmental technology, to win the race between the global industries.

– China Vice Premier Li Keqiang, speech given
at the 11th Five-Year Plan Environmental
Achievement Exhibition, Beijing,
June 7, 2011, cited in CCID 2011b).

Broadly speaking, global competition in renewables techno-innovation and production is intensifying. East Asia's export-oriented economies view RE industries as presenting significant international market opportunities. The South Korean government has set a Green Growth Strategy target of capturing 10 percent of global market for renewable energy and other green technology products by 2020 (Chapter 2). However, this is ultimately a competition between companies rather than countries. Moreover, in a globalising world economy of ever-dense patterns of international business linkages, defining the national winners and losers has become increasingly difficult and often meaningless. We have seen how East Asian and particularly Chinese firms appear to be strengthening their position in equipment manufacturing in many RE sectors. Their growing production capacity in this value-chain element (Table 1.4) can have significant benefits for upstream and downstream firms located both inside and outside the region. For example, China's ability to increasingly mass produce competitively priced solar PV products has created more business demand for polysilicon suppliers upstream and system installers and maintainers downstream worldwide. This explains, for example, why coalitions of solar PV installation and maintenance firms in both the United States (Coalition for Affordable Solar Energy) and European Union (Alliance for Affordable Solar) lobbied against the application of trade barriers on low-priced Chinese imports.[6] The expansion of China's solar PV manufacturing base also helped support the US polysilicon industry, which in the early 2010s supplied a quarter of the world's demand for the material.[7]

The growth of international business linkages in RE industries has often made it difficult to establish the demarcations of competition between firms themselves. These linkages can exist between companies working in the same part of the value chain (e.g. Hanwha Solar One from South Korea and Q-Cells from Germany) or different parts of it (e.g. American firm First Solar's joint venture announced in 2011 with utility-scale systems firm China Power).[8] As Kirkegaard et al. (2010) contended, the global integration of renewable energy industries should create greater economies of scale and scope, thus further reducing unit costs and helping to drive future techno-innovation. Notwithstanding the complications of identifying the exact lines of international competition, rising renewable energy exports from East Asia have caused certain political economic conundrums in importing countries elsewhere. For instance, an important public policy debate in Germany has centred on whether the government's FiT programme for encouraging further installation of solar PV systems is mainly benefitting Chinese manufacturers at the expense of their German rivals, most of which are unable to compete on price (Grau et al. 2012).

Taking the above points into account, international competition in renewables from a government policy perspective is primarily concerned with optimising as much value-added business activity as possible in home territory, whether generated by domestic or foreign firms. Chapters 2 and 3 examine how states around the world have increased their support for promoting R&D in renewable energy to develop stronger techno-innovation capacity. Firms that develop new technological breakthroughs in an industry enjoy first-mover advantages and can capture particular market segments or even create them. The world's most developed economies (European Union, Japan, United States) still enjoy their historic advantages on this aspect of global renewables competition. However, the improving innovation capabilities of China, South Korea, Taiwan, Singapore, and Malaysia also represent an important future competitive challenge to incumbent advanced economy firms on this front, possibly to a similar extent as experienced in manufacturing competition over the last decade.

1.5.2. Trade barriers and disputes

The number of trade disputes over RE products has gradually risen. The disputes have not focused so much on normally applied import tariffs because these have remained relatively low. In the wind energy industry, the 2013 most favoured nation (MFN) duty rates were zero percent for China, Indonesia, Japan, the Philippines, Singapore, and Vietnam; 5 percent for Taiwan and Malaysia; 8 percent for South Korea; and 10 percent for Thailand. These rates can be compared to 3.3 percent for the European Union, 2.6 percent for the United States, 5 percent for Australia, and 7.6 percent for India. A similar pattern is evident for solar PV modules: zero percent MFN duties for Japan, Malaysia, Singapore, South Korea, and Vietnam; 2 percent for Taiwan; 7 percent for the Philippines; and 10 percent for China, Indonesia, and Thailand, which compared to 1.5 percent for the United States, 3.3 percent for the European Union, 5 percent for Australia, and 7.5 percent for India. In addition, virtually all countries worldwide have zero rates for biomass wood pellets and low single-digit rates for both hydropower and biomass power generation equipment. Somewhat higher rates apply in developing countries in the much smaller geothermal sector.[9]

However, non-tariff trade barriers, state subsidies, and foreign investment barriers on renewables have been an issue of contention. Prime examples include domestic-level taxes and charges, obstructive customs procedures, local content requirements in manufacturing, the lack of mutual recognition of certification systems, and soft loans from state-owned banks (Kirkegaard et al. 2010). In 2008, South Korea introduced a 19 percent surcharge on foreign PV goods, although this was removed the following year. Chinese wind turbine producer, A-Power, experienced significant local resistance to its planned investment in Texas in 2009, and it also was subject to 'Buy American' government procurement laws (Kirkegaard et al. 2009). In 2008, China applied a local content requirement on wind turbine manufacturing, thus requiring firms producing in the country to source minimum supply quotients from local firms. At this time, Beijing defended its policy using the infant industry argument often used by developing countries, stating that its aim was to both help establish a scaled-up industry value chain and enhance indigenous investment in wind energy R&D.[10] China removed its local content regulation in February 2011 after increasing pressure from the United States and the European Union. In the late 2000s, the Chinese government also imposed restrictions on foreign firms wishing to enter certain parts of the nation's offshore wind industry (Soares 2009). Meanwhile, Japan undertook a complaint against the Canadian provincial government of Ontario at the World Trade Organisation (WTO) regarding its FiT scheme, which included a 60 percent local content rule. It also was alleged the scheme offered preferential treatment to certain investors, such as Korean firm Samsung, which in January 2011 had announced plans to invest US$6.7 billion for the construction of four new solar and wind energy clusters in Ontario.[11]

After the US government applied trade barriers of alleged state-subsidised exports of Chinese solar PV products in 2012, the Chinese government responded the following year by imposing its own barriers on polysilicon imports from US producers.[12] The EU and China were locked in a similar dispute in 2013, with the EU accusing the Chinese government of breaking WTO rules for allegedly providing cheap state loans, land, interest-free credit lines, and tax breaks to the country's PV manufacturers.[13] Although the level and scope of trade-related disputes over renewable energy products are still small by comparison with many other sectors, such disputes are likely to rise as green energy industries further expand and strategic rivalries within them intensify in the future.

1.5.3 Rare earths elements and resource scarcity issues

Renewable and other green energy technologies are estimated to account for approximately one-fifth of the total world consumption of critical materials, including cobalt, lithium, and the rare earth elements (a group of 17 elements possessing unique properties of a catalytic, magnetic, or optical nature).[14] For example, neodymium is an extremely magnetic rare earth mineral used in the manufacture of advanced wind turbines.[15] Many countries, including the United States, were once very active in rare earth mining but significantly cut back or closed down operations altogether due to the adverse environmental impacts arising from extraction and processing. China, however, has ramped up its production capacity. By the early 2010s, China had a very strong monopoly position in global supplies of rare earth elements, reaching percentage shares in the upper 90s for most elements by 2010. At this time, the country was thought to possess over a third of the world's rare earth deposits. By 2012, China had reportedly fallen to just around a quarter of global known reserves, and its production shares had also correspondingly declined to some degree.[16]

Recent annual rates of growth in the demand for rare earth elements have been between 10 and 15 percent, and consumption is projected to rise further as sectors such as wind energy and solar PV continue to expand. From 2009 onwards, the Chinese government introduced a number of production caps and export restrictions on rare earth elements, explaining that the decision was based on environmental and resource management factors. New pollution controls implemented toward the end of 2010 are thought to have further restricted rare earth mining. Prices for these commodities rose sharply in 2010 as a consequence, by as much as 300–700 percent for certain elements. The restrictions also created supply-chain difficulties in RE industries. In early 2010, China reduced its rare earth exports by 72 percent and an additional 11 percent in early 2011.[17] The export quota amounts were around a third of the capped national production total, leaving two-thirds for domestic consumption. Further restrictions were announced in 2012; in March of that year, the United States, Japan, and the European Union all confronted China at the WTO regarding the matter, claiming that Beijing's actions were a breach of international trade rules. The United States instigated a formal WTO dispute settlement case against China, which was joined by other trade parties.

Policy-makers around the world have responded to these developments in different ways. Large mineral-rich nations, such as Brazil, Canada, Australia, South Africa, Tanzania, Kazakhstan, Russia, and the United States, have increased their rare earth exploration activities. Other nations, including Japan and South Korea, have increased their R&D efforts, focusing on improving material utilisation efficiency, using ferrite magnets as a substitute for neodymium magnets used in wind turbines, developing longer-term nanotechnology options for creating substitute composites for rare earth elements, and additional scientific fields. Japan has also invested in increasing its stockpiles of rare earths and other critical minerals, especially those essential to its solar PV sector, such as lithium. Here also, Japan is investing heavily in recycling key materials and reprocessing used components.

This situation reveals some key issues relating to the future development of renewables. First, it shows the strategic importance now conferred to securing upstream value chains in key industries such as wind and solar energy, and how international disputes over such matters of resource management can be expected to increase in the coming years. It also further highlights the dependency of even RE technologies on what are ultimately finite and depletable materials. Therefore, recycle and reuse practices will need to improve and expand if renewable energy development itself is to maintain its own sustainability over the long term. Greater international cooperation on this matter could prove vitally important in the future.

1.6. Multi-level governance and institutionalism

1.6.1. National level

The governance of renewable energy generally concerns managing, promoting, and optimising its development through various institutionalised means, including strategic planning, public policy, funding mechanisms, interagency coordination and cooperation, and generating shared data and information. At the national level, this can be viewed from two main perspectives. First, state or public governance relates to how renewable energy is managed within the apparatus of national government, as demonstrated by the formation of RE policies (Chapter 3). In East Asia, managing renewable energy policy has been mostly assigned division or section status under the purview of government agencies (e.g. ministries, departments, bureaus, offices) that are more broadly responsible for energy or industry policy, and these may be embedded in turn within even larger economic development agencies. It is hence rare for renewable or green energy to be afforded a distinctive high profile institutionally within East Asian governments. Many of the region's energy research institutes are state or quasi-state institutions, performing important analytical services in the state's governance structures on renewable energy development.

The second main governance perspective at the national level concerns business associations that have grown in number over time as renewable energy development has advanced. In East Asia, sector-based bodies in Japan are especially well established (e.g. the Japan Photovoltaic Energy Association, founded in 1987), and the region's other large economies of China, South Korea and Taiwan now have business associations across a good range of RE sectors that also work closely with their respective national governments. In Southeast Asia, these organisations are still relatively new, such as the Thai Photovoltaic Industries Association, which formed in 2012. Some national associations also generally represent the renewable energy industry, the most prominent being the Japan Council for Renewable Energy, the Chinese Renewable Energy Industry Association, the Korean Society for New and Renewable Energy, and the Sustainable Energy Association of Singapore.

1.6.2. Global level

Most of East Asia's national renewable energy associations are members of their respective global associations, including the Global Wind Energy Council (GWEC), World Wind Energy Association (WWEA), International Solar Energy Society (ISES), International Hydropower Association (IHA), World Bioenergy Association (WBA), and International Geothermal Association (IGA). Each seeks to promote the worldwide development of their sector. From a historic perspective, institutional developments on promoting renewables at the global level date back to the early 1970s. The United Nations (UN) 1972 Stockholm Conference on the Human Environment is seen by many as the first agreement to address global environmental issues, which made reference to how 'new-energy' solutions can alleviate anthropologically caused environmental problems. It took some time, however, for renewable energy to become firmly fixed on multilateral environmental and energy agendas. For example, G7 responses to the 1973–1974 and 1979–1980 oil crises primarily concerned how to mitigate oil dependencies by exploring coal and nuclear power alternatives. The first oil crisis led to the creation of the International Energy Agency (IEA), based on membership of the Organisation of Economic Co-operation and Development (OECD); despite its still somewhat narrow international representation, the IEA remains the world's most important multilateral energy organisation. In the late 2000s, the IEA developed a renewable energy policy database that covers all countries with active RE policies.

The first global endeavour specifically focused on RE development was the 1981 United Nations Conference on New and Renewable Sources of Energy, which convened at Nairobi, Kenya. Although this called for the UN and other multilateral forums to ardently promote renewables, no direct actions followed for some years. The IPCC (established in 1988) and United Nations Framework Convention on Climate Change (UNFCCC, created in 1992) were the next most important milestones in environmental and energy-related multilateralism. The IPCC is essentially an international science committee and a key agency for examining the causes and effects of climate change, as well as advocating measures to address it. This has included comprehensive reports on renewable energy's role in climate change mitigation strategies. The UNFCCC, meanwhile, is an agreement signed by more than 150 nations at the UN Conference for Environment and Development (the first 'Earth Summit') in June 1992 at Rio de Janeiro. The UNFCCC then became international law in 1994, leading also to annual international Conference of the Parties (COP) conferences. There was, however, no explicit role assigned to renewable energy in the UNFCCC text for tackling climate change; rather, renewables were referred to only in the development policy context of the agreement. When Japan hosted COP3 in December 1997, the Kyoto Protocol was added to the UNFCCC text, laying out a set of commitments to reduce carbon emissions over agreed time periods for each signatory to the agreement, as well as the specific instruments used to meet targets set. The Kyoto Protocol documentation has made relatively few and short references to date on how renewable energy can help achieve the agreement's objectives (e.g. UNFCCC 2005, 2011) and has been somewhat criticised for doing so (Hirschl 2009). However, as discussed in Box 1.2, its Clean Development Mechanism scheme, which was formally launched in 2006, had helped fund over 6,000 new renewable energy projects in developing countries worldwide by 2013, with the majority of these being situated in East Asia.

At the 2000 Okinawa G7 summit hosted by the Japanese government, national leaders discussed a British government proposal to create a special task force for promoting renewables worldwide (Kirton 2006). As it transpired, the remit of the subsequently formed task force focused more on development cooperation than energy system transformation. Although its 2001 report was widely commended by the international community, the task force was disbanded soon thereafter, with no momentum on renewables carried forward in subsequent G7 summits. At the 2002 World Summit for Sustainable Development (or second 'Earth Summit') held in Johannesburg in 2002, renewable energy was afforded greater attention than at the 1992 Rio summit. Although these talks failed to meet their main objective of establishing a general consensus on setting sector-specific RE development targets, they helped lay the groundwork for subsequent International Renewable Energy Conferences (IRECs). The first IREC was held in Bonn, Germany, in 2004; this was followed by Beijing, China (2005); Washington, DC, USA (2008); Delhi, India (2010), and Abu Dhabi, United Arab Emirates (2014). The IRECs have brought together between 80 to more than 150 countries to discuss common challenges on developing renewable energy policies, capacities, and techno-innovation, further helping institutionalise global dialogue and cooperation in these fields.

An outcome of the 2004 Bonn IREC was the establishment of the International Renewable Energy Alliance, a new lobby coalition formed between the aforementioned IGA, IHA, ISES, WBA, and WWEA sectoral groups. Soon thereafter, this became a formal organisation and was renamed the Renewable Energy Policy Network for the 21st Century, or REN21, with additional collaborative help from the UN Environmental Programme, the IEA, and the German Technical Assistance agency. REN21's prime aims are to promote policy development, dialogue, and analysis on renewable energy issues based on multi-stakeholder participation. Since 2005, it has published annual global status reports, which are the most important comprehensive analyses on renewable energy development worldwide produced on a regular basis, providing key sources of data intelligence and drawing on the work of its multiple partner agencies. At the ill-fated COP15 conference held at Copenhagen in December 2009, REN21 proposed the creation of a global investment fund for renewables but failed to receive sufficient support. The organisation continues to lobby for this initiative, with a core element of the fund being a possible global FiT scheme. Another new hybrid organisation,

Box 1.2 The Kyoto Protocol and the Clean Development Mechanism

The Kyoto Protocol of the United Nations Framework Convention on Climate Change has made an important contribution to international efforts on renewable energy development, most critically through its Clean Development Mechanism (CDM) scheme. The agreement had been signed by 84 countries by 1999, including all East Asian nation-states, and came into force in February 2005. By this time, most national governments from the region had domestically ratified the Protocol, with the exceptions being Brunei, North Korea, and Singapore; however, quite soon thereafter, these countries completed the process. By early 2014, there were 192 signatory parties. Of these, 42 signatory nations (developed countries and some economies in transition, such as Russia and Ukraine[18]) have legally binding targets on carbon emissions if they had ratified the protocol, forming the Annex I group. The general target set for these countries was to reduce their emissions by an average of 5.2 percent of their 1990 levels by the end of 2012. In addition, an Annex II group, comprised 24 developed countries, are obliged to offer financial and technical aid to developing country parties in their efforts on climate change mitigation. Japan is the only East Asian country in either group, with the rest of the region falling into the developing country category in which there are no legally binding obligations on CO_2 emissions, although they are expected under the agreement to set their own future emissions targets unilaterally. Many East Asian countries have set such targets (see Table 3.2, p. 78). After the first commitment period had passed in December 2012, a second period up to 2020 has been incorporated into an updated and amended Kyoto Protocol, but at the time of writing this had not yet legally entered into force. There were 37 countries by early 2014 that had committed to this second period. This excluded Japan, who in December 2010 announced it would not commit to the Protocol beyond 2012, although of course this may change in the future.

Parties with Protocol commitments are supposed to meet their emission targets primarily by the use of national measures. This has relevance for government RE policy, although Kyoto does not set any formal guidelines or stipulations here. To augment national-level efforts, the agreement introduced three market-based mechanisms with the aim of creating the basis of a global carbon market. The international emissions trading mechanism 'allows countries that have emission [reduction] units to spare—emissions permitted them but not 'used'—to sell this excess capacity to countries that are over their targets.'[19] Under a second mechanism, joint implementation, an

Annex I party can fund an emission reduction project in another party from this group, contributing toward the former's emissions targets. It was expected that most of these projects would occur in the economies in transition sub-group to assist their efforts on replacing dilapidated coal-fired power plants with cleaner and more efficient technology, which in some cases has included renewables.

However, the most relevant mechanism for renewable energy under Kyoto is the CDM, which became fully operational in 2006. The CDM facilitates emission reduction projects funded by Annex I countries to take place in non-Annex I countries, leading to the former earning tradable certified emission reduction (CER) credits (1 tonne of CO_2 per credit) that contribute toward achieving their Protocol targets. One of the key advantages of the CDM is that emission reductions are achieved at a much lower global cost by financing projects in developing countries, where energy generation costs are lower than in the developed world. Similarly, the technical scope for emission reductions in developing countries is much greater given their lower energy-efficiency levels. By use of an illustrative example, the construction of a wind farm to replace an old, inefficient coal-fired power station with outdated emission-control technology located somewhere in developing East Asia would yield a much higher investment return in cost per tonne of emission reduction terms than if a modern coal power plant was being decommissioned somewhere in Japan, Western Europe, or the United States.

The CDM is also the world's first environmental investment and credit scheme to operate on a global scale with 161 countries engaged in the scheme by 2012, helping mobilise up to US$215.4 billion of investments in developing countries by that year. The UNFCCC claims that the mechanism has reduced carbon emissions by over 1 billion tonnes (UNFCCC 2012), or about 3 percent of the annual global emissions total (Table 1.1). By September 2013, a total of 8,822 CDM projects had been either fully registered or in the registration pipeline process. Approximately 70.2 percent of these projects concerned new renewable energy installations (6,192 projects in total), consisting of 30.0 percent wind energy, 25.9 percent hydropower, 9.6 percent bioenergy, 4.6 percent solar, and 0.1 percent for tidal and mixed renewables (Table 1.5). The UNFCCC estimated that, by around this time, the CDM had created an extra 120 GW of installed renewables capacity (UNFCCC 2012); if accurate, this represented 8.2 percent of the global total. Moreover, most of this additional capacity has occurred in East Asia. By September 2013, East Asia accounted for 60.3 percent of the renewable energy CDM project total. These projects were heavily skewed toward China, which alone accounted

Table 1.5 Clean Development Mechanism projects, by September 2013

	Wind	Hydro	Bioenergy	Solar PV	Geothermal	Tidal	Mixed renewable energy	Renewable energy total	Renewable energy % of total	Other projects	Total CDM projects
China	1,517	1,373	163	158	2	0	4	3,217	81.0	755	3,972
Vietnam	5	201	16	0	0	0	0	222	86.0	36	258
Indonesia	0	22	18	1	14	0	0	55	32.4	115	170
Thailand	4	6	33	25	0	0	0	68	40.2	101	169
Malaysia	0	5	45	0	1	0	0	51	31.3	112	163
South Korea	13	16	1	30	1	1	5	67	70.0	29	96
Philippines	6	10	6	1	4	0	0	27	33.3	54	81
Laos	0	8	1	0	0	0	0	9	75.0	3	12
Cambodia	0	4	1	0	0	0	0	5	50.0	5	10
Singapore	0	0	2	0	0	0	0	2	25.0	6	8
North Korea	6	0	0	0	0	0	0	6	100.0	0	6
Mongolia	0	2	1	0	0	0	0	3	75.0	1	4
Myanmar	1	0	0	0	0	0	0	1	100.0	0	1
East Asia total	**1,551**	**1,647**	**287**	**215**	**22**	**1**	**9**	**3,733**	**75.4**	**1,217**	**4,950**
% world total	*59.4*	*72.2*	*33.8*	*53.0*	*62.9*	*100.0*	*90.0*	*60.3*	*–*	*–*	*56.1*
India	805	238	375	139	0	0	0	1,557	73.4	563	2,120
other Asia and Pacific	7	35	12	1	1	0	1	57	54.3	48	105
Latin America	189	318	141	25	7	0	0	680	57.3	506	1,186
Africa	39	21	28	13	5	0	0	106	40.6	155	261
Middle East	3	2	2	13	0	0	0	20	23.8	84	104
Europe and Central Asia	15	21	3	0	0	0	0	39	40.6	57	96
World	**2,609**	**2,282**	**848**	**406**	**35**	**1**	**10**	**6,192**	**70.2**	**2,630**	**8,822**

Source: Clean Development Mechanism (CDM) pipeline projects databank (http://cdmpipeline.org/publications/CDMPipeline.xlsx).

Notes: Figures were derived from CDM 'pipeline' statistics—that is, total projects that are at different stages of the registration process. The vast majority are fully registered. Other categories include requested registration, in review, or possibly still in validation.

for 52.0 percent of the renewable energy CDM projects, as well as 45.0 of all CDM projects by this time. Table 1.5 provides further details on the CDM's contribution to renewable energy development in East Asia.

Vietnam has been another important CDM beneficiary, accounting for the same number of renewable energy project registrations as Africa, the Middle East, East Europe, Central Asia, and other Asia and Pacific (minus India) combined. In most East Asian cases, renewables comprised the majority project total. However, notably for Southeast Asia's largest countries, other clean development technologies (e.g. emission abatement, carbon emission capture, forestation, green transportation systems, energy efficiency) were collectively more important. Japan has funded the fourth largest number of CDM projects (8.3 percent, 578 projects) among Annex I countries, behind Britain, Switzerland, and the Netherlands and slipping from an earlier held third ranking of CDM investors. The scheme has enabled developing East Asian countries to secure renewable energy technology transfers, improve management performance levels, and foster indigenous RE capabilities over the long term, as reportedly demonstrated in China's wind energy sector (GWEC 2010b).

The CDM additionality clause requires that a project must lead to a net emissions reduction greater than what would have occurred in the most plausible alternative scenario to the proposed project activity. Proving additionality obviously becomes harder when developing country governments are offering an ever-wider range of incentives to promote decarbonisation across various sectors, thus presenting many alternative options to CDM projects (Pechak et al. 2011). The mechanism has also come under increasing criticism. Lloyd and Subbarao (2008) were early observers of how the majority of CDM projects appear to be concentrated in larger developing countries such as China, India, and Brazil (the three top-ranked host countries, accounting collectively for three-quarters of the project total by 2013), to the extent that smaller nations lacking the supportive infrastructure have failed to attract CDM investment. On another scale-related issue, high transaction costs associated with the CDM market have tended to favour large, high-CER-volume projects, leading to small community-based project proposals in rural or remote areas–where the socio-economic needs are more often than not the greatest–being deemed less commercially viable and therefore not funded (Bagliani et al. 2010). Furthermore, it is somewhat incongruous for East Asia's relatively well-developed and high-income nations of Singapore and

South Korea to actually host CDM projects, when arguably they should be instead funding them in developing countries.

Finally, CDM activity rose to a sharp peak in 2012 as the first implementation phase of the Kyoto Protocol ended in December of that year, but levels fell back drastically in 2013. The total project number actually declined a little in early 2014 due to aforementioned uncertainty over the implementation of the accord's second phase to 2020. Japan introduced its own Bilateral Offset Crediting Mechanism (BOCM) in 2010–initially as a complement to its CDM scheme, but now effectively as a substitute given the country's decision not to participate in post-2012 Kyoto unless other major carbon emitters (e.g. China, the United States, India) make legally binding targets on emission reductions. The BOCM has similar aims to the CDM but involves a less bureaucratic process and permits more scope for private investment. Japan's new scheme became fully activated in 2013 after a 3-year study period, and four bilateral agreements were first signed with Bangladesh, Mongolia, Ethiopia, and Kenya. If a similar bilateral approach is adopted by other nations due to frustrations with the CDM and UNFCCC multilateralism generally, it could lead to climate change diplomacy, emulating the path that international trade relations has taken from the late 1990s in which the proliferation of bilateral free trade agreements has significantly contributed to the decline of the WTO.

the Renewable Energy and Energy Efficiency Partnership (REEEP), was formed in 2002 and comprises representatives from the regulatory, business, financial, and non-government organisation communities; it advocates improved access to renewable energy and energy efficiency technologies. By 2013, REEEP's membership had expanded to 45 national governments (including Indonesia, Japan, Philippines, Singapore, and South Korea from East Asia) and the organisation had established two regional secretariats, one in China the other in India.

The most important recent institutional development from a global governance perspective was the foundation of IRENA in 2009. It became fully functional in 2011 and has its headquarters in Masdar City, Abu Dhabi. IRENA is the now the world's principal international institution for promoting renewable energy, and its membership had grown to 131 members by June 2014, including some from East Asia: Brunei, China, Japan, Malaysia, Mongolia, Philippines, Singapore, and South Korea.[20] The main tasks of IRENA are to provide advice and support for governments on policy issues, capacity building, finance generation, technology transfer, information sharing, and improved coordinated actions across international institutions that are working to promote renewables. Japan is the second largest contributor to IRENA's core budget. A World Council for Renewable Energy (WCRE) was also formed in 2001, essentially as a lobby group for promoting the sector; it also strongly advocated the creation of IRENA. In Chapter 2, we outline additional efforts to promote renewable energy development through new 'green growth' multilateral agendas and institutions that emerged largely in response to the 2008/09 global financial crisis.

1.6.3 Regional level in East Asia

Regional cooperation and diplomacy on renewable energy issues in East Asia has remained very low key. Although the region hosts a growing number of conferences and business networking events on renewables, inter-governmental dialogue on the subject has been negligible and institution building has been extremely limited. Even the region's most established and fullest functioning regional organisation–the Association of Southeast Asian Nations (ASEAN)–has afforded relatively little attention to the subject. The ASEAN Plan of Action for Energy Cooperation (APAEC) 2004–2009, however, initiated numerous minor development projects with the help of German government assistance for its ten member states to reach a collective 10 percent target on installed renewables power capacity. Under APAEC 2010–2015, this target was raised to 15 percent by 2015. However, as Table 1.1 shows, half of these countries already surpassed this goal by 2009, before the Action Plan became operational[21] (ASEAN 2011). A separate action plan was devised with the aim of realising this target and additional highly generalised goals on introducing clearer policies and programmes on RE development for improving commercialisation, investment, market, and trade potentials of renewable energy technologies. However, APAEC remained largely aspirational on these areas, lacking detail on what exact institutional mechanisms and funding would be deployed to help reach the goals set. Other instances of regional-level dialogue have occurred within the ASEAN Plus Three regional framework (the 'Three' being China, Japan, and South Korea), leading to special forum meetings on renewables but with no real substantive institutional or other kinds of results.

Arguably the most important contribution thus far made at the East Asia regional level toward renewable energy development has been made from the Asian Development Bank (ADB), which originated in the early 2000s under its Promotion of Renewable Energy, Energy Efficiency, and Greenhouse Gas Abatement Project that established loan facilities for investment in RE sectors. The ADB's Energy Efficiency Initiative was superseded in 2009 by its broader Clean Energy Programme, which by 2012 had helped create over US$2 billion of new investments in renewable energy installations and technological advancements (ADB 2012a). Prominent sector-specific programmes also regionally managed by the bank include the Asia Solar Energy Initiative, Quantum Leap in Wind, and the Small Wind Initiative. The ADB is also involved in the Climate Investment Funds programme that comprises a series of financing instruments to promote low-carbon and climate-resilient development projects. The ADB generally has become a key advocate of low carbon development in the East Asia region and the wider Asia continent; it is estimated that over a quarter of its approved loans support projects that are in some way clean-energy related. In the private sector, the only known sectoral business group formed at the regional level relevant to East Asia is the Asian Photovoltaic Industries Association, which was established in 2009 and based in Singapore. However, outside Europe such organisations are quite rare.

Two more recent regional initiatives led by Japan on promoting low carbon development involve efforts on renewable energy cooperation. The first is the East Asia Low Carbon Growth Partnership (EALCGP) established in 2012 as an East Asia Summit[22]-based dialogue process overseen by Japan's Ministry of Foreign Affairs; its inaugural meeting convened in April 2012 in Tokyo. The EALCGP's early work has focused to date on bringing together key decision-makers and policy influencers, as well as knowledge transfer and capacity building for developing countries on renewables and low carbon development. The second EALCGP dialogue meeting was held in Japan in May 2013. Meanwhile, in October 2012, the Low Carbon Asia Research Network (LoCARNet) was created as a research-based initiative running parallel to the EALCGP. As its name suggests, LoCARNet essentially comprises a network of researchers working on low carbon development policy processes in the region. Knowledge and capacity building on renewable energy development and policy aimed at developing countries is also a priority objective of LoCARNet. Global and regional endeavours on strengthening renewable energy governance are augmented by a myriad of bilateral cooperation programmes and projects, which are often oriented toward development cooperation, such as the Japan International Cooperation Agency's work in supporting RE projects in remote, poor areas of Southeast Asia.

1.7. Conclusion

Renewable energy has a vital role to play in the low carbon future of East Asia and the wider world. The energy-intensive nature of East Asia's contemporary economic development, driven by rapid industrialisation, makes renewables especially critical to the region. Renewable energy development in East Asia is also critical to the international community as a whole, given the region's growing impact on global energy security and climate change. China's burgeoning economic development in particular is presenting crucial challenges on both of these fronts, but the country is also making vitally useful contributions to the worldwide expansion of renewables. Other East Asian countries have become important players in certain RE sectors in terms of techno-innovation, production, installed capacity, investment, trade, and/or finance.

Three main inter-related drivers behind renewable energy development in East Asia were proposed, namely environmental and sustainable development imperatives, energy security, and emerging strategic industries. Renewables are being promoted by the region's governments as a means to address the increasingly serious multi-level environmental challenges facing the region, as well as to place its countries on more sustainable low carbon development trajectories. These sources of energy are also essentially indigenous in nature, helping to mitigate the chronic energy import dependencies already confronting East Asia's largest and most developed economies. The region's domestic fossil fuel reserves are depleting quickly; thus, renewables are providing solutions for many kinds of energy supply risks. In addition, many RE sectors are now viewed as emerging industries of the 21st century; hence, they are critical to creating future prosperity, employment, and income.

From historic and global perspectives, it was noted that renewable energy sources have been used for a long time in human society; however, aside from hydropower, the deployment of their modern technological forms was still very limited by the end of the 20th century. Renewables comprise a broad spectrum of different energy technologies, and most recent development patterns have varied significantly. Since the early 2000s, wind, solar, and bioenergy have expanded rapidly; they are now scaled-up and commercialised energy sectors that are increasingly able to compete with fossil fuels on power generation. East Asia has been at the forefront of the global renewable energy development, accounting for around a quarter of RE-generated electricity worldwide. This has corresponded to the region's burgeoning public and private investment in renewables, which by the early 2010s accounted for approximately two-fifths of the world total. However, the future expansion of renewable energy in East Asia and worldwide faces a number of key challenges, constraints, and contentious issues related to intermittent power problems, the non-dispatchable nature of many renewables, infrastructure development, and a number of environmental and sustainability predicaments. There are also important socio-technical and socio-cultural matters to consider if society is to make greater use of renewable energy.

Patterns of renewable energy techno-innovation have varied greatly across the different sectors. The most significant advances have occurred in the two largest non-hydro sectors—wind and solar PV—helping to reduce production and operating costs and raise technical efficiency levels. Firms from East Asia's more developed economies are among the technology leaders in renewables, and the region's developing countries have benefitted

from cross-border technology transfers. East Asian governments are supporting the techno-innovative endeavours of home-based firms for strategic
industry reasons. The region is already the main production hub for RE
products—a position that is likely to strengthen in the future due to both
expected strategic government support and dynamic private entrepreneurship. However, high rates of wind and solar PV production capacity growth,
most notably in China, have caused over-capacity and other problems; they
are part of the significant structural changes that have recently occurred in
both sectors. Key shifts in international trade patterns for renewable energy
products were also examined in this chapter.

East Asia's emergence as the world's primary manufacturing centre for
renewable energy products has been the cause of competitive tension in its
economic relations with the United States, European Union, and other key
trade partners. However, it was noted that deepening global business linkages across the value chains of industries such as wind energy and solar PV
make it increasingly difficult to establish the demarcation of international
competition between companies and between countries. For example, the
expansion of East Asia's productive capacity to make affordable solar PV
cells and modules has helped to create new business for upstream and downstream companies in other parts of the world. Nevertheless, international
disputes in RE industries have been on the rise since the late 2000s as a
number of governments have imposed various types of trade barriers.
Similar tensions have arisen over supply shortages and controls placed on
rare earths and other critical materials that are vital in the production of
renewables and other green energy products.

Finally, this chapter explored the multi-level governance and institutionalism of renewable energy development. At the national level, this concerned
firstly state or public governance relating primarily to government agencies
and policy, and secondly the collective business governance through the
formation of business associations that champion the interests of their renewable energy sector(s). The governance of renewables globally has advanced
slowly but steadily over recent decades, with contributions made from various existing and relatively new institutional elements of the global governance structure, such as the Clean Development Mechanism of the UNFCCC
Kyoto Protocol. It has only been since the early 2010s that a fully functional
global institution for specifically promoting renewables, IRENA, has been
established; it also receives support from other important international
organisations, such as the IEA and the GWEC. The formation of REN21
and WCRE has also established firmer institutional foundations on which to
advance cooperation on renewables in the future. The prospects for success
will be further enhanced if renewable energy issues are more substantially
incorporated into global climate change diplomacy, such as the COP and
other UN-led processes.

Regional-level governance and institutionalism on renewables has meanwhile remained stunted in East Asia for two main reasons. Firstly, parties

from the region are being drawn instead into international arrangements at the global level. Secondly, East Asian governments are primarily interested in their own national-level policies and strategies governing renewable energy development, and they seek to gain competitive advantages over each other. In Chapter 2, we introduce the concept of 'new developmentalism' to the recent expansion of renewable energy in East Asia.

Notes

1 Author's own calculations based on various sources (e.g. REN21, IRENA).
2 The next ranked countries in 2013 were the United States in second place (172 GW), Germany in third place (84 GW), India in fourth place (71 GW), and Italy and Spain jointly at sixth place (49 GW). Even with hydropower excluded, China would still be ranked first globally with 118 GW, with the United States in second place with 93 GW (REN21 2014).
3 However, as we later discuss, the much higher operating efficiency levels of nuclear power plants compared to wind and solar (3–4 times greater) means that nuclear still produced more power generation that year globally.
4 In the early 2000s, renewables enjoyed only an estimated fifth to a quarter of all such total investments worldwide (REN21 2005, 2014).
5 From 2008 to 2012, wind turbine prices fell by up to 20–25 percent in Europe and the United States, and by more 35 percent in China (GWEC 2013).
6 *Green Tech Media News*, January 21, 2012 (http://www.greentechmedia.com, accessed May 29, 2014). *Global Times*, July 17, 2013 (http://www.globaltimes.cn/index.html, accessed May 29, 2014).
7 *Washington Post*, August 13, 2013 (http://www.washingtonpost.com, accessed May 30, 2014).
8 *Wall Street Journal*, May 11, 2011 (http://online.wsj.com/Europe, accessed May 30, 2014).
9 Data available from Duty Calculator (http://www.dutycalculator.com).
10 *Bridges Weekly Trade News Digest*, Vol. 15, No. 21, June 8, 2011 (http://www.ictsd.org/bridges-news/bridges/news, accessed May 29, 2014).
11 *Bridges Weekly Trade News Digest*, Vol. 15, No. 27, June 20, 2011 (http://www.ictsd.org/bridges-news/bridges/news, accessed May 29, 2014).
12 *Washington Post*, August 13, 2013 (http://www.washingtonpost.com, accessed May 30, 2014).
13 *Reuters*, August 27, 2013 (http://uk.reuters.com, accessed May 30, 2014).
14 The 17 rare earth elements are cerium, dysprosium, erbium, europium, gadolinium, holmium, lanthanum, lutetium, neodymium, praseodymium, promethium, samarium, scandium, terbium, thulium, ytterbium, and yttrium.
15 These use permanent magnets instead of traditional gears, and approximately 2 tonnes of neodymium are required in the production of each turbine.
16 *BBC News*, June 20, 2012 (http://www.bbc.co.uk, accessed May 30, 2014).
17 *China Daily News*, July 15, 2011. *Bridges Weekly Trade News Digest*, Vol. 15, No. 27, June 20, 2011 (http://www.ictsd.org/bridges-news/bridges/news, accessed May 29, 2014).
18 The European Union is also a member, making in effect 43 parties to Annex I of the Protocol.
19 Cited from the UNFCCC website: http://unfccc.int/kyoto_protocol/mechanisms/emissions_trading/items/2731.php.
20 Cambodia and East Timor were signatories to the IRENA treaty and in accession talks at this time. Taiwan has observer status.

21 The ASEAN (2011) plan document is not clear about whether renewables includes large-hydro, which may explain the target figure discrepancies.
22 The East Asia Summit is a leader-driven regional framework of cooperation, whose members include the East Asia regional group as defined by this book (minus Taiwan) plus India, New Zealand, Australia, and more recently Russia and the United States.

2 Renewable energy and East Asia's new developmentalism

2.1. Introduction

This chapter introduces the concept of new developmentalism to explain the broader development context in which the recent expansion of renewable energy (RE) in East Asia has occurred. An analytical framework of new developmentalism is presented to synthesise theories of state capacity and ecological modernisation, which links renewable energy to wider ambitious endeavours to achieve low carbon development in the region. The new developmentalism approach reminds us that renewable energy development is essentially embedded in important broader processes, whether part of advancing the larger green energy technology cluster to which it belongs or more generally of efforts to decarbonise economic activity.

2.2. New developmentalism: defining the concept

New developmentalism can be defined as revitalised and refocused forms of state capacity aimed at realising the transformative economic objectives associated with low carbon development (Dent 2012).[1] This especially applies to East Asia, the region with arguably the strongest state capacity tradition that has historically hosted the largest number of successful developmental states (e.g. Japan, South Korea, Taiwan, Singapore, Malaysia) and socialist market economies (e.g. China, Vietnam). In all cases, this relates to the critical role played by the state in guiding the path of economic development; earlier practices of this may be considered as traditional or old developmentalism. The new developmentalism concept thus represents a revisionary update of state capacity thinking and practice in East Asia, and this book argues that the recent expansion of renewable energy in the region can is best understood in this political economy context.

Empirically speaking, East Asian state strategies on RE development are one of three main policy domains of the region's new developmentalism, the other two being climate change strategies and strategic macro-plans[2] on low-carbon development. Table 2.1 shows that many East Asian states have introduced all three elements from the mid-2000s onwards, and a good

Table 2.1 East Asia's new developmentalist plans and carbon mitigation commitments

	Renewable Energy Plan	National Climate Change Action Plan	Macro-Plan on Low Carbon (Green Growth) Development	Kyoto Protocol of UNFCCC				Carbon Emission Reductions Targets
				Ratified	In force	Status	Targets	
China	Medium-to-Long-Term Development Plan for Renewable Energy to 2020	National Climate Change Programme (2007)	12th Five-Year Plan 2011–2015 (elements of)	August 2002	February 2005	Non-Annex member		Energy consumption per GDP unit: –16% (2011–2015). CO$_2$ emission per GDP unit: –17% (2011–2015), –40 to –45% (2011–2020)
Japan	Innovative Strategy for Energy and Environment (2012)		New Growth Strategy (2010)	June 2002	February 2005	Annex I, II	[–6%: 2008–2012]. 2013–2020 not set	3.8% below 2005 emission levels by 2020
Mongolia				December 1999	February 2005	Non-Annex member		
South Korea	New Renewable Energy Medium-Term Plan 2010–2015	[integrated into Green Growth Strategy (2009)]	Green Growth Strategy (2009)	November 2002	February 2005	Non-Annex member		30% reduction compared to 'business as usual' scenario by 2020.
North Korea				April 2005	July 2005	Non-Annex member		
Taiwan	New Energy Policy to 2030	Adaptation Strategy to Climate Change (2012)	Green Energy Industry Sunrise Plan (2009)			Non-signatory		Return to 2005 emission level (2020, 45% business as usual reduction), return to 2000 level (2025)

Northeast Asia

	Country							
Southeast Asia	**Brunei**							
	Cambodia		National Adaptation Programme of Action to Climate Change (2007)	National Strategic Plan on Green Development 2013–2030	August 2009 / August 2002	November 2009 / February 2005	Non-Annex member / Non-Annex member	
	Indonesia	National Energy Blueprint 2005–2025, National Biofuel Roadmap 2006–2025	National Action Plan Addressing Climate Change (2007)	Mid-Term Development Plan 2010–2014	December 2004	March 2005	Non-Annex member	26% from business-as-usual level by 2020 based on domestic funding, or 41% by 2020 based on international funding (US$18.2 billion)
	Laos			7th Five-Year National Socio-economic Development Plan 2011–2015	February 2003	February 2005	Non-Annex member	
	Malaysia	National Renewable Energy Policy and Action Plan, 2011–2015	National Policy on Climate Change (2009)	Green Technology Strategy (2009), 10th Malaysia Plan 2011–2015	September 2002	February 2005	Non-Annex member	
	Myanmar				August 2003	February 2005	Non-Annex member	
	Philippines	National Renewable Energy Programme 2011–2030	National Framework Strategy on Climate Change (2007)		November 2003	February 2005	Non-Annex member	

(continued)

Table 2.1 East Asia's new developmentalist plans and carbon mitigation commitments *(continued)*

| | Renewable Energy Plan | National Climate Change Action Plan | Macro-Plan on Low Carbon (Green Growth) Development | Kyoto Protocol of UNFCCC | | | | Carbon Emission Reduction Targets |
				Ratified	In force	Status	Targets	
Singapore		National Climate Change Strategy (2008)	Sustainable Singapore Blueprint (2007)	April 2006	July 2006	Non-Annex member		16% from projected 2020 business-as-usual level
Thailand	Alternative Energy Development Plan 2012–2021	National Strategy on Climate Change (2008)	11th National Economic and Social Development Plan 2012–2016	August 2002	February 2005	Non-Annex member		30% from 2009 level by 2020, 42 million tonnes CO_2 equivalent in 2020
Vietnam	Renewable Energy Master Plan 2011–2015	National Climate Change Strategy (2011)	Vietnam Green Growth Strategy (2012)	September 2002	February 2005	Non-Annex member		

Southeast Asia

Sources: Asian Development Bank database on national climate change action plans (http://www.aric.adb.org/climate-change.php), UNFCCC website on Kyoto Protocol (http://unfccc.int/essential_background/kyoto_protocol/items/6034.php), authors' research.

majority introduced at least two elements. There are close inter-linkages among them. It is normally the case that a strong explicit connection between how renewable energy strategies contribute to meeting national climate change strategy targets on carbon emission reduction targets where these are set. Renewable energy strategies also tend to be embedded within strategic macro-plans on low carbon development.

An analytical framework of new developmentalism is presented later, which synthesises theories of state capacity with those of ecological modernisation; the latter have been highly influential in shaping environment-related government policies and business activity since the early 1990s. Briefly defined, ecological modernisation prescribes the reform, modification, and improvement of existing economic, business, and social structures to realise environmental objectives through the application of new technologies, policies, and practices. It may be considered an evolved form of economic modernisation that seeks to reconcile the conventional aims of economic growth with sustainable development and essentially prescribes gradual, adaptive change rather than radical or revolutionary change to economic and societal structures. The scaling-up and growth of renewable energy industries in East Asia is hence compatible with ecological modernisation thinking. This chapter shows how ideas on state capacity and ecological modernisation share common theoretical and empirical ground regarding East Asia's efforts on RE development, set within the context of the region's broader plans on low carbon development.

2.3. State capacity theory and practice

Various ideas have been offered to explain the profound transformation of East Asia during the latter half of the 20th century into one of the world's three dominant economic regions, the other two being Europe and North America. However, state capacity theories—which centre on the crucial role played by the state in shaping the paths of economic development—are generally recognised as providing the most complete and convincing explanations. Of these, developmental state theory has gained most traction, having its roots in both late industrialisation theory and the German Historic School (Gerschenkron 1962, Polyani 1944, Weber 1947), being first directly applied to Japan (Johnson 1982) and thereafter to other East Asian economies, especially South Korea, Taiwan, and Singapore (Amsden 1989, Evans 1995, Gold 1986, Low 2001, Pereira 2008, Rodan 1989, Woo-Cumings 1999, World Bank 1993).

Developmental states are fundamentally capitalist economies with three general attributes:

- *State institutions* with the capacity to strategise and plan a transformative economic development path, with institutional and planning coordination led by a pilot agency, such as Japan's Ministry of International

Trade and Industry, South Korea's Economic Planning Board, and Singapore's Economic Development Board.

- *Strategic policies* to operationalise these plans and realise their transformative economic goals, which change over time in accordance with developmental progress. Multiple-year economic plans have been adopted across the whole East Asia region.
- *Developmental partnerships* forged by the state primarily with the business sector but also with society, whereby non-state actors are co-opted into transformative economic projects through various institutionalised consultative mechanisms.

Meanwhile, the socialist market concept has been applied to China, Vietnam, and other countries where the state also has strategic economic planning institutions and policies but where the state plays a substantial interventionist role in the economy through more direct control of markets and the means of production, where state-owned enterprises are highly active (Suliman 1998). At the same time, these states still harness the power of private enterprise to help realise strategic economic objectives. The state's capacity to bring about prosperity-generating transformative change is a core issue that links both concepts, and the notion of state capacity has been deployed more commonly to examine the effective role played by governments in economic management (Beeson and Pham 2012; Weiss 1998, 2003). Conversely, predatory statism concerns situations arising in East Asia and elsewhere, where structural and chronic abuses of state official power for wholly personal gain have caused persistently poor economic governance outcomes (Evans 1995).

Strategic industry policy is a key aspect of state capacity practice. It principally concerns promoting industries deemed essential to the nation's long-term economic security, welfare, and prosperity. Chapter 1 examined how renewable energy sectors are generally viewed in East Asia as important and still largely emerging strategic industries. Over time, strategic industry policies have been a core component of the macro-development plans of East Asian states; by international comparison, these plans have been largely successful at meeting their transformative economic objectives. Japan was the first in the region to practice strategic industry policy, dating back to the late 19th century and based on then formative developmental state practices (Johnson 1982). These days, Japan's Ministry of Economy, Trade and Industry (2010a: 2) defines the policy's main purpose as 'the act of policy intervention to cope with market failures in resource distribution under the pricing system,' as well as to 'use various measures to influence resource distributions among industries and to control, restrain and promote certain economic activities of private sector companies.' It can involve a number of policy instruments, including direct financial support for research and development, tax incentives, state coordination of complementary and competing investments, regulation of technology imports and foreign direct investments, leveraging private venture capital to help incubate new technologies, and exporting promotion measures (Chang 2010).

Theoretically, the rationale for state support of strategic industry development is mainly premised on public good and externality arguments. Although the market mechanism can account for the private and purely price-determined costs and benefits arising from economic transactions, it can fail to capture their external social costs and benefits (therein negative and positive externalities, respectively) that affect society. Public actions undertaken by the state are thus required in such circumstances to minimise negative externalities and optimise positive externality outcomes. For example, state support for developing new renewable energy installations and technologies will reduce carbon emissions, lead to cleaner air, and mitigate society's energy supply risk dependency on exhaustible fossil fuel resources. These state actions constitute the provision of public goods where the market (i.e. private enterprise) alone is unable to independently and sufficiently deliver in these areas, and more broadly place the economy and society on track to a lower carbon future. Left purely to the market, many RE sectors would not become technologically developed enough or adequately commercialised to compete on price with fossil fuel energies. Drawing closely on public good and externality concepts, strategic trade theory provides a more specific intellectual basis for strategic industry policy. This theory argues that the state should significantly contribute toward the proportionately high initial capital costs (e.g. infrastructure, new technology research) arising at the early development stage of emerging strategic industries, with significant futures based on the expectation of substantial long-term economic and public welfare returns on this state investment (Lall 2003; Rodrik 2004, 2007; Schmitz 2007).

Following related discussions in Chapter 1, strategic industry policy has become less concerned with protecting domestic firms from foreign competition in a globalising world economy and more on developing locally embedded, high value-added activity with particular emphasis on industry or technology cluster formation, where clustering may be spatially concentrated in certain geographic zones. Multinational enterprises tend to anchor their core activities in such areas, which can be considered locational centres of competitive advantage due to infrastructural, human capital, value-chain capacity, and other kinds of assets situated there (Dicken 2011). Clustering may also concern closely related and mutually reinforcing industry- and technology-based activities and competences (often also arising in particular geographic zones), and it is common for states to facilitate both aspects of cluster development. This is relevant to a later debate on long-wave techno-economic development theory and renewable energy's core role in a green-energy technology cluster, which is expected to become an important foundation of the future global economy.

From the late 1980s onwards, globalisation and ascendant neo-liberalism created critical pressures and challenging conditions for both developmental state and strategic industry policy practice in East Asia. The combination of globalising forces, technological developments, and strengthening neo-liberal orthodoxy made national economic spaces more permeable through the growth of transnational systemic linkages (e.g. production networks and supply chains) and more liberalised and integrated national markets. This presented certain challenges for governments when defining and controlling national development projects at both macro-plan (i.e. whole economy) and industry-sector levels. Moreover, the stronger adoption of neo-liberal-oriented policies in East Asia led to less proactive state economic manage-ment and a corresponding greater faith placed in purely market mechanisms (Radice 2008). The concurrent rise in the influence of business and civil society over economic policy formation also compelled the region's governments to adapt and rebalance their developmental partnerships accordingly (Holden and Derneritt 2008, Lee and Park 2009, Pirie 2008, Weiss 2010).

As the 2000s progressed, however, the increasingly evident risks associ-ated with globalisation and neo-liberalism—culminating in the 2008–2009 global financial crisis, its prime cause being generally attributed to misman-aged deregulation—combined with the emergent global challenges of climate change and deepening energy security predicaments worldwide. This resulted in reinvigorations of state capacity in East Asia and elsewhere, or at least refocused state capacity in response to these risks (Dittmer 2007, Gills 2010, Holden and Derneritt 2008, Karagiannis and Madjd-Sadjadi 2007, Weiss 2010). Emphasis was placed on the stronger institutionalisation of market order in the Weberian and Polyanian sense, where states seek to make markets work better rather than supplant them (Khan and Christiansen 2010, Mok and Yep 2008). This was, in many senses, a market-oriented developmental partnership between government and business, but addi-tionally where the state was engaged in developing new or emerging markets and industries through the implementation of strategic plans and their associated policy instruments. Both the substance of these plans and the proportionate state funding to support them were generally greater in East Asia than in any other region. The chapter's later debate on the 'green growth' strategies that emerged after the 2008–2009 financial crisis provides empirical evidence of this; it is also noted how this created an important impetus across many East Asian states for renewable energy development.

More generally, we later examine how the post-crisis green growth response brought state capacity and ecological modernisation thinking closer together than ever, creating a firmer underlying basis of East Asia's new developmentalism and a new evolutionary phase of the region's strate-gic industry policy (Hayashi 2010, Lim 2010, Stubbs 2009). Given that the socialist market states of China and Vietnam have continued to devise their development plans more or less as usual from a governance perspective, this change especially applies to the region's developmental states, where new state institutions have been formed and new strategic economic plans devised to pursue new transformative goals focused on addressing the closely inter-related challenges of climate change, low carbon development, and energy security. Renewable energy has been a core element of these plans in both East Asia's developmental states and socialist market economies. In sum, from a state capacity theory (SCT) perspective, East Asia's new developmentalism is based on adapted technical and relational methods, as well as refocused (and in many cases, reinvigorated) state capacity set on achieving low carbon development goals and other new transformative objectives.

2.4. Ecological modernisation theory and practice

2.4.1. East Asia's economic development and modernisation

As an important pretext to this chapter's discussion on ecological modernisation and renewable energy, it is first necessary to explore how East Asia's contemporary economic development can be understood from the viewpoint of modernisation generally. Renewables are often referred to as modern technologies and modern industries; here, *modern* implies that they are at the current or new frontiers of economic development. This raises the question of whether the expansion of RE systems may be considered as part of a long-term modernisation process or alternatively helping establish a new kind of post-modern development paradigm or philosophy in the early 21st century. This point is important when examining the relationship between renewable energy and ecological modernisation theory. Before we turn to that, let us first consider modernisation's relevance to shaping the path of East Asia's contemporary economic development.

The ideas and principles of modernisation are inherently humanist or anthropological in orientation, and mainly derive from European Enlightenment thinking.[3] In essence, modernisation proposes that the 'progress' of humankind is achieved through scientific and technological improvements, leading to the 'advancement' of society. Economic modernisation more specifically entails the application of new science and technology in subjugating nature's power and resources to the material and other economic needs of humanity. As Alexander (1990: 16) commented, 'To be modern is to believe that the masterful transformation of the world is possible, indeed that it is likely.' Modern economic development could thus be reduced to the simple equation of harnessing natural resources, human endeavours (labour, enterprise, science, technology), and capital to deliver measurable higher material standards of living (Arat 1988, Bernstein 1971, Hoselitz 1952, Tominaga 1991). Theories on the progressive economic modernisation of society were first popularised by Rostow (1960), whereby 'traditional societies' were raised out of their low income cycle and transformed into prosperous modern societies through linear-path stages of development, primarily aligned with capitalist forms of economic organisation and based on Western social norms and philosophies, such as the primacy of individual liberty and rights.

East Asia's own economic modernisation has involved exploiting the region's substantial human and natural resources, combined with considerable levels of entrepreneurial energy allied, in many instances, to foreign capital to deliver sustained high rates of economic growth over long periods, depending on the country in question. Taking China as an example, at the outset of the economic reforms in the late 1970s, Deng Xiaoping revived Zhou Enlai's idea of the 'four modernisations' (agriculture, industry, science and technology, and national defence), as first articulated in 1963. In the 1990s, the four modernisations were revised to economy, society, politics, and culture, thus including non-material domains. Then, at the National People's Congress in 2001, Premier Zhu Rongji announced that the government would continue its mission—as envisaged by Deng decades earlier—to make China a fully modernised country by the mid-21st century. Both national and local governments in China have devised various quantifiable criteria for charting progress toward this vision of national modernity (Zhang et al. 2007).

The emphasis of quantifiable, material-based economic advancement—such as gross domestic product (GDP) growth and income per capita—has been the underlying objectives of most East Asian nations for decades, with these being seen as the ultimate indicators of their modernisation, and moreover which shaped both the aims and design of past government strategic development plans and other forms of state capacity practice. Perpetual techno-industrial upgrading of the economy is still seen by many in the region as the prime route to delivering national prosperity and societal welfare enhancement, with strategic industry policy being highly relevant here. However, for some time now, East Asia's most developed economies (Japan, South Korea, Taiwan, and Singapore) have afforded greater priority to *qualitative* economic, social, and environment improvements. Evidence also indicates that the region's developing economies are gradually moving toward this position. Most significantly, as we later discuss in more detail, China has looked to modify its socialist market model and thinking on development per se, primarily because the country's 'modernisation' has created critical social, health, economic, and environmental problems (Li et al. 2012, National Development and Reform Commission 2006, 2007b, 2011). The nation's burgeoning middle class have correspondingly become more vocal about quality-of-life issues, such as pollution abatement. This is an emerging regionwide trend; however, in most East Asian countries, civil society still lacks significant political power to contest carbon-intensive development policies. Further discussion on these issues is continued in the debate on ecological modernisation and its application to East Asia's economic development and renewable energy.

2.4.2. Ecological modernisation theory

2.4.2.1. Introduction

Theories and thinking on ecological modernisation date back to the early 1980s. By this time, there were mounting concerns over the past cumulative and future expected impacts of modern industrial activities on the environment, leading to calls to move as quickly as possible toward sustainable development policies and practices. The 1987 Brundtland Commission report on the global environment and development laid down a foundational definition of sustainable development as 'development that meets the needs of the present without compromising the ability of future generations to meet their own needs' (43). This built on the work of earlier studies on the mounting environmental problems caused by human economic activity from the local to global scales. For example, Meadows et al. (1972) argued the case for scaling back material-intensive economic development given the planet's finite natural resources.

Sustainable development became a mainstream policy discourse in developed countries, especially from the late 1980s onwards; it was the core theme of discussion at the first Earth Summit held in Rio de Janeiro in 1992 (see Chapter 1). There was, however, a wide spectrum of views on how sustainable development was to be achieved as a long-term goal, and therefore what policies and practices should be accordingly employed (Brand 2010, Langhelle 2000, Redclift 2005). At the more radical end were those calling for a deep paradigm shift in human economic activity that envisaged an end to the modernist fixation with material-intensive economic growth and progress, instead prescribing revolutionary change in economic, social, and business structures, with some even advocating a return to certain sustainability practices of pre-modern society (Garcia 2012, Heinberg 2004, Jackson 2009).

2.4.2.2. The ascendance of ecological modernisation

What prevailed, however, was a different line of thought, which argued that current structures could deliver sustainable development through continual reform, modification, and adaptation of the existing order. This became known as ecological modernisation theory (EMT), the origins of which date back to the works of Huber (1982, 1985), Jänicke (1984, 1988), Weale (1992), Hajer (1995), Mol (1996), Christoff (1996), and Dryzek (1997). It is essentially an evolved form of modernisation that maintains an inherent faith in science and technology to deal with the twin challenges of climate change and energy security. As York and Rosa (2003: 274) succinctly summarised:

> Central to EMT is the view that the era of late modernity offers promise that industrialization, technological development, economic growth, and capitalism are not only potentially compatible with ecological

sustainability but also may be key drivers of environmental reform... It argues for the potential of attaining sustainability from within–a greening of business as usual–thereby avoiding such challenging alternatives as radical structural or value changes in society.

From this perspective, renewable energy is viewed as providing important technical solutions to the key problems facing modern human society.

Most ecological modernists ascribe importance to processes rather than structures–more specifically, to processes that circumvent structural impediments (Hajer 1995, Warner 2010). Its axiomatic belief in incremental, perpetual reform of existing systems links it closely with ideas on reflexive modernisation and late modernity (Beck 1992, Beck et al. 1994). These associated theories contrast with the rather amorphous and contested concept of post-modernity, which has been used generally to explain the condition of society *after* modernity. Thinking on post-modernity has been most commonly deployed as a critique on modernisation, highlighting the dangers of modern development 'progress' that fails to take into account the ecological contexts in which it occurs, and thereby longer-term sustainability constraints. A core idea of post-modernity is that there is no one supreme scientific method that can or should be applied universally, either at a singular point in time or over time; moreover, structures and methods need to change as reality changes. Contemporary environmentalism is often viewed in a post-modernist context (Soule and Lease 1995, White 1997), and we may associate postmodernity with the aforementioned more radical views on sustainable development. While there is still little consensus as to what a post-modern society should look like, post-modern approaches to economic development tend to prescribe a more holistic understanding of humankind's relationship with nature, which takes more into account factors of social and cultural diversity and eschews what it sees as ecological modernisation's somewhat reductionist emphasis on scientific and technological determinism. More generally, EMT's technical and technocratic approach to sustainable development has been criticised for lacking a robust socio-cultural dimension (Buttel 2000, Hajer 1995, Murphy and Gouldson 2000).

Despite EMT's apparent weaknesses, it remains the dominant discourse among government and business decision-makers on environmental sustainability issues. An important reason for this is the appealing economic case EMT makes on how companies can increase profitability by pursuing green corporate strategies, for instance, by improving material efficiency and waste management and exploiting the emerging market potential of environmentally friendly products, such as renewables (Christoff 1996, Jänicke 1988). Ecological modernisation is thus not viewed as a threat to existing business structures and interests, but rather as a commercial opportunity. It hence postulates that economic and business growth is reconcilable with resolving environmental problems, and firms can generate financial gains by adopting new environmental technologies and practices (Buttel 2000,

Dryzek 1997, Langhelle 2000, Weale 1992). Unlike later evolved understandings of sustainable development—which took on a stronger social dimension by tacking issues of global economic inequity between rich developed and poor developing nations—ecological modernisation has remained largely preoccupied with economic, business, and technologically related challenges.

Huber (2000: 269) neatly summarised the traditional contrast between the revolutionary 'deep green' view generally championed by civil society on sustainable development and the 'light green' ecological modernisation approach adopted by business, which is still relevant today:

> [Non-governmental organisations'] understanding of sustainable development has been formulated by themselves as an anti-industrial and anti-modernist strategy of 'sufficiency,' meaning self-limitation of material needs, combined with 'industrial disarmament,' withdrawal from the free world-market economy, and an egalitarian distribution of the remaining scarce resources. Contrary to that, the industry's understanding of sustainable development is the 'efficiency revolution.' Industry and business are looking for a strategy that would allow for further economic growth *and* ecological adaptation of industrial production at the same time. The means for achieving this goal is seen in the introduction of environmental management systems aimed at improving the environmental performance, i.e. improving the efficient use of material and energy, thus increasing resource productivity in addition to labour and capital productivity.

One of the most important technical sub-fields of ecological modernisation is the concept of industrial ecology. Learning from the processes and operations of natural ecosystems, industrial ecology studies the flow of materials and energy through industrial systems. It looks to create closed loops within those systems where any produced waste is recycled or fed back as new inputs by utilising so-called roundput techniques (Huber 2000, Jensen et al. 2011, Korhonen 2004). It has relevance for certain renewable energy technologies, especially material-based forms such as bioenergy and hydropower (Ayres and Ayres 1996), and its principles have been more generally applied in eco-industrial park or zone design (McManus and Gibbs 2008). Industrial ecology partly originated from research conducted by the Japan Industrial Policy Research Institute and the government's Ecology Research Group from the early 1970s, exploring how to reduce the Japanese economy's dependence on natural resources. The work of Chihiro Watanabe (1972, 1993, 1995; see also Chapter 3) was especially seminal at this time, as well as decades after, in the study of Japan's RE development in this context. Other scholars have also applied industrial ecology principles to modelling clean development in East Asia where renewables plays a role (Chiu and Yong 2004, Rock et al. 2000).

Ecological modernisation's main protagonists have long stressed its applications to policy, and more generally contended that the state and other institutions have a crucial role to play in achieving ecological modernisation aims (Andersen and Massa 2000, Buttel 2000, Gibbs 2000, Huber 1982, Jänicke 1985, Mol and Spaargaren 1993, Murphy and Gouldson 2000, Seippel 2000). As we later explore, this constitutes important common ground with state capacity theory. Ecological modernisation thinking began to significantly influence environmental policy-making in developed countries from the late 1980s onwards, especially in Europe. The landmark from the Commission of the European Communities (1993), *White Paper on Growth, Competitiveness and Employment*, was among the first major policy documents to use EMT as its underlying analytical approach. Other international organisations, such as the World Bank, soon followed suit, although ecological modernisation has been comparatively slow to penetrate policy thinking in the United States (Massa and Andersen 2000, Schlosberg and Rinfret 2008). This, it could be argued, is indicative of why it has been Europe and not the United States that has taken the Western lead on renewables, other green energy sectors, and climate change.

Over time, EMT thinking has spread from advanced industrial countries to developing economy regions, such as East Asia. Of connected relevance here is the idea of the environmental Kuznets curve that hypothesises the relationship between economic development and the environment will change over time, depending on which stage of development a country or economy has reached. Having resonance for East Asia's own economic development story, the theory contends that environmental quality will fall at first during the initial development phase of mass material utilisation, increased industrial output, but low attention paid to economic efficiency and emissions control. However, environmental quality will increase at higher development stages when efficiency and technological capacities improve, incomes rise, and socio-economic priorities change. The evidence for this is mixed. For example, South Korea has become far more energy efficient over recent decades of rapid economic modernisation, but its CO_2 emission levels in absolute total and per capita terms have continually increased (see Table 1.1, p. 4). If, in fact, the advanced industrial economies have failed to 'dematerialise' their development, reduce waste levels, and scale up clean energy generation since adopting ecological modernisation practices and policies, then they obviously set a poor example to developing countries (Rock et al. 2000). Furthermore, the pursuit of greening industry in developed countries has, to a notable degree, involved simply displacing carbon-intensive business activities overseas to lower-cost developing nations, especially in East Asia's competitive production hubs (Redclift 2005). This has been clearly evident in China's case, where approximately 60–70 percent of the nation's exports originate from foreign invested enterprises. From this perspective, one could make the case that China and East

Asia's high carbon emission levels are as much a product of global capitalism as national economic development.

Chinese views on ecological modernisation are largely representative of, as well as crucially important to, how developing East Asia has approached its ideas and principles. Like many other emerging economies from the region, it began to examine the applicability of EMT to its national context in the early 2000s, and subsequently incorporated it into policy discourses and planning thereafter (Huan 2007, Lai et al. 2012, Li and Lang 2010, Yee et al. 2013, Zhang et al. 2007). The scientific development concept, first formally proclaimed by then President Hu Jintao around this time, became the official ideological basis for China's future economic and social development when it was ratified into the national constitution in October 2007. Guided by this concept, China was to foster a more resource-saving and environmentally friendly 'harmonious society,' and '[develop] the economy in a more balanced manner, paying less attention to GDP growth per se and more attention to such things as the ecological costs of the headlong rush for development' (Fewsmith 2008: 88). In the same year, the Chinese Academy of Sciences published the *China Modernization Report 2007: Study on Ecological Modernization,* when the government also launched its Medium- and Long-Term Development Plan for Renewable Energy as well as its National Climate Change Strategy, where the EMT thinking was clearly evident in each case (National Development and Reform Commission 2007a, 2007b).

The scientific development concept was more broadly and systematically applied to China's five-year plans (FYPs) and associated strategic development programmes, such as on energy. The 11th FYP (2006–2010) accordingly reflected a shift toward more qualitative development objectives, and this relative change in thinking was evidently stronger in the design of the 12th FYP (2011–2015). Ecological modernisation has had a clear influence on these FYPs and related development policies, with renewable energy forming an integral part of the state's recent low(er) carbon, 'clean industry' strategies. Yet, as in other developing East Asia, EMT's influence on development has manifested itself in a rather top-down and technocratic manner, primarily through state policies and strategies. Most elements of East Asia society have yet to be strongly imbued with ideas of ecological modernisation. In China's case, although certain elements of the domestic business community are aligned with EMT thinking, Chinese society generally is only just beginning to assimilate EMT's ideas and afford more serious priority to environmental goals (Huan 2007, Yee et al. 2013).

2.4.2.3. Application to renewable energy policy

As noted in Chapter 1 and examined in further detail in Chapter 3, modern renewable energy policy (excluding hydropower) and RE sector development generally originated in the early 1970s; a major catalyst was the impact of the 1973–1974 oil crisis on global energy security and emerging concerns over industrialism's long-term adverse environmental effects. Efforts thereafter to develop renewable energy technologies could be viewed as an ecological modernisation process that pre-dated EMT thinking. The ensuing slow pace of renewables development up to the end of the 20th century is certainly consistent with ecological modernisation's emphasis on a gradual process rather than dynamic structural change—on steady reform rather than systemic revolution. Even after a period of rapid growth in the 2000s decade, renewables still only make a relatively minor contribution to the energy mix of most countries.

Despite this, there has been growing speculation on whether we are on the verge of a new green or eco-industry paradigm shift, where renewables will play a major role. This connects closely with Kondratiev's (1935) long-wave theory on the relationship between techno-innovation and economic development. According to Kondratiev and other long-wave theorists, such as Perez (2002, 2010) and Warner (2010), world economic development has been driven forward in spurts based on the emergence of certain technology clusters, which arise every few decades and lead to the creation of new techno-economic paradigms that have transformative effects across industry and society.[4] For example, according to this theory, the Industrial Revolution— the foundational stone of economic modernisation—was kickstarted in the late 18th and early 19th centuries by water mill and steam power technologies, as well as innovations in the textile, metal, railway, and other civil engineering sectors. A new techno-economic paradigm arose in the late 19th century based on electrical power and chemical industries. Another paradigm shift occurred in the early 20th century, founded on mass production techniques and petrochemicals, and then later in the late 20th century by the information technology revolution. Notwithstanding the main criticism of Kondratiev's long-wave theory being rather over-generalised, there is nevertheless an observed historic regularity of normally around 50–60 years of new techno-economic paradigms arising. Green energy, with renewables at its core, is certainly among the strongest contenders as one of the new technology clusters on which the next 'long wave' of world development in the early 21st century could be based, with the others being biotechnology and biomedicine, artificial intelligence systems, new material science, and advanced aerospace. There tend to be strong links among different technology clusters within the same long-wave paradigm. By way of illustration, Chapter 1 noted wind energy's techno-innovatory connections with aerospace and new material sciences. The connections between technology cluster development and strategic industry policy were also noted earlier in the chapter.

In addition, Schumpeter's (1911, 1939) work on economic development and business cycle dynamics put new technology and techno-entrepreneurship at the heart of paradigmatic economic change. By applying his concept of creative destruction, green energy technologies such as renewables would completely displace fossil fuel and other pollutive energy types, establishing a new energy system. Again, though, the question is how long this structural change would take. For most countries, the phasing in of a total change in energy system or paradigm based on near or absolute complete dependence on renewables would normally take decades to achieve for technical, financial, and other reasons. Furthermore, like Kondratiev, Schumpeter believed that paradigmatic change first required a critical mass of new technology clusters to form and become mainstream (Bowen and Fankhauser 2011, Perez 2010, Schumpeter 1939). Ecological modernisation theory would suggest that these formations do not occur suddenly, but rather tend to slowly evolve; very few new and singular technology breakthroughs on their own transform the economy and society.

However, one strand of EMT thinking contends that constant incremental reforms and modifications will eventually amount to a long-term revolution retrospectively in human society's relationship with the environment (Mol 1995). This would be achieved, for example, by the gradual scaling up and expansion of RE technologies across the planet. However, a very long transition to a global clean energy system could prove to be too slow, leading to humanity ultimately failing to avoid the worst disastrous impacts of climate change.

Finally, in an EMT context, one could argue that the scaling up of RE sectors as hydropower, wind, and solar has been to not only deliver cleaner methods of power generation but also to meet the growing energy demands of existing economic structures, as well as the continuity of material lifestyles to which high-income societies have become accustomed and to which low income societies (in East Asia and elsewhere) aspire to achieve in the future. By expanding renewables only for this purpose, they therefore serve only to 'ecologically modernise' these incumbent structures of high per capita energy consumption and thus essentially maintain the structural status quo.

2.5. New developmentalism and renewable energy

2.5.1. The green growth impetus

To thus far summarise, exercises of state capacity were on the decline in East Asia at the same time (i.e. the late 1980s and early 1990s) that ecological modernisation theories and ideas were starting to shape environmental policy-making in the European Union. By the 2000s, EMT began to influence East Asian policy-making as well, contributing to revitalised state capacity on green economy development in the region. In the aftermath of the 2008–2009 global financial crisis, ecological modernisation and state capacity approaches to economic development further converged around the introduction of various 'green growth' initiatives and strategies in different parts of the world. This has been substantively evident in East Asia and has provided an important impetus to the region's new developmentalism, and renewable energy development more specifically.

Although originating ideas and debates on green growth pre-dated 2008, the global financial crisis brought them into the mainstream policy arena. According to Jacobs (2013), the new green growth discourse has succeeded for similar reasons that the discourse on ecological modernisation did, namely by attracting support from powerful economic interests. Indeed, green growth policies and strategies may be viewed as the latest phase of ecological modernisation as well as a major basis for reinvigorated state capacity in East Asia in particular, thus establishing stronger common empirical ground between SCT and EMT. After observing that there appears to an analytical consensus that green growth refers to 'environmentally sustainable, biodiverse, low carbon and climate-resilient growth in human prosperity,' Bowen and Fankhauser (2011: 1157) went on to claim a strong association with EMT: 'From a strategic point of view, green growth allows environmental protection to be cast as a question of opportunity and reward, rather than costly restraint' (ibid). Jacobs (2013) also summarised that this recent trend can be understood in the following inter-related terms:

- *Countercyclical fiscal stimulus*: Applying the principles of Keynesian theory by increasing public investments to compensate for significant contractions in private sector activity was a consequence of the 2008–2009 global financial crisis. Because this is countercyclical in nature, it is essentially a short-term solution to the problem. The fiscal package incorporated into South Korea's later-examined Green Growth Strategy, implemented from 2009 onwards and accounting for an estimated 79 percent of the country's total stimulus measures, was one of the largest in national GDP proportional terms (2 percent, or US$87 billion). Around a third of China's own fiscal boost measures (US$219 billion) were categorised as green. By comparison, figures were 60 percent for the European Union (US$23 billion) and 12 percent for the United States (US$118 billion; Barbier 2010a, 2010b; Robins et al. 2010).
- *Correcting market failure*: This is understood with regard to the above in terms of correcting macro-economic-scale market failure (i.e. dealing with a deep financial crisis). However, more importantly, the aforementioned positive and negative externality arguments for the rationale of state support for renewable energy are based on the idea that market forces fail to account for the environmental benefits that RE sectors bring to society. New green growth measures have been aimed at 'correcting' the supply and price issues in renewable energy markets (Bowen and Fankhauser 2011, World Bank 2011).
- *Strategic industry development*: This is connected closely to previously made arguments on the rationale for strategic industry policy, and in the context of recent green growth measures to foster RE industries deemed to be of future strategic importance. Also relevant here are the motivation

to develop a comparative advantage in emerging export-trade sectors and the long-wave theory on new technology clusters driving forward future economic development.

In sum, the above could be considered exercises of state capacity oriented by ecological modernisation thinking. Green growth has moreover now established itself as an international economic policy discourse and core element of climate change strategy and diplomacy, with these twin developments being consolidated by various incumbent and new multilateral institutions promoting their own green growth agendas and recommendations. This has augmented more general global-level governance efforts to promote renewable energy development, as discussed in Chapter 1. The United Nations Economic and Social Commission for Asia and the Pacific (2005, 2008, 2012a, 2012b) was one of the first international organisations to take ideas on green growth seriously, from 2005 onwards and thus pre-dating the 2008–2009 crisis. The Organisation of Economic Co-operation and Development (OECD 2009, 2010, 2011) was also a relatively early mover and active advocate, promulgating its Green Growth Declaration at the June 2009 OECD Ministerial Council and in 2012 implementing a green growth strategy of research and publications on the subject (OECD 2012). Other organisations followed suit. The World Bank and five other multilateral development banks, including the Asian Development Bank (2012b; World Bank 2011, 2012a, 2012b), produced key publications in the early 2010s, and the United Nations Environment Programme (2011) published its own report (entitled *Towards a Green Economy*) around the same time.

The Global Green Growth Institute, an international body established in 2010 to promote the green growth agenda worldwide, was a South Korean government initiative led by the Lee Myun-bak administration. Working together in unison, the World Bank, OECD, United Nations Environment Programme, and Global Green Growth Institute formed the Green Growth Knowledge Platform in 2012 to help coordinate international efforts in this area. On the elite networking front, the Global Green Growth Forum and Green Growth Leaders held their inaugural meetings in 2011, with again South Korea playing a key formative role. Green growth was also a priority theme of G20 summits held in 2011 and 2012, and the green economy was a major focus of the Rio+20 United Nations summit convened in June 2012. More specifically in relation to East Asia, in 2011 the Association of Academies of Sciences in Asia (2011a, 2011b) launched a new study programme entitled *Towards a Sustainable Asia*, which is based on green growth aims and acknowledges ecological modernisation arguments that green legislation and policies will have a strong positive effect on green innovation (Jacobs 2013, Jänicke 2012). It will be perhaps some years, though, before we can ascertain how strong a platform these green growth diplomacy initiatives will provide for further advancing new developmentalist thinking and practice in East Asia, and thereby give additional impetus to renewable energy development. One outcome may be a more substantive level of future regional cooperation on renewables, which has hitherto been rather limited (see Chapter 1).

Relatedly, Martinelli and Midttun (2012) contended that the emergence of the green growth discourse in international diplomacy has created new layers of multi-stakeholder governance in that the formation of governments' low carbon development policies has necessitated closer involvement of business, international organisations, and civil society. Thus, global efforts on climate change do not now just involve inter-governmental treaties but a wide range of state and non-state actors. This has implications for the relational aspects of state capacity, as well as how policies on renewables and other green economy sectors are formulated. A similar multi-stakeholder dynamic applies in an intra-government sense, both among various relevant national government agencies as well as among national, provincial, and city government levels. Growing international diplomacy amongst city and other local governments that effectively bypass national government (e.g. Cities for Climate Protection, the C-40 Large Cities Climate Leadership Group) represents another layer of stakeholder interaction on green economy issues. This is relevant to the general governance of East Asia's new developmentalism and renewable energy development generally, as later discussed in this and subsequent chapters.

2.5.2. An analytical framework of new developmentalism

In this section, we develop an analytical framework of new developmentalism that is founded on the synthesis of state capacity and ecological modernisation theories (Figure 2.1). Previous connections made between them have been extremely rare, such as an EMT work of Buttel (2000) that cites Evans' (1995) embedded autonomy concept of the state's collaborative (rather than arms-length, antagonistic) relations with strong societal actors in pursuing a common transformative project. With specific reference to fellow EMT thinkers Leroy and van Tatenhove (1999), Mol (1995) and others, Buttel (2000: 62) further observed that 'the core thinkers of ecological modernization share very similar ideas about state effectiveness and state-civil society ties.'

We have seen more broadly how SCT is principally concerned with the state playing an effective positive role in transforming the economy toward more advanced development ends. It was contended that both past exercises of state capacity as well as recent revitalised forms of it have been especially evident in East Asia compared to other regions. Ecological modernisation meanwhile highlights the importance of state policies and institutions in helping shape markets and paths of economic development toward more environmentally friendly outcomes, and EMT's influence on East Asian policy-makers has grown, as is evident in their ambitious renewable energy plans and strategies. Both SCT and EMT are thus essentially concerned with

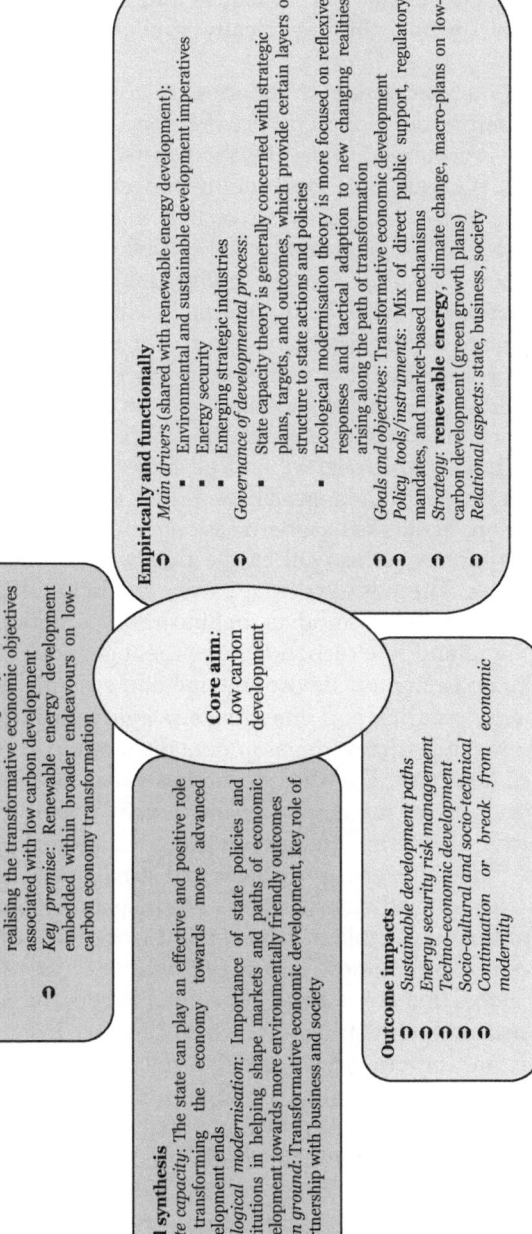

Conceptually

⚬ *New developmentalism defined:* Revitalised and refocused forms of state capacity aimed at realising the transformative economic objectives associated with low carbon development

⚬ *Key premise:* Renewable energy development embedded within broader endeavours on low-carbon economy transformation

Empirically and functionally

Main drivers (shared with renewable energy development):
• Environmental and sustainable development imperatives
• Energy security
• Emerging strategic industries

Governance of developmental process:
▪ State capacity theory is generally concerned with strategic plans, targets, and outcomes, which provide certain layers of structure to state actions and policies
▪ Ecological modernisation theory is more focused on reflexive responses and tactical adaption to new changing realities arising along the path of transformation

Goals and objectives: Transformative economic development
Policy tools/instruments: Mix of direct public support, regulatory mandates, and market-based mechanisms
Strategy: **renewable energy**, climate change, macro-plans on low-carbon development (green growth plans)
Relational aspects: state, business, society

Core aim:
Low carbon development

Theoretical synthesis

⚬ *State capacity:* The state can play an effective and positive role in transforming the economy towards more advanced development ends

⚬ *Ecological modernisation:* Importance of state policies and institutions in helping shape markets and paths of economic development towards more environmentally friendly outcomes

Core common ground: Transformative economic development, key role of the state, partnership with business and society

Outcome impacts

⚬ *Sustainable development paths*
⚬ *Energy security risk management*
⚬ *Techno-economic development*
⚬ *Socio-cultural and socio-technical*
⚬ *Continuation or break from economic modernity*

Figure 2.1 East Asia's new developmentalism: analytical framework.

transformative economic development, where the state has a key contributing role to play in partnership with business and society, and new developmentalism focuses on how this specifically applies to realising low carbon development.

It can be further argued that both theories are concerned with governance of this transformative development process, and the case for the state's engagement here is grounded in previously discussed public good and externality arguments. Whereas SCT is generally concerned with strategic plans, targets, and outcomes–which provide certain layers of structure to state actions and policies–EMT is more focused on reflexive responses and tactical adaption to new changing realities arising along the path of transformation. At this generalised level, these apparently contrasting 'structured' and 'evolutionary' approaches to state governance of low carbon development can be viewed as complementary to the other. For example, as later noted in other chapters, certain East Asian state strategies on renewables development have been significantly revised (e.g. in Thailand's case), targets have been amended (e.g. China's RE policies), and all have incorporated new policy mechanisms over time into their structures as part of the evolutionary policy process. In addition, ecological modernisation thinking has influenced both the nature of the targets set as well as the policy instruments deployed in these state strategies. There is also an apparent conflation of SCT and EMT ideas on the main drivers or motives behind the promotion of renewable energy development and low carbon development generally in East Asia (as outlined in Chapter 1), namely environmental and sustainable development imperatives, energy security, and emerging strategic industries.

The technical practice of new developmentalism outlined here has important relational dimensions. The shared emphasis that SCT and EMT place on the state–business partnership was previously noted. Following from the earlier analysis of international green growth diplomacy, good governance of national renewable energy policies and strategies depend on responsible state agencies and other institutions working in accordance with the principles of multiple stakeholder involvement from business, civil society, and other relevant communities. This is not least because of the significant diversity of the actors that are normally involved in renewable energy development processes and their complexity, which can vary from installing small off-grid systems in remote areas and consultations with utilising local communities to the construction of large utility-scale plants involving numerous firms engaged in different functional areas, such as installation operators, grid companies, equipment supply, and logistical services. In both examples, there will be additionally several levels of mediation between national and local government agencies, and often between policy-makers and the state-owned enterprises that can dominate renewables power generation in East Asia.

Notwithstanding the aforementioned criticism made of EMT especially (and also SCT, to a lesser extent) for being somewhat technocratic and technical, studies from both theoretical fields have highlighted the growing influential role played by civil society and social movements as developmental partners to the state. For instance, Toke's (2011a, 2011b) work on ecological modernisation and renewable energy emphasises how social movements were pivotal through various stages of the wind energy industry's development in Denmark and other RE sectors worldwide. Yet, there is rather limited evidence of this bottom-up approach arising in East Asian societies. Analysis on the relational dimension of the region's new developmentalism has to include the extent to which conducive socio-cultural and socio-technical conditions exist for advancing the further deployment of renewable energy and other green energy technologies. To summarise, there are a number of *empirical and functional elements* to the analytical framework, namely:

- Governance of developmental process
- Goals and objectives
- Policy instruments
- Strategy
- Relational aspects.

We may identify certain *outcome impacts* that we may expect from East Asia's new developmentalism. This element of the analytical framework draws attention to the future possible consequences and results of new developmentalism practice, with specific reference to renewable energy development. The outcome impact areas outlined below relate back closely to the main drivers behind new developmentalism and the region's recent expansion of renewables, and they link to other parts of the framework. They have been devised to bring greater analytical focus when thinking about the various kinds of impacts arising from new developmentalism and RE development in East Asia. In addition, we may pose more specific research questions under each outcome impact area as part of this analytical and evaluative process. The main outcome impact areas are as follows:

- *Sustainable development paths*: Is new developmentalism placing East Asia on more viable paths to long-term sustainable development?
- *Energy security risk management*: To what extent will renewables and other aspects of green energy development help to mitigate energy supply and environmental risks in East Asia?
- *Techno-economic development*: For example, how will the region's renewable energy development help build an increasingly strong foundation for future green energy technology cluster development? Also, will new developmentalism prevail over the carbon-intensive industries (e.g. steel, ship-building) of old developmentalism? More specifically, will coal and other fossil fuel systems continue to dominate over renewables?
- *Socio-cultural and socio-technical change*: For instance, will the deepening use of renewables in East Asia bring about important socio-cultural and

socio-technical change in relation to energy relationships and economic development practice generally?

- *Continuation or break from economic modernity*: Will new developmentalism prove to be just a continuation of economic modernity in another guise (i.e. a kind of neo-modernisation) in which its aims and practices are more conventionally economistic than environmentalist in orientation?

Investigations within these outcome impact areas could be undertaken in the short-term to very long-term. Moreover, some degree of overlap exists between them, as is evident from the specific research questions provided here. To conclude, it has been argued that renewable energy development in East Asia is embedded within the broader context of the region's new developmentalism, and Figure 2.1 outlines the analytical framework of the new developmentalism concept. This is based this section's discussion of its main elements:

- Theoretical synthesis of SCT and EMT
- Core aim of low carbon development, and its main drivers
- Empirical and functional aspects
- Outcome impact areas.

The framework informs the approach taken to the empirical studies that follow in subsequent chapters and is applied more directly in the book's concluding chapter, which draws together the main arguments of this study.

2.5.3. New developmentalism: a critique

It is evident that while low carbon development forms an important and often core goal of East Asia's new developmentalism, it also vies with other strategic imperatives, such as strengthening national economic competitiveness. Furthermore, the region's new strategic macro-development plans contain programmes or promote industries that are strongly associated with 'old developmentalism.' For example, the expansion of physical infrastructures, export-oriented manufacturing, and an emphasis on state direct financial support remain defining characteristics of China's 12th FYP, while its promotion of higher levels of domestic consumption somewhat contradict the plan's sustainable development objectives. It should too be noted that in the 11th FYP, the increase in coal-fired power station capacity was greater than that for wind energy, although the 2011–2015 plan has a target for reducing the nation's electricity generation dependence on coal from 70 percent to 62 percent.

Meanwhile, conventional energy and chemicals are amongst the 20 strategic areas that comprise the Singapore Economic Development Board's industrial master plan, which includes further expansion of the city-state's already huge petrochemical complex and the construction of a large new liquefied natural gas terminal. In South Korea, the Green Growth Strategy must simultaneously contend with the Ministry of Trade, Industry and Energy's strategic industry policy for upgrading the traditional flagship industries, including the energy-intensive sectors of shipbuilding and steel.[5] Evidence of old-style strategic industry policy can be seen in certain aspects of RE policy, such as China's local content rule (now rescinded) on wind turbine production and export controls on rare earths, and rules constraining foreign firm eligibility on competitive bidding, as found in the Philippines' 2007 Biofuels Act. Indeed, some have contended that China's new innovation policies are based on traditional protectionism, thus marking no real departure from the mercantilist past, which in their view may not bode that well for its future RE policy (US–China Economic and Security Review Commission 2010).

More generally, East Asia's national energy strategies and systems are still heavily dependent on securing fossil fuel resources. This is indicative of the broader problem of how many of East Asia's core industrial structures, policies, and practices remain geared toward carbon-intensive development. In Indonesia, for instance, the continued heavy subsidisation of fossil fuel prices make it difficult for the nation's geothermal sector to seriously compete in the electricity generation market. Furthermore, many countries persist with old industrial policies involving mass-scale ecological damage (e.g. Indonesia and Malaysia's palm oil industries) whilst rolling out low carbon development strategies (Gunningham 2011). Despite the March/April 2011 Fukushima disaster, many East Asian countries may still maintain their ambitious plans on nuclear power sector development, and moreover include nuclear under the policy rubric of green or clean energy. While the inclusion of nuclear power in low carbon strategies may be defensible, it remains highly contentious given the very serious environmental risks associated with nuclear fission's byproducts, as clearly demonstrated at Fukushima in 2011.

East Asia's 'modernist' industrialised development is likely to remain entrenched for some considerable time given the enormity of the structural change required to establish the broad foundations of low carbon development. As discussed earlier, the influence of ecological modernisation ideas and theories is growing within East Asia and has evidently had a positive effect on greening state policies and exercises of state capacity studied in this chapter, yet this influence still remains somewhat limited in both depth and scale. Furthermore, EMT itself has limitations in bringing about the profound structural changes required to secure humanity's low carbon future. In addition, many Southeast Asian governments still lack sufficient state capacity to implement coherent low carbon development and renewable energy policies technocratically and institutionally, as shown by the lack of RE policy definition and detail, as well as poor coordination of RE policy across their state bureaucracies (IEA 2013, Olz and Beerepoot 2010, REEEP 2014).

East Asia's developing countries face especially difficult challenges moving toward low carbon development, given both the high proportionate GDP costs of replacing old or rudimentary equipment and infrastructure and their relative lack of indigenous techno-innovatory capacity. Furthermore, there are domestic political pressures to meet more immediate socio-economic needs (e.g. poverty alleviation, provision of basic welfare and utility services) than to prioritise environmental-related goals. This being said, socio-economic and environmental problems are increasingly conflating in developing country regions. Many of East Asia's major cities (most notably China's) are now subject to acute levels of pollution, causing chronic health problems on a mass scale and palpably deteriorating societal welfare. Even in relatively poor and authoritarian states, there has been mounting civil unrest regarding pollution and a growing acknowledgement of problems caused by pursuing economic growth, whatever the costs. Worsening environmental conditions and tensions at the local, national, regional, and global levels will only create stronger imperatives for expanding and developing the use of renewable energy technologies in East Asia and worldwide.

2.6. Conclusion

This chapter has explored how renewable energy development in East Asia is embedded within the region's new developmentalism–this being defined as revitalised and refocused forms of state capacity aimed at realising the transformative economic objectives associated with low carbon development. It was argued that the expansion of renewables in the region has become closely linked to evolving forms of state capacity. The introduction of strategic master plans specifically for renewables and their programmatic integration with broader strategic macro-plans on development and climate change strategies is a generally unique feature of East Asia's approach to promoting RE sector development, helping create the conditions for substantial future public and private investments in renewable energy. Yet, new developmentalism in East Asia coexists with the continued promotion of carbon-intensive industrial activities, including conventional fossil fuel–based sectors. Such apparent contradictions are likely to persist for some time given the structural dependencies of East Asian economies on carbon-intensive industries for delivering material growth and prosperity, and that, like most others around the world, East Asian states have only just embarked on the very long transition to meaningful low carbon development–a process that most expect to take many decades, if indeed it is to be achieved.

Furthermore, East Asia's new developmentalism has thus far tended generally to afford primacy to economistic objectives rather than environmentalism, to economic growth rather than sustainable development per se. While the promoted growth of the RE sector under new developmentalism naturally involves supplanting fossil fuel energy practices and consequently bringing about important systemic changes to address climate change, most

East Asian states concurrently view this as an export growth opportunity in a new emerging strategic industry. This raises the question of just how far East Asia's new developmentalism marks a departure from the economic modernisation-oriented policies of the past, despite being influenced by ecological modernisation ideas and thinking. It is perhaps more realistic to expect this initial current phase of new developmentalism and its corresponding early expansion of renewable energy to deliver, at best, *relatively lower* carbon development over the medium-term. This chapter also discussed how achieving low carbon development requires deep socio-cultural and socio-technical transformations, especially changing people's mindsets toward creating a green society that cannot be realised solely through a top-down and purely technical policy process. Relational dimensions to East Asia's new developmentalism are critically important to the region's future renewable energy development.

Notes

1 Bresser-Pereira (2011) also used the term 'new developmentalism,' but to explain new state-active policies in Latin America in the conventional neo-Keynesian macro-economic sense with no explicit linkage to low carbon development, the promotion of renewable energy, or other eco-industries. As deployed here in this book, the new developmentalism concept is based on a distinctly different theoretical approach and empirical foundation.
2 These may also be referred to as macro-development plans.
3 Most scholars date this development from the mid-18th century onwards; however, some, such as Giddens (1990), have argued that ideas of modernity originate earlier, from the 17th century.
4 Connections here exist with Kuhn's (1962) seminal work on paradigm shifts and the structure of scientific revolutions. The term 'techno-economic' reflects the conflation between techno-innovation and economic development, when new technology clusters provide the basis for an emerging development paradigm.
5 For details, see http://www.motie.go.kr/language/eng/policy/Ipolicies_04.jsp (accessed June 1, 2014).

3 Renewable energy policy in East Asia

3.1. Introduction

This chapter provides an overview of the general development of renewable energy (RE) policy in East Asia set in its global context. It first examines the methods and approaches to RE policy-making and conducts a study of recent international policy trends. Thereafter, it explores how RE policy has evolved over time in China, Japan, South Korea, Taiwan, and Southeast Asia, as well as how RE policy-making has become increasingly framed on new developmentalist strategic plans. The development context of each part of East Asia is first considered, taking into account pertinent energy security and environmental risk imperatives. Key policy agencies and stakeholders are then introduced, which is followed by the core analysis of policy developments. The main challenges for future renewable energy development in different parts of the region are also studied.

3.2. Renewable energy policy: an overview

3.2.1. Origins and principles

Germany, Denmark, and Finland were the first countries to initiate modern multi-sector renewable energy policies (beyond large-hydro) from 1970 onwards (IEA 2004). Most other IEA/Organisation for Economic Co-operation and Development (OECD) countries followed suit a few years later, including Japan, which was the first East Asian country to initiate a multi-sector RE policy programme (formally commencing in 1974). Other nations from the region followed later that decade and into the early 1980s (Philippines in 1978, Singapore in 1979, Malaysia in 1981, South Korea in 1985). China, Indonesia, and Thailand began their programmes in the 1990s, and Vietnam in 2001. By 2013, 138 countries worldwide had introduced RE policy support mechanisms (up from just 55 countries in 2005) and 144 countries had set defined renewable energy targets. The general expansion of renewable energy policy has occurred in developed and developing countries alike, and East Asian countries have been at the forefront of this trend (Hirschl 2009, REN21 2014). As Western governments have cut back on their renewables development budgets in the aftermath of the 2008–2009 global financial crisis, their lesser crisis-affected East Asian counterparts have continued to increase their investments, augmented by resilient private sector investments (see Chapter 1).

Although national governments remain the most important formulators of RE policy, local governments (e.g. cities, provinces, sub-national states) are increasingly becoming players in this field. In Japan, Tokyo city government has a well-established renewables and climate change strategy. After the nuclear disaster of 2011, the Fukushima Prefecture set a target of 100 percent total energy self-sufficiency on renewables by 2040, while Nagano Prefecture is committed to reaching 70 percent by 2050. Shanghai has a target of generating 12 percent of its electricity from renewable energy by 2015 and has set technology-specific installations targets, including 150 MW of solar photovoltaic (PV). In Beijing, solar water heater devices must be used for all new buildings and swimming pools. Many provincial governments in China have established their installed wind energy capacity targets. In South Korea, the Seoul city government is working toward generating 20 percent of electricity from renewable sources by 2020, and solar PV systems will be installed in 1,000 schools by 2014. Taipei City in Taiwan has a 12 percent by 2020 target (REN21 2014). Local governments in East Asia and worldwide are furthermore networking together to share the best renewable energy and low carbon development policy practices, on both national and international levels. An example on the national level is the China Low Carbon City Programme. On the international level, the C40 Cities Climate Leadership Group comprises a network of 63 cities with 14 members from East Asia.

Chapters 1 and 2 outlined the motives and rationale (e.g. public good and externality arguments) for state support of renewable energy development, as well as government policy as a mechanism for facilitating this development by removing barriers impeding it. Such barriers may be thought of in the following terms:

- *Costs and prices*: Renewables are not sufficiently cost or price competitive in relation to incumbent or conventional energy fuel systems, especially fossil fuels.
- *Market development*: Market development is linked closely to the above. On the supply side, this may arise due to firms' unwillingness to enter renewable energy industries, a lack of firms able to do so, or the unwillingness of incumbent firms to invest further. There may also be limited credit and finance options to support investments in renewables. Demand-side issues include the lack of a consumer base, especially if alternative fuels are cheaper and energy fuel consumption represents a high proportion of average incomes (particularly relevant to low-income

countries). Risk and long-term market certainty issues may also impede market development.

- *Legal and regulatory*: There may be deficiencies in or lack of legislative and regulatory frameworks to support production, consumption, and new technology development. Existing rules and laws in place (e.g. planning restrictions, subsidies for fossil fuels) may also hinder development.
- *Infrastructural*: Gaps in infrastructure development make it difficult to expand production capacity and consumption. This can relate to electricity grid connectivity and technical load capacity issues.
- *Socio-cultural and socio-technical issues*: Poor public awareness concerning the benefits of renewables, and society's possible current and future possible RE utilisation options, should be addressed. A low-level 'sustainability culture' and the prioritisation of more immediate socio-economic problems (e.g. poverty alleviation; see Chapter 1) may also impede renewable energy development. The workforce may also lack the technical education and skills to work with RE technologies.

With the exception of hydropower, RE sectors are relatively new technology industries that mostly first emerged from the 1960s and 1970s onwards. Policies aimed at supporting their development have generally progressed in accordance to the following pattern (Wu and Huang 2006):

- *Research and development (R&D) stage*: This can involve high costs and investment risks. Hence, government support has been necessary to help mitigate these risks for private-sector investors.
- *Demonstration stage*: This stage involves installing new renewable energy technology devices or systems to prove their technical and potential commercial viability, helping build investor confidence, and deepening public awareness of the technology concerned.
- *Cost-reduction stage*: Once a renewable technology has been proven, financial support and other incentive measures are needed to help scale-up production to achieve cost reductions and efficiencies in manufacturing, installation, operating, and maintenance. Further R&D support may be required for this.
- *Widespread deployment stage*: A renewables sector will enter this stage once it has become commercially competitive in its particular market or end use. Progressing to this stage is the ultimate aim of RE policy.

As Chapter 1 noted, renewables can be used for electrical power generation, transportation, and thermal heat purposes. The primary focus of most governments' RE policies is electrical power generation. Achieving the widespread deployment of renewables in power generation is invariably based on the long-term objective of *grid parity*. This relates to the ability to generate electricity without subsidised or other support at the same levelised cost of energy (LCOE) equal to the usual grid purchasing price (i.e. based on conventional fuels, such as coal and gas). In turn, the LCOE refers to the cost of production distributed over an installation's entire lifetime, thus factoring in all investment and operational costs, and allows for cost comparisons between producing a kilowatt hour (kWh) of electricity between different generation technologies.

3.2.2. Methods and instruments

Renewable energy policy instruments fall broadly into one of three follow-ing main categories. *Direct financial support* concerns financial instruments used by governments (state subsidies, grants, loans, and capital investment) for renewables development generally, including R&D support and new institution creation. By early 2013, approximately 65 countries were engaged in direct financial support for renewable energy development (REN21 2013). This policy approach has remained a particularly important feature of RE policy for most East Asian governments, which can be broadly explained by their traditionally strong strategic industry policies and exercises of state capacity generally (Chapter 2).

The second policy instrument category is *regulatory mandates*, which estab-lish legally binding requirements for firms and other organisations on deploying renewables. For example, renewable portfolio standards (RPS, which have been gradually introduced into East Asia) and more ad hoc mandatory obligations oblige energy producers to source minimum quanti-ties of power generation from renewables, or work on a similar principle in building construction and biofuel processing. By early 2013, six East Asian national governments operated an RPS system—South Korea (from 2001), Japan (2003), Thailand (2004), China (2006), the Philippines (2011), and Indonesia (2011)—out of a global total of just 18 national governments (REN21 2013, 2014).

The third policy instrument category is *market-based instruments*, which adapt or use the market mechanism to provide a variety of different finan-cial incentive measures. For example, competitive bidding involves firms contesting for state contracts on RE development. Tradable permits or certificates are another means of controlling the supply of renewables power to energy systems by tracking and registering it. These are exchanged between producers on a voluntary market basis, and they are frequently used in RPS and other quota systems to document compliance. Tax incen-tives include tax rate reductions, exemptions, holidays, and credits, and they could arguably be placed under direct financial support. Lastly, feed-in tariff (FiT) schemes involve paying premium tariff rates[1] to suppliers of renewables to electricity grids under long-term contracts. Some policy instruments cut across these three categories. For example, power purchas-ing agreements between sellers (e.g. wind farm operators) and buyers (e.g. grid operators) are often the contractual basis of regulatory mandates and FiT schemes, providing investor certainty to renewable energy producers of a secure market for their output. Another example is when generic renewable energy laws lay down the legislative basis for a whole new RE policy framework.

Consistent with the policy development process outlined earlier, initial RE policies of developed countries in the early 1970s focused primarily on R&D support measures, with demonstration project policies following soon after. These same countries then began cost-reduction stage policies in the late 1970s and early 1980s, mainly through the use of direct financial support and other financial incentive policy instruments to promote the scaling up and initial commercialisation of renewable energy technologies. Regulatory mandates and policy instruments started to appear in the early 1990s, and tradable permits and certificates from the mid-1990s; East Asian countries were not present at the cutting edge then but rather lagged in comparison. The same applies to FiT schemes: the United States introduced the world's first in 1978 and mostly European nations followed suit in the 1990s. By 2013, a total of 71 countries and 27 sub-national states around the world had enacted FiT legislation; those from East Asia comprised, in chronological order, South Korea (in 2001), China (2003), Taiwan (2003), Thailand (2007), Indonesia (2008), Philippines (2008 legislated, 2012 implemented), Japan (2009: surplus solar PV only; from 2012 all sectors), and Malaysia (2009; IEA 2013a, REN21 2014). Table 3.1 shows important RE policy milestones in East Asia, marking when countries from the region first introduced key types of policy measures.

Most recently, many Western governments have curbed their policy support for various aspects of renewables development. This has been mainly due to the longer-term fiscal fallout from the 2008–2009 global financial crisis. Chapter 2 discussed how many countries adopted countercyclical 'green growth' policies in response to the crisis, with the subsequent increasing of renewable energy support budgets that followed, especially on new or revised FiT schemes. However, not only did numerous governments worldwide underestimate the public popularity of such schemes, but certain countries experiencing fiscal difficulties decided to reverse positions and began to tax rather than subsidise the use of renewables. In 2012, Bulgaria applied temporary retroactive taxes on solar, wind, hydro, and bioenergy; Greece introduced a new levy on renewable energy consumers that was later raised in early 2013; and Spain set a flat rate tax on renewables and other power generation forms (REN21 2013). East Asian countries, however, continued to maintain high levels of RE policy support throughout the period. The region experienced resilient growth in public and private investments in renewables in 2012 and 2013, during a time when most other parts of the world experienced a decline (Table 1.3). The 2008–2009 global financial crisis mostly affected East Asia indirectly through sharp falls in export demand for its manufactured products. After learning the lessons of the 1997–1998 East Asian financial crisis a decade earlier, its national governments had reregulated financial systems, built up foreign exchange reserves

Table 3.1 Renewable energy policy milestones in East Asia

Renewable energy policy measure	Year first introduced	Scheme
Research and development (R&D) support		
Japan	1974	Sunshine Project
Taiwan	1980	Wind Turbine (150 kW) Prototype Development
South Korea	1985	Special Accounts for Energy and Resources
Singapore	1991	Joint Research with Universities on Environment Related Research
Thailand	1992	Energy Conservation Programme
China	2007	National Climate Change Programme
Malaysia	2010	National Renewable Energy Policy and Action Plan
Installed capacity expansion support		
Philippines	1978	Act to Promote the Exploration and Development of Geothermal
Japan	1979	Project for Geothermal Power Generation
Singapore	1979	Energy Recovery from Biomass in Municipal Waste
South Korea	1985	Special Accounts for Energy and Resources
Taiwan	1986	Solar Water Heater subsidy
Indonesia	1990	Micro-Hydro Project Programme
Thailand	1992	Energy Conservation Programme
China	1996	Brightness Programme
Malaysia	2001	Five Fuel Policy
Vietnam	2001	Renewable Energy Action Plan
Target-setting		
Japan	1996	New Renewable Energy Target
China	1996	Brightness Programme
South Korea	1997	10-Year Energy Technology Development Plan
Taiwan	1998	First National Energy Conference
Malaysia	2001	Small Renewable Energy Programme
Vietnam	2001	Renewable Energy Action Plan
Philippines	2003	Renewable Energy Policy Framework
Thailand	2004	Strategic Plan for Renewable Energy Development
Indonesia	2005	National Energy Blueprint
Feed-in tariffs (FiT) scheme		
South Korea	2001	
China	2003	
Taiwan	2003	
Thailand	2007	
Indonesia	2008	
Philippines	2008	
Japan	2009	
Malaysia	2011	
Renewable portfolio standards (RPS) scheme		
South Korea	2001	
Japan	2003	
Thailand	2004	
China	2007	
Philippines	2008	
Indonesia	2011	

(continued)

Table 3.1 Renewable energy policy milestones in East Asia *(continued)*

Renewable energy policy measure	Year first introduced	Scheme
Strategic development plan		
South Korea	2001	Basic Plan for New and Renewable Energy Development
Malaysia	2001	Small Renewable Energy Programme
Vietnam	2001	Renewable Energy Action Plan
Philippines	2003	Renewable Energy Policy Framework
Thailand	2004	Strategic Plan for Renewable Energy Development
Indonesia	2006	National Biofuel Roadmap
China	2007	Medium and Long-Term Development Plan for Renewable Energy
Singapore	2007	Comprehensive Blueprint for Clean Energy
Taiwan	2008	Green Energy Industry Sunrise Plan
Japan	2008	Cool Earth 50 Energy Innovative Technology Plan

Sources: IEA (2014), REN21 (2014).

and other forms of savings, and strengthened regional financial cooperation (Dent 2013b).

East Asian states' predilection for strategic planning on renewables and low(er) carbon development provides a further explanation for their relatively stronger promotion of renewable energy (Chapter 2). Strategic plans are more than just target-setting: they entail structured, multi-layered policy frameworks that comprise a series of instruments and programmes backed up by public finance and often mechanisms for leveraging private investments. They also normally set staged development steps that progress toward defined medium- to long-term goals or targets on installed capacity and production, advances in techno-innovation, and industrial capacity-building. Strategic planning locks in policy commitment to developing renewables and in principle brings greater coherence to RE policy overall. Plans may be revised or superseded by others, especially when new governments come into office. The broader the plan, the greater the number of policy-making agencies that are usually involved and–depending on the political system or culture–the greater the level of stakeholder engagement is from business and society. Table 2.1 shows that most East Asian countries have introduced renewable energy strategies in recent years, which are closely linked to low carbon development and climate change strategies (Chapter 2).

Although East Asian nations do not possess the world's highest-level targets on renewables' share of the total energy mix, their level of strategic ambition is very high when compared to similar income-level countries worldwide (REN21 2011, 2012, 2013, 2014). Furthermore, most countries from the region have set a broad range of multi-sector targets, especially on raising installed power capacity levels (Table 3.2). The earliest targets set by most East Asian governments are into the mid-2010s and beyond; therefore, we cannot yet evaluate whether they were met. Japan, South Korea, and China did, however, set targets prior to 2011, and only mixed success was achieved. Japan and South Korea failed to meet more or less all of their targets in the 2000s. China also failed to realise its general 11th Five-Year Plan (FYP) non-fossil fuel energy target of 10 percent of primary energy consumption by 2010 (achieving only 8.4 percent, which included nuclear), although it had well surpassed its 2015 targets on specific RE sectors, such as wind and solar, by 2010.

Chapter 1 first noted how East Asia's developed economy governments (Japan, South Korea, Singapore, and Taiwan) have adopted a more *technology-oriented* approach to RE policy, where the main emphasis is on R&D support, new advances in techno-innovation, and improving technical efficacies. This may be considered as much an innovation policy as an industrial policy (Noland 2007). In contrast, the region's developing nations (China and most of Southeast Asia) have hitherto adhered to a more *installed capacity-oriented* approach on renewables, in which realising domestic installation and production-based goals are the priorities. These countries have lower levels of techno-innovatory capacity, although China, Malaysia, and Thailand are looking to significantly raise their game in this area. China's push toward becoming an 'innovation hub' economy has special relevance in this regard (Segal 2010).

3.3. China

3.3.1 Development context

Pressures to promote renewable energy development are arguably far greater on China than on any other East Asian nation. Environmental and energy security imperatives are especially relevant. China's burgeoning energy demand is a consequence of sustained high rates of economic growth achieved since the late 1970s, when the country's economic reform process began. The country's future energy demand will be primarily and functionally driven by realising key domestic goals. Many communities in remote interior provinces either have inadequate electricity supply or none at all. Addressing this issue is integral to achieving the government's twin overarching socio-economic objectives of reducing income gaps that have grown in recent years within China's society and lifting further hundreds of millions of people out of poverty. The provision of key welfare services (health, education, utilities such as freshwater supply) all depend on electricity power generation.

China's power generation capacity has increased almost 10-fold from 1990 (126.6 GW) to 2012 (1,150.5 GW), and accounts for approximately 70 percent of East Asia's total. Coal, the most carbon-intensive fossil fuel, was responsible for 758.0 GW (66.0 percent) of Chinese generation capacity in 2012.

Table 3.2 East Asia's renewable energy and carbon emission targets

		General renewables	Hydropower	Wind	Solar
Northeast Asia	**China**	Non-fossil fuel, primary energy (inc. nuclear): 11.4% (2015), 15% (2020)	325 GW, including 41 GW pumped storage (2015); 430 GW (2020)	100 GW on-grid, including 5 GW offshore (2015); 200 GW, including 30 GW offshore (2020)	40 GW, including 1 GW CSP (2015); 50 GW, including 3 GW CSP (2020)
	Japan	Primary energy: 10% (2020) Non-hydro electricity: 1.63% (2014), including hydro electricity: 10% (2014)		8.03 GW offshore (2030)	4.8 GW (2010), 28 GW (2020), 53 GW (2030)
	Mongolia	Electricity: 20–25% (2020)			
	South Korea	Primary energy: 4.3% (2015), 6.1% (2020), 11.5% (2030). Power generation output: 13,016 GWh, 2.9% (2015); 21,977 GWh, 4.7%, (2020); 39,517 GWh, 7.7% (2030)	Large: 3,860 GWh (2030); Small: 1,926 GWh (2030)	23 GW and 50,000 GWh (2030), 2 GW offshore (2020)	PV: 1.2 GW (2015), 1,971 GWh (2030); Thermal 2,046 GWh (2030)
	Taiwan	Electricity: 9,952 MW, 14.8% total (2025), new: 6,600 MW (2025); 12,502 MW, 16.1% total (2030)	2,502 MW (2030) excluding pumped storage	1,200 MW onshore (2030), 3,000 MW offshore (2030)	610 MW (2015), 3,100 MW (2030)
Southeast Asia	**Indonesia**	Primary energy: 25% (2025); Electricity output: 26% (2025)	1.3 GW new (2015), 2 GW (2025)	300 MW (2014), 970 MW (2025)	870 MW (2025)
	Malaysia	Primary energy: 11% (2020), 14% (2030), 36% (2050)			1,250 MW (2020)
	Philippines	Electricity output 40% (2020), capacity 15,236.3MW (2030, triple 2010 level)	8,729.1 MW (2030)	2,378 MW (2030)	285 MW (2030)
	Singapore				
	Thailand	Primary energy: 25% (2021), 9,201 MW electricity (2021)	1,608 MW new (2021)	1.2 GW (2021)	PV: 2 GW (2021); Thermal: 100 ktoe (2021)
	Vietnam	Primary energy: 5% (2020), 8% (2025), 11% (2050); Non-hydro electricity: 4.5% (2020), 6.0% (2030), 241 MW p/year average new (2006–2015), 160 MW p/year (2016–2025), 4,050 MW new (2025)	17.4 GW (2020); pumped storage 1.8 GW (2020), 5.7 GW (2030)	1,000 MW (2020), 6,200 MW (2030)	

Sources: EPIA (2014), GWEC (2014), IEA (2014), REN21 (2011, 2012, 2013, 2014).

Biomass power	Biofuels	Geothermal	Ocean	Fuel cells	Carbon emission reduction targets
13 GW (2015), 30 GW (2020)	4 million tonnes bioethanol, 1 million tonnes biodiesel used (2011–2015)	110–120 MW (2015) geothermal and tidal; ocean 50 MW (2015); tidal 100 MW (2020)			Energy consumption per GDP unit: −16% (2011–2015) CO2 emission per GDP unit: −17% (2011–2015), −40 to −45% (2011–2020)
6 GW (2030)		3.88 GW (2030)	1.5 GW – tidal (2030)		3.8% below 2005 emission levels by 2020
Forest: 2,628 GWh (2030)	Biogas: 161 GWh (2030)	2,803 GWh (2030)	6,159 GWh (2030)		30% reduction compared to 'business as usual' scenario by 2020, or 4% below 2005 levels by 2020.
1,400 MW (2030)		200 MW (2030)	600 MW (2030)	500 MW (2030)	Return to 2005 emission level (2020, 45% business as usual reduction), return to 2000 level (2025)
400 MW new (2015), 810 MW (2025)	Transportation: 5% (2025); Biodiesel: 20% diesel (2025); Bioethanol: 15% gasoline (2025)	12.6 GW (2025)			26% from business-as-usual level by 2020 based on domestic funding, or 41% by 2020 based on international funding (US$18.2 billion)
1,065 MW (2020)					
306.7 MW (2030)		3,467 MW (2030)	70.5 MW (2030)		
					16% from projected 2020 'business-as-usual' level
3,630 MW (2021); Thermal: 8,200 ktoe (2021). MSW 160 MW (2021); Thermal 35 ktoe 500 MW (2020), 2,000 MW (2030)	Biogas: 600 MW; Bioethanol: 9 m ltrs/day (2021); Biodiesel: 6 m ltrs/day (2021) Various for bioethanol and biodiesel (2015, 2020)	1 MW (2021)	2 MW (2021)		30% from 2009 level by 2020, 42 million tonnes CO_2 equivalent in 2020

Rising fossil fuel consumption has caused significant environmental and welfare degradation in China with increasingly adverse effects (CCICED 2009, 2012; Wang, Y. 2010; World Bank 2001; World Bank/China State Environmental Protection Administration 2007). China overtook the United States in 2006 as the world's biggest carbon dioxide (CO_2) emitter, and by 2010 accounted for around a quarter of global carbon emissions (Table 1.1). Increasing levels of energy emissions are creating acute health problems across the country. The CCICED (2012) reported that, in the year 2000, almost 300 people per million population were prematurely dying in China due to high emission particulate levels–the world's highest rate–and this was expected to increase to almost 900 per million by 2030.[2] China's Ministry of Environmental Protection reportedly estimated the cost of environmental degradation in the country at around US$230 billion in 2010, roughly three times the 2004 level.[3] Many parts of China are also highly vulnerable to the climate change risks of extreme weather and rising sea levels.

China's coal reserves (approximately 40 percent of the global total) will completely run out by the early 22nd century, its oil by around 2025, and gas by 2040 (Zhao 2008). Despite the government's longer-term aim of reducing coal's share of total power generation, the most optimistic estimates suggest this level is unlikely to fall below 50 percent by 2030 (Li 2010, Yu et al. 2011). The government's deregulation of electricity and coal prices as well as the removal of coal transport subsidies in 2013 helped other fuels to compete on a more level playing field, and China is investing more than any other country in carbon capture and storage technology (Garnaut 2013). At the same time, the 12th FYP (2011–2015) includes a programme to develop 16 large coal-power bases in north and northwest provinces, where the bulk of China's coal deposits are situated, and raises the target for coal-power capacity to 960 GW by 2015, a 45.5 percent increase from the 2010 level. According to Yang and Cui (2012), Chinese state-owned energy companies are planning to collectively add a further 558 GW of new coal-power capacity (363 new plants) in coming years, roughly equating to 40 percent of total worldwide planned growth. Given that these new plants will have an expected operating life of 35–40 years, it still unclear whether China's energy policy strategists view coal as a transition or destination fuel source (You and Xu 2010). The 12th FYP also sets a target to more than double national gas-power capacity, from 26.4 GW in 2010 to 56.0 GW by 2015, and raise total generation capacity for all sectors to 1,490 GW by the same year–a 53.6 percent increase over the 5-year period.

There are also energy supply security risks confronting China from its growing dependency on foreign fuel sources. The country became a net oil importer in 1993, and it imports gas in growing volumes through Russian and Central Asian pipelines. Despite being the world's dominant coal producer, China now buys in ever larger quantities from Australia, Indonesia, and other nations. China is the world's second largest oil and coal importer, and the fourth largest importer of liquefied natural gas (BP 2013, Yang and Cui 2012). Concerning nuclear, the country possesses only an estimated 1 percent of global uranium reserves, currently maintains a 65 percent import dependency on the mineral, and saw its uranium import levels triple from 2009 to 2011. Analysts expect China to overtake the United States as the world's largest uranium importer by 2020 (Massot and Chen 2013). Table 3.3 indicates the recent development paths of China's renewable energy sectors. Hydropower remains dominant, which is consistent with international norms. However, the country's installed wind energy capacity continues to increase quickly, while solar PV has recorded very dynamic recent growth from a relatively low starting base. The table further shows that the government has set ambitious future growth targets for almost all RE sectors.

3.3.2 China's renewable energy policy agencies and stakeholders

The Chinese government is a large and complex political entity with multiple competing interests. China's energy policy was for some years pulled in many opposing directions due to the different agendas of various energy-related state agencies (Herberg 2009, Xu 2007). This was despite the creation of two centralised organisations–the National Energy Bureau in 2003 and the National Energy Administration (NEA) department in 2008, both within the National Development and Reform Commission (NDRC)–which were supposed to coordinate Chinese energy policy overall. These bodies had been vested with limited authority and resources to exercise meaningful control.

In an acknowledgement of growing coordination problems, the Chinese government set up a new National Energy Commission (NEC) in 2010, originally under the leadership of Premier Wen Jiabao and subsequently by his successor Li Keqiang. The NEC's membership comprises broad high-level ministerial representation, including from the NRDC, Ministry of Finance, Ministry of Foreign Affairs, Ministry of State Security, and the People's Liberation Army. The NEC is reportedly responsible for national energy development strategy and strategic oversight generally, energy security reviews, coordination on major energy issues, international cooperation on energy affairs, and therefore in principle on all such matters concerning China's renewable energy policy. However, it remains unclear just how functional and effective the body has proved since its creation. The executive power of the NEA is still evidently strong and remains involved in renewable energy policy design, analysis, and coordination.

The NDRC thus remains the key overarching actor in China's renewables policy-making, given its central role in formulating FYPs and development policies generally as well as parenting the NEA and the China National Renewable Energy Centre, which was established as a joint Sino-Danish

Table 3.3 China's renewable energy profile

Power generation	Installed capacity (GW)						Sector targets
	2004	2006	2008	2010	2012	2013	
Hydropower	105.2	128.6	171.5	199.1	229.1	260.2	284 GW, plus 41 GW pumped storage (2015); 350 GW + 80 GW PS (2020)
Wind	0.8	2.6	12.2	44.7	75.3	91.4	100 GW, including 5 GW offshore (2015); 200 GW, including 30 GW offshore (2020)
Solar PV	0.0	0.1	0.1	0.9	8.3	18.8	39 GW (2015); 47 GW (2020)
Biomass power	0.0	2.5	3.3	5.5	8.0	8.5	13 GW (2015); 30 GW (2020)
Geothermal	0.0	0.0	0.0	0.0	0.0	0.0	110–120MW (2015) geothermal and tidal; 50 MW ocean (2015), tidal 100 MW (2020)
Ocean	0.0	0.0	0.0	0.0	0.0	0.0	
Concentrating solar power (CSP)	0.0	0.0	0.0	0.0	0.0	0.0	1 GW (2015); 3 GW (2020)
Total	**106.0**	**133.8**	**187.1**	**250.2**	**320.7**	**378.9**	Non-fossil fuel (inc. nuclear): primary energy 11.4% (2015), 15% (2020); electricity generation capacity 30% (2015)
% share total installed capacity	*23.6*	*21.2*	*23.2*	*25.3*	*27.9*	*30.5*	
% share total power output generated	*16.2*	*15.0*	*17.0*	*18.0*	*20.0*	*20.0*	
Nuclear	7.0	7.0	8.9	10.8	13.8	14.6	40 GW (2015); 80 GW (2020)
% share total installed capacity	*1.6*	*1.1*	*1.1*	*1.1*	*1.2*	*1.2*	
% share total power output generated	*2.3*	*1.9*	*2.0*	*1.8*	*1.8*	*2.1*	
Other energy production	2004	2006	2008	2010	2012	2013	
Solar water heaters (million m²)	–	90.0	136.0	185.0	257.7	317.0	400 million m² (2015); 800 million m² (2020)
Biogas users (million households)	–	21.8	30.5	39.0	40.8	41.2	50 million households (2015)
Biogas (billion m³)	–	9.0	12.0	14.0	15.7	15.7	22 billion m³ (2015); 30 billion m³ (2020)
Solid biomass (1,000 tonnes)	–	0	1,000	3,000	6,000	6,000	10 million tonnes (2015); 50 million tonnes (2020)
Bioethanol (1,000 tonnes)	–	1,020	1,580	1,860	2,020	2,100	4 million tonnes (2015); 10 million tonnes (2020)
Biodiesel (1,000 tonnes)	–	50	300	500	500	400	1 million tonnes (2015); 2 million tonnes (2020)

Sources: CNREC (2013), GWEC (2014), EPIA (2014), IEA (2014), IGA (2014), BP (2013).

Notes: Hydropower capacity figures do not include pumped storage. Biomass includes waste incineration. Data source figures differ for China's installed wind, solar PV, and geothermal energy capacity. The author has chosen to use international sourced data from international sectoral and other organisations for these sectors for international comparative reasons over Chinese national data, which is generally believed to underestimates capacity levels.

venture in 2012. The Ministry of Commerce, Ministry of Foreign Affairs, Ministry of Finance, Ministry of Water Resources (involved in the country's hydropower sector), Ministry of Science and Technology, Ministry of Industry and Information Technology, Ministry of National Land and Resources, and Ministry of Environmental Protection are other relevant RE policy-making agencies. After the 2002 reform of the energy market, two state-owned grid companies now monopolise China's electricity generation grid system–the State Grid Corporation of China and the China Southern Power Grid, which are responsible for power transmission. In actual physical terms, the country's electricity grid is divided into four synchronous regional grids: the northeast-central, east, and northwest, which are operated by the State Grid Corporation of China, and the south, which is operated by the China Southern Power Grid[4] (Kahrl *et al* 2011). There are five large power generation companies–Guodian Corporation, Huaneng Group, Datang Corporation, Huadian Corporation, and China Power Investment Group–which are also state-owned enterprises (SOEs). The state-owned Assets Supervision and Administration Commission–an agency under the State Council's purview–exercises national government representation across these companies, with local government representation present. The State Electricity Regulatory Commission regulates the electricity generation sector on behalf of the State Council, with regional branches across the country, and in 2013 was merged with the NEA.

A number of quasi-state and non-state agencies in China have stakeholder interests in renewables policy. The Energy Research Institute of the NDRC is the country's main think tank on energy issues. The Chinese Renewable Energy Industry Association is the nation's most important private business organisation on renewables. Sector-based groups include the Chinese Wind Energy Association and the China Photovoltaic Industry Alliance. Additionally, there are various renewables-related networks, including the China Renewable Energy Information Network, China Renewable Energy Scale-Up Programme,[5] China Sustainable Development Programme, and China New Energy Network. While China's RE associations and networks lobby the government on policy issues, they are generally more preoccupied with strengthening business relationships and fostering international cooperative ventures. Overall, the political economy of China's renewable energy development is strongly state-centric (see Box 3.1).

3.3.3. The evolution of China's renewable energy policy

China's FYPs have provided the main framework for shaping the nation's energy policy-making. The FYPs have also been the principal mechanism through which the Chinese government has channelled the world's largest public investment programme in renewable energy. The country's first hydropower dams were built in the 1910s, and the government began to sponsor experimental tests in renewable energy technologies in the 1950s.

Box 3.1 Business and China's renewable energy industry

The Chinese economy exhibits features of dynamic entrepreneurism and strong competition. This is not only evident in the private sector but also among the country's SOEs, which play a dominant role in China's renewable energy development. These enterprises can be generally categorised as large-scale, national government-administered corporations or smaller local government-owned companies serving the energy needs of city municipalities and individual provinces. Examples of the former include the two state-owned grid companies that monopolise China's electricity grids, the government's five power generation companies (e.g. Guodian, Huadian), and national energy companies (e.g. Sinopec, SinoHydro). Naturally, both SOE types have close relationships with the governing state agencies that are responsible for national and local government RE policy, although they can exercise high degrees of corporate autonomy and moreover compete among each other. Just as renewable energy sectors vary significantly from each other, so are there many differences in the types of enterprises and markets that exist across them. Broadly speaking, an energy-industry value chain can be divided into the following elements, listed from upstream to downstream:

• Fuel source explorers or surveyors
• Equipment manufacturers (including upstream suppliers)
• Source or plant developers
• Energy infrastructure providers
• Plant or system maintainers
• Retail market suppliers.

Figure 3.1 provides a comparative overview of how these differ across fossil fuel, nuclear, and renewable energy sectors, while additionally indicating the mix of Chinese public and private sector enterprises that exist in each part of the matrix. As shown, SOEs dominate along the whole value chain of China's 'incumbent' fossil fuel, nuclear, and hydropower sectors, as well as many aspects of RE industries. Li (2013) reported that, by 2011, there were approximately 700 SOEs involved in wind farm plant development, for example. The country's five main power generation companies alone were responsible for installing 57.0 percent of wind energy generation and local SOEs for another 22.4 percent. This contrasted with just 4.6 percent for China's private wind farm developers, 1.3 percent for foreign firms, and 14.7 percent for foreign joint venture projects. A similar situation exists at the plant development level in the biomass power generation industry.

	Coal, oil, gas	Nuclear	Hydropower	Wind	Solar PV	Biomass
Fuel source explorers or surveyors	Public	Public	Public	Public/private	Public/private	Public/private
Equipment manufacturers (plus upstream suppliers)	Public/private	Public/private	Public/private	Private	Private	Private
Source or plant developers	Public	Public	Public	Public	Public/private	Public
Energy infrastructure providers	Public	Public	Public	Public	Public	Public
Plant or system maintainers	Public	Public	Public	Public	Public/private	Public/private
Retail market suppliers	Public	Public	Public	Public	Public, n/a prosumers	Public

Figure 3.1 Energy enterprises in China's power generation (public/private sector mix).

For the public/private sector mix in China status for each segment, *Public / Private* indicates a situation where neither sector dominates. *Public* or *Private* indicates a situation where either correspondingly dominates. Private can also include foreign firms, such as hydropower turbine manufacturers.

However, the solar PV industry is different due to its highly distributed nature (i.e. a myriad of building-integrated installations), which has allowed a growing number of private sector firms to enter various aspects of the value chain. The wider deployment of micro-scale RE technology applications generally (e.g. rooftop PVs, small wind turbines, biomass boilers, pico-hydro devices) has created a growing number of 'prosumer' individuals and organisations (including firms) that both produce and consume their own generated electricity and thermal power by and large independently, thus diminishing the need for energy supply companies (Chapter 1). However, despite the rapid growth of PV prosumers in China, utility-scale solar park capacity has grown even faster and is now presenting opportunities for energy SOEs to become market players here also (IEA 2013a). In general, the country's established energy SOEs with roots in fossil fuels, nuclear, and hydropower have simply diversified into wind, solar, biomass, and other renewables by applying their economies of scope advantages of existing technical expertise and assets to these new emerging power sectors. China's public and private sector enterprises have also become more vertically integrated up and down industry value chains. The following are some illustrative examples of the above points:

- National oil company CNOOC formed a new energy subsidiary in 2007 that used its engineering expertise to develop offshore wind projects, onshore wind farms, biomass generation plants, and biofuel production. China's other major state-owned oil companies, Sinopec and CNPC, have also become scaled-up biofuel producers.
- Hydropower SOEs SinoHydro Group, HydroChina Corporation, and China Three Gorges Corporation have engaged in wind farm development. The latter two enterprises are involved in overseas projects, such as in Pakistan. Shenhua Group, the world's largest coal company, and China Guangdong Nuclear Power are also among China's major wind farm developers.
- China's largest wind farm developer is Longyuan Power Group, a subsidiary of power generation SOE China Guodian Corporation. Longyuan is also involved in developing solar, bioenergy, geothermal, and tidal energy projects. In August 2013, Guodian announced it would develop its first overseas wind farm, a 100 MW plant in Ontario, Canada, becoming operational in 2014.
- Ningxia Yinxing Energy is a local SOE with multi-sector plant development (wind, solar PV, coal) and equipment manufacturing interests.

- Sinovel, an SOE that then turned into a private enterprise in 2011, is China's largest wind turbine manufacturer and additionally has operations in wind-field design and planning, equipment transportation and installation, plant maintenance, and remote data analysis services.
- Mingyang Wind Power, a subsidiary of the Mingyang Energy Group conglomerate, is a large private wind turbine manufacturer that is diversifying downstream to wind farm development. In June 2013, it formed a joint venture with the state-owned China National Nuclear Corporation to develop up to 300 MW of wind projects in Henan province.
- GCL Poly is a Hong Kong–based private company and world's largest producer of polysilicon, the base material for first-generation solar PV. The company has recently ventured downstream into wafer manufacture and is also developing the 300 MW Datong solar park (see Chapter 5).
- The solar PV subsidiary of China Huaneng Group, one of the big power generation SOEs that historically focused on coal-power stations, is the developer of the 100 MW Shilin solar park. Another big state-owned power generation company, Huadian, developed the 100 MW Gansu Jiayuguan solar park. These are among China's largest PV installations (see Table 5.3).

In many other countries, fossil fuel companies are also major players in developing clean energy systems. However, it is the scale on which this business diversification is occurring in China's energy companies that is rather exceptional. Their evolutionary transformation into eventually green energy businesses, if this indeed transpires, is likely to be a very slow process through gradual adaptations in response to changing technological, market, and policy conditions. This is consistent with the ecological modernisation approach (Chapter 2). Government regulatory mandates and other policy measures will, in the meantime, require Chinese enterprises to source more power generation from renewables. In general, the country's larger incumbent companies (mainly SOEs) have most successfully diversified and thrived in most aspects of China's RE industry, primarily due to both economies of scale and scope advantages. They are additionally well positioned to extract state support due to close connections with national and local government (Zhang et al. 2013). The combination of these factors explains why incumbent power generation SOEs have been able to secure wind farm development contracts with low bids.

Furthermore, there are some interesting market dynamics at play for policy-makers to consider. For instance, wind power generators often do not compete against coal-power generators because they can be the same company. However, there may be intra-company competition between old and new energy divisions, China's hydropower companies may take a competition position against their fossil fuel rivals, and the country's five large state-owned power generating companies may compete against their local-level counterparts (but as they are all SOEs, this may be considered an intra-state competition of sorts).

Indeed, intra-state competition and tensions have had a profound impact on China's renewable energy business on various levels, proceeding from governance and policy coordination challenges involving national government, local government, SOEs, and private companies. On the retail side, competing SOEs provide the main bulk of electricity sold to consumers (Figure 3.1) at prices regulated locally but in accordance to centralised NDRC guidelines; however, a significant lack of transparency persists on both price formulation and the allocation of retail market contracts among enterprises (Lin and Purra 2012). Furthermore, Shi (2013: 8) observed that local governments in China were often encouraging the development of RE plants to help meet their economic growth targets: 'The construction of power plants is mainly for increasing GDP rather than meeting the demands for electrical power, leading to serious blind construction of power plants in various areas.' Applying this to the wind energy industry, plant developer companies such as Longyuan have built numerous wind farms that may take a considerably long time before they are connected to the grid, thus explaining why by the early 2010s an estimated fifth of China's wind power installations were in effect dormant (Chapter 4). Contestations over grid connectivity and market demarcations among mainly power generation SOEs, grid infrastructure SOEs, local energy SOEs, local governments, and national governments have created a difficult business environment in many aspects of this industry (Global Wind Energy Council 2014, Li 2012), and also to a lesser degree in the smaller solar PV power generation sector (China Photovoltaic Industry Alliance 2013). Notwithstanding these problems, these industries continue to demonstrate robust growth.

Figure 3.1 shows that the only element of the value chain where the private sector dominates in China's RE industries (except hydropower) is in equipment manufacture. Wind energy companies such as Goldwind, Sinovel, Mingyang, United Power, and Dongfang have all become key global players in their industry, as have Yingli, JA Solar,

ReneSola, and Jinko in the PV sector. Although some of these firms once had SOE origins (e.g. Sinovel, Goldwind), the rapid expansion of this value-chain element is a critically important part of China's renewable energy development story (Chapter 5). In 2004, the country had just six wind turbine firms, rising to 40 by 2007 and to around 90 by 2011 (Li 2013). Best estimates suggest that more than 400 solar PV manufacturers now operate in China (IEA 2013a). There has thus been a dynamic entrepreneurial response generally from the nation's public and private business sectors to the government's strengthening policy support and ambitious strategic planning on renewables.

Greater efforts on renewables R&D came in the 1980s under the '863' State High-Tech Development programme. However, it was not until the 1990s that a holistic, multi-sector approach on RE policy in China emerged. After the United Nation's 1992 Earth Summit, the Chinese government's 'Agenda of China for the 21st Century' white paper outlined its inaugural sustainable development strategy, noting a key role for renewables in future energy planning (Yang et al. 2003). The 1995 China Electric Power Act thereafter introduced general legislation to promote renewables development (Han et al. 2009). This paved the way for the landmark 1996 Brightness Programme, which was outlined in the 9th FYP (1996–2000) and based on approximately US$1.2 billion public investment for small-scale installations in rural communities. An early sectoral emphasis on wind energy was evident (the 1997 Ride the Wind and 2000 National Debt Wind Power programmes), along with renewable energy's contribution to rural electrification schemes, such as the 2001 Township Electrification and 2006 Village Electrification Programmes, which were launched at the beginning of the 10th and 11th FYPs, respectively (NDRC 2001, 2006; REN21 2009a).

The 10th FYP also included plans to ardently promote hydropower and wind energy. New large dam projects were linked to the government's ambitious Western Development Strategy launched in 2001–2002 (Chapter 6). Under the 2003 Wind Power Concession Programme, China introduced its first FiT scheme and the goal of creating 20 farms of 100–200 MW capacity each and 20 GW of national capacity by 2020. At the start of the 11th FYP, the government's REL introduced in 2006 laid another landmark foundation for China's RE policy by providing a much firmer legal framework to foster renewables development. The REL is based on five main regulations: a revised and more comprehensive FiT scheme, price categorisation mechanisms, cost-sharing mechanisms, special fund mechanisms, and RE sector targets (National Energy Bureau 2008). It also established national standards for renewable energy technologies and production. Although the implementation of the REL has been somewhat patchy in parts, it helped create a much stronger investment environment for renewable energy industries. The increase in enterprises entering the wind energy sector after 2006 was particularly noteworthy, and investment levels in all RE sectors rose significantly after the law's introduction. The REL was soon complemented by the Medium and Long-Term Development Plan for Renewable Energy, introduced in September 2007, which created a new and highly defined strategic policy framework to foster RE development. The plan, covering the 2007–2020 period and anchored in the 11th FYP, set out China's ambitious targets for renewables, backed up with approximately US$263 billion of public investments. A series of sector-specific targets were also set, although these were all later revised substantially upwards under the 12th FYP (2011–2015) due to higher than expected subsequent growth rates.

Chapter 2 noted how the 11th FYP and Medium and Long-Term Development Plan on Renewable Energy reflected how China's leaders had become more influenced by ecological modernisation when devising new approaches to and thinking about the country's future development. This was aligned with the strategic intent of fostering renewables as new growth industries. Chapter 2 also commented on how renewable energy was strategically linked to China's National Climate Change Programme, introduced in July 2007. Thus, by that year China's RE policy had a stronger legal and strategic development framework, and it had become firmly integrated into the country's broader development strategy and emerging environmental policy framework. This was more broadly the principal basis for China's new developmentalism.

Further developments in China's renewable energy policy over the remainder of the 11th FYP term of 2006–2010 concentrated mainly on revisions and extensions to existing policy measures. This included changes to FiT schemes, strategic targets, and planning, funding, and coordinating mechanisms. The 11th FYP also set the target of renewables contributing 10 percent of total primary energy consumption by 2010. However, as the deadline approached, the Chinese government revised the energy category from renewables to non-fossil fuels, thus including nuclear power. Even with this adjustment, non-fossil fuel energy only reached the 8.4 percent mark by 2010 (Delman and Odgaard 2011). The government's current targets are to raise this share further to 11.4 percent by 2015 and 15 percent by 2020, and to reduce CO_2 emissions per gross domestic product (GDP) unit by between 40 and 45 percent by 2020, as embodied in both the 11th FYP and the National Climate Change Programme.[6] Coupling nuclear with renewables in a 'new' or 'alternative' energy policy framework remains controversial but not uncommon, as we later discuss in the Japan and South Korea sections of this chapter.

In May 2011, the State Council launched China's New Energy Industry Development Plan, in which the 'new energy' classification grouped renewables with clean coal technology, smart grid systems, coal bed methane, natural gas hydrate, and other unconventional oil and natural gas resources.[7] The 12th FYP selected 'new energy' (primarily renewables but also nuclear and energy efficiency) to be part of the RMB4 trillion (US$610 billion) funded programme to promote seven strategic emerging industries for 'clean' development and a new industry base. In addition, the 12th FYP introduced sector-specific strategic development plans for solar PV, wind, hydropower, and biomass, as announced in 2012. In sum, disregarding the country's well-established hydropower sector, China's RE policy evolved from being mainly focused on realising the socio-economic objective of delivering off-grid electricity to remote communities to an increasingly comprehensive policy framework strongly integrated with strategic development planning by the mid-2000s.

3.3.4 China's renewable energy development: main future challenges

Despite the remarkable expansion of RE development in China over the last decade or so, renewables faces stiff competition from other energy sectors. Coal and nuclear are particularly noteworthy in this regard. Like renewables, they present the option for high-growth indigenous energy supply development, notwithstanding the significant environmental costs and risks they pose (Singapore International Energy Week 2012). In 2013, China had 18 operational nuclear reactors with a total combined capacity of 13.8 GW and plans to construct another 28 by 2018, which accounts for 42 percent of the total number of new reactors planned for construction globally (Schneider and Froggatt 2013).

Chapters 4 and 5 on wind and solar energy, respectively, highlight the various policy-related challenges concerning grid connectivity to have recently arisen. China's grid companies have not expanded their transmission infrastructures fast enough to connect with the rapidly growing number of new renewable energy installations, especially wind farms and now solar parks. This is a planning policy as much as an infrastructure construction issue. Expanding grid connectivity is vital: investment growth in the country's RE plants will only be sustained if there is a corresponding investment in electricity grid infrastructure. Closely linked to this are policy implementation and enforcement challenges concerning aspects of the Chinese state apparatus. For example, there is no guarantee that policies formulated by the national government will be enacted by provincial and municipal authorities (Delman and Odgaard 2011). Local governments enjoy relatively high degrees of autonomy in China's policy-making structures, especially in terms of the interpretation, implementation, and enforcement of policy. In addition, the country's RE laws are still relatively young and many stakeholders lack awareness of China's regulatory and wider policy environment on renewables. Another future key challenge for Chinese RE policy-makers will be fostering stronger indigenous techno-innovation capacity. There are limits as to what policy can achieve in this respect because it largely depends on the entrepreneurial response from domestic energy companies.

Finally, despite the considerable financial resources of the Chinese state, the country's growing debt ratios–especially in the SOE sector, and thus the energy industry–are a cause for concern, with potentially significant ramifications for the future funding of renewable energy projects. While the green fiscal stimulus funding of the late 2000s and early 2010s channelled considerable sums of money into China's energy SOEs (see Box 3.1), huge losses experienced from investments in coal-fired power stations led to the National Asset Management Bureau (NAMB) imposing debt-to-equity ratio controls on the big power companies in 2012. At the end of the year, the NAMB provided new loans to the power companies to stabilise investment levels. Whether China can maintain its relatively high debt structures is a matter that affects the entire economy, but it is particularly relevant to industries such as renewables, where the government has strategic plans to progressively increase investment levels over time. Attracting more private sector investment, especially from foreign companies, into the country's SOE-dominated energy sectors could make an important contribution to addressing this problem.

3.4. Japan

3.4.1. Development context

Japan is East Asia's most economically and technologically advanced nation, the world's third largest economy, and the fifth largest carbon emitter. It was the first Asian country to industrialise, founded on the developmental state policies introduced by the Meiji government from the late 19th century onwards. Japanese companies are among the largest and most sophisticated in the global economy, as well as often being technology leaders in their respective fields. The country's international economic status and power are historically founded on rapidly expanding energy-intensive manufacturing industries. In many ways, the environmental and energy-related problems experienced by China in the last 20 or so years are similar to those that faced Japan in the 1950s to 1970s. Then, Japanese cities were among the world's most polluted due to high industrial and energy emissions. The so-called four big pollution diseases–Minamata, Niigata Minamata, Yokkaichi asthma, and Itai-itai–caused thousands of deaths, consequently leading the government to initiate its first substantive environmental policies (Broadbent 2002). As noted in Chapter 1, Japan has long maintained an acute energy import dependency: it remains the world's largest coal and liquefied natural gas (LNG) importer and third largest oil and uranium importer. The 1973–1974 oil crisis more starkly revealed the energy supply and price security risks confronting the country. This, combined with high pollution levels, spurred the Japanese government to initiate a comprehensive renewable energy policy–the first East Asian nation to do so.

Japan passed a series of new pollution abatement laws in November 1970. A few months later, the Japanese government created an Ecological Research Group in May 1971 (thus, two full years before the 1973–1974 oil crisis), which, together with the Japan Industrial Policy Research Institute, proposed the industrial ecology concept discussed in Chapter 2 (Watanabe 1972, 1995). This established the ideational foundation of the country's new renewable energy policy in the early 1970s. However, far more investment and policy priority were directed toward nuclear power; thus, Japan became a leading nuclear energy generator by the 1980s. Given its energy security vulnerabilities, the country's long-term energy strategy had been based on nuclear's growing share in power generation (Toichi 2003). However, the Fukushima disaster of March 2011 led to a major re-evaluation of this strategy. At the time, there were 54 operating nuclear reactors in Japan spread over 18 plants with a combined installed capacity of 49 GW (ranked third globally behind the United States and France), which were generating around a quarter of the nation's electricity output (Ushiyama 2012). These were all shut down by May 2012 for safety testing, and many are likely to become decommissioned altogether in the coming years.

Over time, companies in Japan's energy-intensive industries began to relocate operations overseas for cost- and market-related reasons, which helped to lower domestic emission levels. New technological advances in energy efficiency by Japanese firms also had a similar positive impact. The country has become a world leader in this field and one of the world's most energy-efficient industrial nations (ARNE 2010). Indeed, Japan's green or new energy policy has hitherto been focused far more on nuclear and energy-efficient technologies than renewables (Moe 2012, Takase and Suzuki 2011). The country's many other green energy and industry sectors (e.g. fuel cells and electric vehicles) are also strong. In terms of policy and economic priority, renewables are thus competing not just against fossil fuels and nuclear power but also a broad range of technologically strong green energy sectors in Japan (Huenteler et al. 2012).

Nevertheless, with the future of nuclear power looking very uncertain after the 2011 Fukushima disaster, renewable energy offers Japan the only safe long-term solution for supplying clean energy and decarbonising economic development. Table 3.4 shows that hydropower continues to dominate the country's renewables power generation profile, but a recent surge in installed solar PV capacity could lead to this position being challenged in the coming years. Meanwhile, Japan's wind, biomass, and geothermal sectors have achieved only modest growth in the 21st century.

3.4.2. Japan's renewable energy policy agencies and stakeholders

Established in 1973, the Agency for Natural Resources and Energy (ARNE)–a division of Ministry of Economy, Trade and Industry (METI)[8]–is the

Table 3.4 Japan's renewable energy profile

	Installed capacity (GW)						Sector targets
	2004	**2006**	**2008**	**2010**	**2012**	**2013**	
Hydropower	20.0	22.2	21.9	22.4	22.1	22.3	
Wind	1.0	1.2	1.9	2.3	2.6	2.7	8.03 GW offshore (2030)
Solar PV	1.5	1.7	2.1	3.6	6.9	13.6	28.0 GW (2020), 53.0 GW (2030)
Biomass power	3.0	3.0	3.0	3.3	3.3	3.4	6.0 GW (2030)
Geothermal	0.5	0.5	0.5	0.5	0.5	0.5	3.88 GW (2030)
Ocean	0.0	0.0	0.0	0.0	0.0	0.0	1.5 GW – tidal (2030)
Total	**26.0**	**28.6**	**29.4**	**32.1**	**35.4**	**42.5**	*Primary energy: 10% (2020). Non-hydro renewables electricity output generated: 1.63% (2014) and 190,000 GWh by 2030. All renewables: 10% (2014), 300,000 GWh by 2030.*
% share total installed capacity	*9.4*	*10.3*	*10.5*	*11.2*	*12.3*		
% share total power output generated	*10.1*	*9.5*	*8.4*	*9.0*	*9.9*		
Nuclear	45.0	48.0	48.0	48.0	38.0		
% share total installed capacity	*16.3*	*17.2*	*17.1*	*16.7*	*13.2*		
% share total power output generated	*24.0*	*27.8*	*24.0*	*25.8*	*2.4*		

Sources: GWEC (2014), EPIA (2014), EIA (2013), IEA (2014), BP (2013), REN21 (2014).

Notes: All data for power generation. Hydropower capacity figures do not include pumped storage. Biomass includes waste incineration.

Japanese government's main body responsible for renewable energy policy. It is supported by three technology-oriented public agencies, the New Energy and Industrial Technology Development Organisation, the National Institute for Advanced Industrial Science and Technology, and the Council for Science and Technology Policy. In addition, the Institute of Energy Economics, Japan and the Institute of Applied Energy perform supporting research and analytical functions regarding renewables policy-making. Other government agencies are occasionally co-opted into the formation of Japan's RE policy. For example, in 2008, the Ministry of Education, Culture, Sports, Science and Technology; the Ministry of Land, Infrastructure, Transport and Tourism; and the Ministry of the Environment worked with ANRE and METI in the design and implementation of a solar energy action plan (REEEP 2013). The Ministry of Foreign Affairs also works closely with ANRE on the later discussed Cool Earth 2050 plan. Japan's electricity grid is divided into separate eastern and western parts; the historic origins of this division date back to the 19th century, when two different foreign technology systems were developed by Tokyo- and Osaka-based companies. Ten regional power companies (Chugoku, Chubu, Hokuriku, Hokkaido, Kyushu, Kansai, Okinawa, Tokyo, Tohoku, and Shikoku) supply the electricity generation market, collectively represented by the Federation of Electric Power Companies of Japan (*Denjiren*). There is a considerable interchange of personnel between ARNE/METI and *Denjiren* companies and a well-established alignment of shared vested interests between them around the nuclear sector relative to renewables (Moe 2012).

Japan has various business and civil society organisations working in RE industries with relatively strong contesting influence over policy-making. Important examples include the New Energy Foundation (NEF) of Japan, formed in 1980, and the Institute for Sustainable Energy Policies, which is a non-profit research organization created in 2000 that, like the NEF, produces reports and recommendations on renewables, energy efficiency, and energy market issues. The Japan Council for Renewable Energy was established in June 2007 as a semi-independent forum involving academics, business, and government officials, supported by the New Energy and Industrial Technology Development Organisation, the National Institute for Advanced Industrial Science and Technology, and the NEF. The Japan Renewable Energy Policy Platform, launched in 2008, is a coalition of eight green energy groups[9] that is coordinated through the Institute for Sustainable Energy Policies (which performs a secretariat function) to both study and recommend proposals for Japan's renewable energy policies. In addition, the Japan Photovoltaic Energy Association was formed in 1987. In the aftermath of the 2011 Fukushima nuclear disaster, the Japan Renewable Energy Foundation was founded in the same year by telecommunications entrepreneur Masayoshi Son.

3.4.3. The evolution of Japan's renewable energy policy

The Sunshine Project was the first landmark programme of Japan's multi-sector renewable energy policy. First conceived in 1973 and introduced in 1974, it provided financial incentives for R&D activity, where the response on solar energy was most significant (Watanabe 1995). A series of sector-specific policy initiatives on demonstration start-up projects then followed: the 1979 Project for Geothermal Power Generation, the 1980 Project for Developing Small and Medium-sized Hydro Power Plants, and the 1994 Subsidy Programme for Residential PV Systems. However, only the latter had any real impact, consolidating solar PV's position as the only non-hydro renewables sector with a substantive foundation. More generic policy measures were introduced from the mid-1990s onwards, bringing greater coherence to Japan's RE policy (Ushiyama 1999). The New Renewable Energy Policy Target initiative of 1996 introduced multi-sector strategic planning and targets (on solar PV, wind, and bioenergy) for the first time, with a long-term goal for renewables (excluding hydro and geothermal) to account for 3.1 percent of total primary energy by 2010. The 1997 New and Renewable Energy R&D programme meanwhile provided incentives for investment in advanced green energy technologies and facilities. Over the next 10 years, Japan's total spending on renewables R&D was to double, but it still remained very small compared to nuclear and other energy sectors.

In 1999, members from the national legislative assembly (Diet) formed the Federation of Diet Members for Promoting Natural Energy as a political coalition, whose aim was to introduce the country's first FiT law. However, this failed due to opposition from METI bureaucrats and the *Denjiren* power companies. Instead, the government opted a couple of years later for a METI initiative on establishing a green power RPS scheme using tradable permit and green energy certificate instruments, which entered into force in April 2003 (Maruyama et al. 2007). This was Japan's first major RE policy initiative since the 1997–1998 East Asia financial crisis. Under the new scheme, the original target set for the power companies was to increase their national collective procurement of renewables production from 7.3 TWh in 2003 to 12.2 TWh by 2010, equivalent to around 1.35 percent of projected electricity generation. This was updated in 2007, when a further target of 16.0 TWh and 1.63 percent of total generation by 2014 was set.[10] These targets, though, were among the lowest set by any country at the time, and according to Moe (2012) reflected pressures exerted by the *Denjiren* companies on the government to minimise their RPS obligations.

In the meantime, Japan's RE policy was shaped by more general developments in Japan's energy policy. The 2002 Basic Law on Energy Policy placed greater emphasis on market-based mechanisms rather than direct financial support, as well as taking more account of the environmental risks of energy consumption. Japan's first strategic development plan on solar (PV Roadmap

2030) was launched in 2004, based largely on R&D support for reducing costs and improving efficiency levels with multiple technical targets set. Later, the New National Energy Strategy, which was launched in 2006, strengthened the country's commitment to expanding the green energy sector, although energy efficiency and nuclear power were afforded much greater priority (REEEP 2014). This pattern of sectoral emphasis continued, as revealed in the 2008 Action Plan for Achieving a Low Carbon Society and the Cool Earth Energy Innovative Technology Plan 2050 (or Cool Earth 2050 plan) introduced in the same year, with nuclear power placed at the centre of both strategies. The former plan included a programme of constructing 13 new nuclear power plants (nine by 2017), whereas the latter sought to extend Japan's green diplomacy through international nuclear cooperation. The Cool Earth 2050 plan additionally aspired to strengthen international collaboration on renewables R&D, although this was deemed to be a lower priority.

Due to slower than expected progress toward the 2010 targets on renewables set in 1996 and 2001, in 2008 the Japanese government revised these slightly downwards, just two years away from the deadline year. In the last two years of the Liberal Democratic Party (LDP) government, new measures promoting solar PV were introduced. The 2008 Subsidy for Residential PV Systems superseded the 1994 scheme and helped to double solar PV modules the following year (Japan Renewable Energy Policy Platform 2010). The 2009 New Purchase System for Surplus Solar Power Energy introduced Japan's first FiT scheme, but this was directed at installed surplus-generated solar PV only and therefore had limited application. The same year, the Use of Non-Fossil Energy Sources Act and the Promotion of the Development and Introduction of Alternative Energy Act laid down co-joined legislation promoting renewables, nuclear, and other non-fossil fuel energy technologies. In addition, new tax credits were awarded for new solar PV installations.

The manifesto promises of the Democratic Party of Japan (DPJ), which came into power in September 2009, stated a broader promotion of renewables beyond the LDP government's past emphasis on solar PV, including a pledge to introduce a multi-sector FiT scheme. A month later in October 2009, the *Denjiren* announced its opposition this policy, and the DPJ's attempts to also sidestep similar resistance within METI failed (Moe 2012). The political compromise that ensued was a narrow-scope FiT scheme restricted to just residential solar PV. With the launch of its 2010 New Growth Strategy (NGS) and Industrial Structure Vision (ISV), the DPJ government sought other means of trying to diversify and promote the country's renewable energy development. The main goal of the NGS is the 'creation of a low carbon society through a comprehensive policy package including new systems design, systems changes, new regulations, and regulatory reform, and to support the rapid spread and expansion of environmental technologies and products' (METI 2010b: 20). This transformation would include 'measures to support the spread and expansion of renewable energies (solar, wind, small-scale hydropower, biomass, geothermal, etc.) by expanding the electric power feed-in tariff system' (METI 2010b: 21–22). A core element of the NGS strategy is to strengthen Japan's industrial-technological capabilities in seven strategic areas up to 2030, with environment and energy ('green innovation') being the first listed; the other six areas are medical/health care, finance, science and technology, tourism/local revitalisation, human resources, and Asian regional economic integration (Jones and Yoo 2011). These are underpinned by 21 national strategic projects, one of which (listed under green innovation) is the expansion of renewables sector into a JPY10 trillion (US$110 billion) market by 2020. The NGS is augmented by the ISV, which sets broad aspirational objectives for diversifying Japan's industrial base toward a multi-sector low carbon economy that includes a relatively prominent role for renewables (METI 2010c). The NGS and associated ISV thus formed an important foundation of Japan's new developmentalism.

In June 2011, the DPJ government announced a plan to increase renewable energy's contribution to 20 percent of electricity output by 2020, over double the then-current level (Table 3.4). To achieve this, Kan argued that the 2009 FiT scheme would need to be substantively revised and made as comprehensive as first envisaged, having wide sectoral coverage. He made this a condition for his resignation as premier (in September 2011) in the political aftermath of the Fukushima disaster (Huenteler et al. 2012). His DPJ colleague and successor, Yoshihiko Noda, succeeded in legislating the new FiT scheme in July 2012, representing probably the most significant political victory against the pro-nuclear vested interests of the *Denjiren*, METI and relevant parts of Japanese industry, and the new centrepiece of Japan's RE policy. This was soon followed up by two other announcements. In August 2012, the Ministry of the Environment launched the new Innovative Strategy for Energy and Environment for expanding six-fold the combined national installed capacity on offshore wind, geothermal, biomass, and tidal power by 2030 (see Table 3.2, p. 78). New plans were announced in September 2012 for new offshore wind projects in the Tohoku region, where Fukushima is located. Meanwhile, the new FiT scheme had reportedly experienced a considerable positive response from business and the public, helping stimulate significant new growth in installed solar PV capacity (Table 3.4).

After the LDP election victory, new Prime Minister Shinzo Abe took office in December 2012. A few months later, his administration announced that nuclear would return as a core foundation of Japan's energy system. Given the regularity of new elected premiers in Japan, it remains uncertain what priority future governments will afford to renewable energy policy. On a more recent positive note, the Japanese Cabinet approved a plan in April 2013 to fully liberalise the electricity market and bring about a structural separation of the transmission and distribution from generation that aimed to break the regional monopoly position of the 10 *Denjiren* power companies (Hughes 2013). If successively implemented, this would allow renewable energy greater scope to compete in a more flexible and open electricity market.

3.4.4 Japan's renewable energy development: main future challenges

Long-term competition with nuclear power is arguably the most important challenge for renewable energy development in Japan. Notwithstanding the cataclysmic events of the March 2011 Fukushima disaster and a number of safety scandals in the early 2000s that pre-dated it (Masuda and Komiyama 2012, Takase and Suzuki 2011), there are important reasons why we may expect the country's preference for nuclear over renewable in years to come. There persist strong entrenched interests between the energy policy-making hub of ARNE/METI and the *Denjiren* power companies to maintain their entrenched financial, technological, industrial, and political investments in Japan's nuclear industry (Aldwich 2011, DeWit and Iida 2011, Huenteler et al. 2012). The regularity of new governments presents notable continuity and coherence problems for the nation's energy policy and strategy-making per se and only serves to strengthen ARNE/METI's position as the energy policy-making hub. State bureaucrats rather than elected politicians are likely to wield most influence over the nation's renewables policy.

Only solar PV has managed to secure a strong political support within Japan due to first-mover advantages developed from the 1970s onwards and its close techno-industrial linkages with traditional Japanese strengths in electronics and high-tech manufacturing. As Moe (2012) contended, solar PV has been able to receive policy support due to its 'insider' position in Japan's state–business nexus, whereas the wind energy industry has long been considered an outsider, owing to lack of advocacy within METI, the *Denjiren*, and the big business (*keiretsu*) community. The broader diversification of non-hydro renewables beyond solar PV presents other policy, industrial, and power-generation challenges. The advantage of nuclear over renewables also extends to geographic and spatial constraint issues. Scaling up many types of RE installations (e.g. onshore wind, solar parks) in Japan's densely populated and mountainous territory will prove difficult. Japan too has a fragmented electricity grid infrastructure with limited inter-grid connections among the regionalised monopolies of the 10 power companies and different frequencies between the southwestern and northeastern halves of the system. This somewhat limits the scope for large RE plants to feed power across many parts of the grid simultaneously.

3.5. South Korea

3.5.1. Development context

South Korea is East Asia's third largest economy and the world's fifteenth largest. Over the last six decades, it has transformed itself into a modern industrial state, primarily through energy-intensive rapid industrialisation. In a famous speech made by President Park Chung-hee in 1962 at the opening of the then new Ulsan Industrial Complex, he stated that 'dark smoke rising from the chimneys is the symbol of our nation's growth and prosperity' (cited in Yoon 2006: 76). This was indicative of South Korea's approach to economic development, energy, and the environment for many decades thereafter. From 1980 to 2010, the country's primary energy consumption increased over five-fold, from 46 million tonnes of oil equivalent (mtoe) to 250 mtoe, and installed electricity grid capacity has grown over eight-fold over this period. Carbon emissions have risen from 136 million tonnes annually to 509 million, and emissions per head have tripled from 3.5 tonnes to 10.4 tonnes per year during this time but have stabilised around this level. In 1971, the country's energy intensity rate (GDP per unit of energy used) was 42 percent below the OECD average, but by 2008 it was 25 percent higher than the average and fourth highest in the OECD group (Jones and Yoo 2011).

Manufacturing and construction still have a large share of the South Korean economy, which explains why the nation's energy consumption rate is still relatively high. Since 1998, industry's share of South Korea's total energy consumption has stayed in the 55–60 percent band but up from its 1971 level of 44 percent (Kim et al. 2011). The country's petrochemicals sector alone accounted for around a fifth of national primary energy consumption in 2009, and the metals sector around an eighth. In the meantime, South Korea's fossil fuel import dependency rose from just over 60 percent in the early 1980s (when domestic coal played a significant role in energy supply) to 97 percent by 2010. By this year, the country was the world's seventh largest oil importer, third largest coal importer, and second largest LNG importer. In 2010, South Korea was the world's eighth largest CO_2 emitter, and its air quality rating was one of the worst in the OECD group. In South Korea's increasingly prosperous society, there is growing public support for low carbon development (Cho 2004, Watson 2012). Table 3.5 indicates that solar PV is challenging hydropower to become South Korea's dominant installed capacity RE sector on power generation, but renewables are dwarfed by nuclear power in this regard. More generally, the country compares poorly to other similarly developed economies in terms of renewable energy's share of the national energy mix.

3.5.2. South Korea's renewable energy policy agencies and stakeholders

The Office of Energy and Resources within the Ministry of Trade, Industry and Energy[11] is South Korea's main government agency responsible for renewables development. However, there is only one division within the office dedicated to promoting 'new and renewable energy,' whereas there are three divisions working specifically on nuclear. Formed in 1980, the Korea Energy Management Corporation's (KEMCO) main remit is to change South Korea's energy culture generally toward stronger sustainable

Table 3.5 South Korea's renewable energy profile

	Installed capacity (GW)						Sector targets
	2004	2006	2008	2010	2012	2013	
Hydropower	1.6	1.6	1.6	1.6	1.6	1.6	Large: 3,860 GWh (2030); Small: 1,926 GWh (2030)
Wind	0.0	0.2	0.3	0.4	0.5	0.6	23 GW and 50,000 GWh (2030), 2 GW offshore (2020)
Solar PV	0.0	0.0	0.4	0.7	1.1	1.5	1,971 GWh (2030)
Biomass power	0.0	0.1	0.1	0.2	0.3	0.3	2,628 GWh (2030)
Geothermal	0.0	0.0	0.0	0.0	0.0	0.0	2,803 GWh (2030)
Ocean (tidal and wave)	0.0	0.0	0.0	0.0	0.3	0.3	6,159 GWh (2030)
Total	**1.6**	**1.9**	**2.4**	**2.9**	**3.8**	**4.3**	
% share total installed capacity	2.5	2.7	3.0	3.4	4.2		*Primary energy: 4.3% (2015), 6.1% (2020), 11.5% (2030). Power generation output: 13,016 GWh, 2.9% (2015); 21,977 GWh, 4.7%, (2020); 39,517 GWh, 7.7% (2030)*
% share total power output generated	1.7	1.5	1.6	1.9	2.1		
Nuclear	16.7	17.7	17.7	17.7	20.7		32% total power generation (2020), 59% (2030)
% share total installed capacity	25.9	25.2	22.2	20.9	22.8		
% share total power output generated	35.5	36.8	33.8	29.7	28.1		

Sources: GWEC (2014), EPIA (2014), EIA (2014), IEA (2014), BP (2013), REN21 (2014).

Notes: All data for power generation. Hydropower capacity figures do not include pumped storage. Biomass includes waste incineration.

energy practices. KEMCO provides educational, R&D, and technical support services on renewables and energy efficiency, administrates policy instrument support (e.g. subsidies on equipment purchase), and hosts a New Renewable Energy Centre. The Korea Institute of Energy Research and Korea Energy Economics Institute are two state-related think tanks that perform analytical functions for South Korea's renewables policy.

The Korea Electric Power Corporation (KEPCO) is the majority state-owned enterprise[12] responsible for supplying 93 percent of national electricity through six power generation subsidiaries (one dominating the nuclear and hydropower sector and the other five being regional power companies)[13] and whole ownership of the national grid. KEPCO thus dominates South Korea's electricity generation, transmission, and distribution, and it operates many of the nation's utility-scale RE installations. Numerous independent wind farm and solar park operator firms exist, but these are not that powerful or influential. The large and highly diversified *chaebol* companies such as Samsung, Hyundai, Doosang, and Hyosung are major RE equipment producers and have close ties with energy policy-makers and government generally due to the legacy of developmental statism, revitalised recently under the Lee Myun-bak administration (Chapter 2). The industry is regulated by the Korea Electricity Commission (KEC) established in 2001. The only two relevant business associations are the Korean Wind Energy Industry Association and the Korea Photovoltaic Industry Association, which primarily champion the interests of their respective sector manufacturers. Green Korea United and the Korea Federation of Environmental Movements are the nation's most important civil society organisations promoting renewables and sustainable development.

3.5.3. The evolution of South Korea's renewable energy policy

South Korea was a relatively slow starter in developing a multi-sector RE policy, with the first landmark initiative being the 1985 Special Accounts for Energy and Resources programme that incentivised investment in renewables R&D and the construction of bioenergy combined heat and power installations. The 1987 New and Renewable Energy Development and Promotion Law then introduced the first major legislation supporting renewables development, comprising measures for combined heat and power stations, solar thermal, and small hydro. A number of small-scale R&D and demonstration stage policies (e.g. National Photovoltaic Project, 1989–2001) were implemented in the late 1980s, but then a hiatus followed until the 10-Year Energy Technology Development Plan launched in 1997. This too maintained the focus on multi-sector R&D support and included a renewables target of 2 percent of national electricity output by 2006. There was then another dry policy period as the country came to terms with the 1997–1998 financial crisis. As the economy recovered and the state's fiscal position strengthened, in 2001 the government introduced its Basic Plan for New and Renewable Energy Development, Utilisation, and Promotion. This was one of East Asia's first strategic development plans on renewables (Table 3.1). Initially set to run from 2001 to 2006, it updated and superseded the 1987 legislation and included financial incentive and regulatory measures, including an RPS system, which was East Asia's first such scheme. South Korea was additionally the first country in the region to introduce a FiT scheme, which also was enacted in 2001 and later amended in 2008.

However, the 2001 Plan was superseded four years early by the 2003 Second Basic Plan, which raised the 1997 set target for renewables from 2 percent to 3 percent by 2006, and to 5 percent by 2011. Neither of these targets, though, were realised (Table 3.5). A total budget of US$7.6 billion was allocated to the Second Plan over its duration (2004–2011), which was the most comprehensive policy framework on renewables development then yet devised in East Asia. Overall, South Korea was emerging as the region's RE policy pioneer by the early 2000s (Table 3.1), although the government's level of ambition for future RE sector development was still low by most international comparisons.

Various relatively minor policy initiatives followed during the mid-2000s period under the presidency of Roh Moo-hyun (2003–2007). A far more ambitious approach emerged after new president Lee Myun-bak took office in 2008. In the same year, his government devised a new national energy plan as an integral part of the broader Green Growth Strategy (GGS; 2009–2030) on lower carbon development (Chapter 2). This was another example of South Korea pioneering East Asia's new developmentalism. The GGS aimed to increase renewable and new energy's share of primary energy consumption to 4.3 percent by 2015, 6.1 percent by 2020, and 11.5 percent by 2030; for electricity power output, these shares were 2.9 percent, 4.7 percent, and 7.7 percent, respectively, for the same years. It targeted nine key sectors for green energy investment, including solar, wind, light-emitting diodes (LEDs), hydrogen fuel cells, gas-to-liquid energy, integrated gasification combined cycle, and energy storage (Korea Environmental Industry and Technology Institute 2009, Presidential Committee on Green Growth 2009, United Nations Environment Programme 2010). The New Renewable Energy Medium-Term Plan (2010–2015) was embedded in the GGS and superseded the Second Basic Plan (2004–2011), based on US$6 billion of public investments that aimed to leverage a further US$29 billion of private sector investments on renewables development. It additionally set sector-specific 2015 targets for installed solar PV and eco-vehicle production.

The South Korean government more recently set sector-specific targets expressed in power output and production (GWh and mtoe) terms rather than installed power capacity. Table 3.2 shows this is quite unique in East Asia. Given how installed capacity figures can put an overly positive spin on actual contributions that many RE sectors make to total power output due to relatively low capacity factor ratings (see Chapter 1), South Korea's approach on target-setting is arguably more transparent and useful. These also extend beyond the aforementioned Medium-Term Plan (2011–2015) sector targets up to 2030, but with intermediate installed capacity targets solely assigned to wind energy. In January 2012, the existing FiT scheme was replaced by a new RPS policy that required power companies to source at least 10 percent of their energy from renewables by 2022. As Kim et al. (2013) observed, the FiT-to-RPS switch represented in effect the government shifting more responsibility for funding onto the power companies and their consumers. Although the deregulation of the country's electricity market started in the 1990s, this has not entailed much market liberalisation because KEPCO's dominate power companies retain strong regional monopolistic positions, meaning they are able to more or less set uncontested high prices if Korea Electricity Commission regulation is weak. Finally, it remained unclear from some time whether the future of South Korea's new developmentalism would continue to rest with the GGS after new President Park Geun-hye took office in February 2013. However, by the end of the year, it appeared that her government would continue to take the nation's low-carbon development strategy in the same direction as her predecessor.[14]

3.5.4. South Korea's renewable energy development: main future challenges

In many ways, renewables face even more serious long-term competition from nuclear in South Korea than in Japan because it is the perceived 'clean energy' solution to national energy security. For some years now, nuclear power has accounted for about one-third of the national electricity output, compared to around a quarter for Japan (Tables 3.4 and 3.5). Despite recent scandals over poor-quality safety standards, South Korea's government has plans to increase this to 59 percent by 2030 based on 40 operational reactors, up from the 23 in 2013. China is the only other country worldwide with plans to install more new nuclear capacity in the coming years (Schneider and Froggatt 2013). The classification of nuclear under the government's green energy rubric is a significant problem for RE development in South Korea. Not only are the future targets for nuclear's expansion far higher, but it is also viewed essentially in the same sustainable energy terms as renewables.

Although the GGS and South Korea's new developmentalism generally has certainly conferred greater strategic importance to renewable energy development, turning around the country's hitherto poor record on this front will prove to be a very tough challenge. Plan targets to more than double power generation output from renewables from 1.9 percent in 2010 to 4.7 percent by 2020 and then to 7.7 percent by 2030 may appear impressive in terms of proportional increase, but the country is starting from an extremely low base by international comparison. Green Korea United and the Korea Federation of Environmental Movements have been very critical of the government's relatively weak RE policy to date and its perceived lack of ambition on renewables (Moon 2010). Although business in South Korea has taken a generally positive view on the GGS, some firms reportedly believe that even these and other low carbon development targets are unrealistic (Mathews 2012). However, Korean companies are renowned for taking long-term strategic views on developing new emerging industries and technologies, so–given the right policy incentives–targets for RE sector development may well be exceeded. Yet, overall, far greater strategic ambition on and policy prioritisation of renewables will be required if South Korea is to make serious progress on low carbon development and be viewed in the international community as a true 'green power.'

3.6. Taiwan

3.6.1 Development context

Like Japan and South Korea, Taiwan–East Asia's fourth largest economy and one of its most advanced–is heavily dependent on foreign energy supplies, importing 98 percent of its used energy in the early 2010s. By this time, it was the world's fourteenth largest oil importer, seventh largest LNG importer, and fifth largest coal importer. The vast majority of these imported fuels are transported through the maritime security chokepoint straits of Southeast Asia. Renewables are therefore a vital element in mitigating energy-supply security risks. As a first-generation 'tiger economy,' Taiwan has also long faced the environmental hazards and challenges that accompany rapid techno-industrial and energy-intensive development (Huang et al. 2011, Williams and Chang 2012). Taiwan has developed a particularly high dependence on coal, accounting for half of the island economy's power generation (Bureau of Energy 2012). The island was ranked as the world's 21st largest carbon emitter in 2010. Taiwan is itself highly susceptible to climate change risks, being prone to extreme weather and rising sea levels.

Taiwan also has a strong tradition of developmental state practice. However, unlike Japan and South Korea, its economy is predominately structured around small- and medium-sized enterprises (SMEs) rather than big business. This has created different kinds of state–business relationships in the pursuit of transformative economic goals, where inter-firm networks and business associations have played a vitally important role. Taiwan is furthermore renowned for having a strong techno-entrepreneurial culture and impressive industrial competences in key fields underpinning the development of many RE sectors, such as electronics, engineering, material sciences, fuel cells, and aerospace. Many Taiwanese companies have developed leading positions in solar energy and other green technology fields, and a strong technology-oriented approach has persisted in the government's renewables policy. Table 3.6 reveals that both wind energy and solar PV have achieved steady but not spectacular progress over recent years in raising their contributions to Taiwan's power generation, but the somewhat static hydropower sector still dominates its renewable energy portfolio.

Table 3.6 Taiwan's renewable energy profile

	Installed capacity (GW)						Sector targets
	2004	2006	2008	2010	2012	2013	
Hydropower	1.9	1.9	1.9	2.0	2.0	2.0	2.502 GW (2030) excluding pumped storage
Wind	0.0	0.2	0.3	0.4	0.6	0.6	1.2 GW onshore, 3.0 GW offshore (2030)
Solar PV	0.0	0.0	0.0	0.0	0.2	0.4	0.61 GW (2015), 3.1 GW (2030)
Biomass power	0.6	0.7	0.7	0.7	0.7	0.7	1.4 GW (2030)
Geothermal	0.0	0.0	0.0	0.0	0.0	0.0	0.2 GW (2030)
Ocean	0.0	0.0	0.0	0.0	0.0	0.0	0.6W (2030)
Total	**2.5**	**2.8**	**2.9**	**3.1**	**3.5**	**3.7**	*Electricity: 9.952 GW, 14.8% total (2025), new added 6.6 GW (2025); new added 12.502 GW, 16.1% total (2030)*
% share total installed capacity	6.0	6.2	6.3	6.3	7.2		
% share total power output generated	4.5	4.9	5.0	4.8	5.4		
Nuclear	5.1	5.1	5.1	5.1	5.1	5.1	
% share total installed capacity	12.1	11.3	11.0	10.4	10.5		
% share total power output generated	18.1	16.9	17.1	16.9	16.1		

Sources: GWEC (2014), EPIA (2014), EIA (2014), IEA (2014), BP (2013), REN21 (2014).

Notes: All data are for power generation. Hydropower capacity figures do not include pumped storage. Biomass includes waste incineration.

3.6.2 Taiwan's renewable energy policy agencies and stakeholders

Created in 2004, the Bureau of Energy, located in the Ministry of Economic Affairs (MOEA), is Taiwan's main state agency responsible for renewable energy policy and development. Its antecedent, the Energy Commission (also located in the MOEA), was established in 1979. The Council for Economic Planning and Development is the Taiwan state's principal agency for strategic economic planning and often is involved in energy policy planning matters. The Industrial Development Bureau (IDB) is involved in industrial policy aspects of renewables development, working closely with the Industrial Technology Research Institute and the National Science Council on relevant R&D, science, and technology policy matters. The state-owned Taiwan Power Company, or Taipower, has dominated the electricity industry since its establishment in 1946, being responsible for approximately 80 percent of power generation, transmission, distribution, and sales. After liberalisation of the market in 1995, nine independent power-generating companies were established, which account for the remaining 20 percent. Taipower also performs certain regulatory functions in the power industry. Since the late 2000s onwards, a number of renewable energy business associations have formed, most notably the Taiwan Wind Energy Association, Taiwan Wind Turbine Industry Association, Taiwan Offshore Wind Power Alliance, Taiwan Photovoltaic Industry Association, Photonics Industry and Technology Development Association, and Taiwan Biomass Energy Industry Association. The most important civil society organisation for promoting renewables is the Taiwan Institute for Sustainable Energy.

3.6.3. The evolution of Taiwan's renewable energy policy

In the East Asian context, Taiwan was an early starter in certain RE policy aspects but a relative latecomer in others (Table 3.1). Its first non-hydro renewable energy policy measures date back to the early 1980s (R&D support for wind and geothermal prototype equipment), but most studies on the subject reference the foundation of Taiwan's RE policy to measures arising from the 1998 First National Energy Conference (Chen et al. 2010a, 2010b; Chiang 2004; Huang and Wu 2007, 2008, 2011; Hwang 2010; Liou 2010a, 2011; Wu and Huang 2006). The conference, held primarily in response to the Kyoto Protocol (see Chapter 1), signalled the government's new priority afforded to renewables development and set a target of 3 percent of total energy consumption by 2020. A government report entitled *New and Clean Energies Utilization Potential in Taiwan* was published soon thereafter, accompanied by around US$300 million public funding for R&D activity (Huang and Wu 2007).

Leading up to the Second National Energy Conference convened in 2005, the government introduced a series of sector-specific measures, mostly direct financial support, under the broader Statute for Upgrading Industries framework aimed at promoting R&D activity, demonstration projects, and installation deployment. A Renewable Energy Promotion Plan launched in 2002 endeavoured to bring greater coherence to Taiwan's emerging RE policy framework. Of the plan's eight key elements, arguably the most important were a new FiT scheme and a comprehensive Renewable Energy Development Act. However, the former proved largely unsuccessful because tariff rates were too low to attract significant investment levels (Liou 2010a). Meanwhile, goals on renewables development were revised in 2003 (New Renewable Energy Development Targets) and again in 2005 at the Second National Energy Conference. Later revisions were made in 2009 and 2011 (see below), making five phases of target-setting in total from 1998, among the highest target revision rates in East Asia.

After 7 years of first being proposed–and mostly delayed due to opposing positions of the major political parties over nuclear power–the landmark Renewable Energy Development Act (REDA) was finally passed in Taiwan's legislative Yuan in 2009. This was a comprehensive general policy framework with multiple key elements spread over 23 constituent articles, with broad stated aims on reducing CO_2 emissions, improving energy diversification, mitigating environmental pollution, enhancing sustainable development, and promoting green energy industries. REDA contained a completely revamped FiT scheme and upward revised targets on renewables, initially set at reaching a total of 9,952 MW or newly added 6,600 MW by 2030; this deadline was subsequently brought forward 5 years to 2025 by the 2011 New Energy Policy, which also laid down the additional goal of 12,502 MW installed by 2030. Although REDA may appear in many respects to be one of the most comprehensive such legislative acts passed in East Asia, it has been criticised for apparent gaps and weaknesses compared to similar legislation in force in other countries. This especially applied to levels of financial support, regulatory clarity, strength of market mechanisms, enforceable standards, and the extent to which REDA improved upon existing laws and measures (Chen et al. 2010b, Hwang 2010, Liou 2010a).

The Taiwanese government began to develop its climate change strategy in parallel to preparations for REDA (Valentine 2010, Young and Huang 2012). In 2008, it introduced its Framework of Taiwan's Sustainable Energy Policy, where renewables played an explicit role in meeting new carbon emission reduction targets by providing a 'cleaner energy supply.' Emission targets were revised in 2010 (see Table 3.2, p. 78) and embodied within the Adaptation Strategy to Climate Change (Council for Economic Planning and Development 2012) strategy document. Furthermore, the strategic industry orientation in Taiwan's RE policy became more evident from the mid-2000s onwards, where expanding production capacity was prioritised over installed capacity. In 2005, the IDB and other MOEA agencies instigated a 3-year Renewable Energy Equipment Industry Developing Plan, which aimed to strengthen foundations for wind energy and solar PV industry development, promote foreign technology transfers, and engage more effectively in international production networks (IDB 2005, Liou 2011). In this sense, it sought to further internationalise Taiwan's growing wind turbine and solar PV manufacturing companies, such as TECO and Motech.

In 2007, the Executive Yuan's Emerging Industrial Technology Strategy Review Board Meeting identified RE technologies as strategically important future industries for investment (Liou 2010b). This was followed a year later by the 5-year Green Energy Industry Sunrise Plan (GEISP), which set targets for production value, job creation, and exports, backed up with substantial public funding support and ambitious goals on leveraging around the equivalent of US$7 billion. The GEISP followed the work of the 2005 Renewable Energy Equipment Industry Developing Plan, which expired in 2008 (Lee and Shih 2011). This new 2009 Plan built on Taiwan's already well-established solar PV and LED lighting technology industries, as well as the lesser established 'five potential growth industries' of wind energy, bioenergy, hydrogen and fuel cells, energy information and communication technology, and electric vehicles. Hydropower, geothermal, and ocean energy are covered under the GEISP's clean energy development component (Chen 2009). The plan was complementary to REDA, promoting renewables as part of a broader green industrial policy, which in turn was constituent with the international trend of 'new green deals' arising in response to the 2008–2009 global financial crisis (Chapter 2).

The GEISP strategic industry planning was augmented by the government's 2011 New Energy Policy. This introduced revised sectoral targets up to 2030, with again special emphasis on solar PV and wind energy. Whereas its Northeast Asian neighbours of China, South Korea, and even possibly Japan are looking to nuclear power as a long-term solution to energy security, Taiwan's New Energy Policy seeks to gradually reduce its dependence on nuclear to zero, moving toward a future 'nuclear-free homeland,' a policy aim first introduced in 2003 (Chiang 2004) and reconfirmed here in response to the March 2011 Fukushima disaster. Thus, in Taiwan's policy parlance, 'new energy' omits nuclear and focuses on renewables and energy-efficiency technologies. More recently, the government introduced new policies on promoting solar PV exports, offshore wind projects, and bioenergy power generation, such as the 2012 Solar PV Overseas Market Expansion Action Plan.

3.6.5 Taiwan's renewable energy development: main future challenges

Taiwan faces challenges on renewables development that are similar to those of Japan and South Korea. Approximately 70 percent of the island's territory is mountainous and the vast majority of its 23 million people are densely concentrated in the western plains. This presents comparable spatial constraint issues for the scaling up of land-intensive installations. The government's recent push on developing offshore wind and buildings-integrated solar installations provides some solutions for this problem. Taiwan's power generation capacity is the fourth highest in East Asia, and its per-capita power needs are also high. Thus far, however, renewables have only made a minor contribution to meeting them. The goal of decarbonising Taiwan's energy structure over forthcoming decades and simultaneously phasing out nuclear power also puts far more responsibility on renewables to deliver a 'clean energy supply' than anywhere else in Northeast Asia.

In further similarity with Japan and South Korea, as well as to some extent China, renewable energy in Taiwan faces competition from other green energy sectors in the strategic industry development context. It was previously noted that the 2009 GEISP included LED lighting, hydrogen energy and fuel cells, energy information and communications technology, and electric vehicles in addition to three RE industries (wind, solar PV, bioenergy), thus omitting solar thermal, hydropower, geothermal, and ocean energy. Although the 2011 New Energy Policy includes plans for all RE sectors, there are nevertheless questions concerning the extent to which renewables will be prioritised within Taiwan's future green energy spectrum.

3.7. Southeast Asia

3.7.1. Development context

Southeast Asia is a highly diverse region (or sub-region, in the East Asia context), comprising a set of relatively small national economies and states in comparison with their Northeast Asian counterparts. The ten Association of Southeast Asian Nations (ASEAN) account for only 12.5 percent of East Asia's total GDP. In addition, the sub-region's largest economy, Indonesia, is around 80 percent of the size of South Korea's (Table 1.1). This has various implications for renewable energy development in Southeast Asia. For example, micro-states such as Singapore and Brunei face chronic spatial constraints on scaling up many renewables, such as wind energy. Poorer nations such as Cambodia, Laos, and Myanmar generally have insufficient infrastructure, economic resources, and development capacity to effectively exploit their renewable energy potentials. The sub-region's RE markets are also relatively small and underdeveloped. Similarly, Southeast Asian firms are not able to capture the same level of scale economies as their Northeast Asian counterparts. Nevertheless, renewables are one of Southeast Asia's fastest growing industries and are a priority sector in state strategies and plans on low carbon development. This section particularly focuses on the sub-region's six largest economies (known as the ASEAN-6): Indonesia, Malaysia, the Philippines, Singapore, Thailand, and Vietnam.

Like China, Southeast Asia is relatively well endowed with fossil fuel resources. Indonesia, Malaysia, Thailand, and Vietnam are all positioned in the world's top-40 oil producers, and these four plus Brunei and Myanmar are additionally in the world's top-40 natural gas producers. In 2012, Indonesia was ranked the world's fourth largest coal producer, while Vietnam, Thailand, and the Philippines were other global top-30 producers. Energy exports have traditionally been key sectors in these countries' economies, indicative of how many Southeast Asian nations have been hitherto able to maintain relatively high levels of energy self-sufficiency. Thus, supply security pressures to find new alternative energy sources, such as renewables, have been less acute in Southeast Asia compared to Northeast Asia. However, Southeast Asia's oil, gas, and coal reserves are fast depleting, and most of its nations are moving toward net energy-importing positions. Most national governments in the region are increasingly looking to generally diversify their energy supply sources. For example, Singapore's decision to develop its first LNG terminal, completed in 2013, was largely due to its 80 percent dependence on Malaysia and Indonesia for its gas, which is the principal fuel used by its electricity grid.[15]

In addition, the sub-region faces similar environmental risks of fossil fuel dependence as China. Many Southeast Asian cities are among the world's most polluted. Current projections indicate that the sub-region's carbon emissions will continue to rise significantly due to urbanisation and industrialisation. Southeast Asia contributed 3.6 percent of global CO_2 emissions in 2010 (Table 1.1), and this is expected to rise to 5 percent by 2030 (IEA 2009). It has been estimated that more than 130 million people in sub-region—over a fifth of the total—remain unconnected to electricity grids (IEA 2013b). Therefore, small-scale renewables can play a crucial role in providing them with power, thereby improving the provision of key welfare services, such as health and education. Only 65 percent of Indonesia's population has access to grid electricity and 86 percent in the Philippines, and these are the region's two most populous nations.

Furthermore, many of the region's grid-connected urban populations (e.g. in Java, Indonesia) are susceptible to intermittent electricity blackouts due to overload demand pressures on the existing transmission infrastructure. Rising or relatively high electricity prices are already a concern in some countries, such as Singapore and the Philippines. With the demand for electricity in most Southeast Asian nations expected to increase substantially into the foreseeable future, and the long-term trend for fossil fuel prices anticipated to gradually rise, renewables offer useful options in mitigating domestic supply security risks. Table 3.7 reveals that, with the exception of Thailand, the ASEAN-6 nations have developed a relatively limited range of renewables for power generation. Hydropower remains the clearly dominant sector, with well over 80 percent of Southeast Asia's total. The only renewables sector where Southeast Asia has a global position is geothermal.

Table 3.7 Southeast Asia's renewable energy profile

	Installed capacity (GW)						Sector targets
	2004	2006	2008	2010	2012	2013	
Indonesia							
Hydropower	4.6	4.9	4.9	4.9	4.9	4.9	1.3 GW new (2015), 2 GW (2025)
Geothermal	0.8	0.9	1.0	1.2	1.3	1.3	12.6 GW (2025)
Total	**5.4**	**5.8**	**5.9**	**6.1**	**6.2**	**6.2**	*Primary energy: 25% (2025); Electricity output: 26% (2025)*
% share total installed capacity	*20.4*	*18.6*	*19.1*	*17.9*	*16.7*		
% share total power output generated	*13.6*	*12.3*	*13.3*	*16.0*	*11.4*		
Malaysia							
Hydropower	2.1	2.1	2.1	2.1	2.1	2.1	–
Total	**2.1**	**2.1**	**2.1**	**2.1**	**2.1**	**2.1**	*Primary energy: 11% (2020), 14% (2030), 36% (2050)*
% share total installed capacity	*8.6*	*9.3*	*9.0*	*8.3*	*8.0*		
% share total power output generated	*7.1*	*7.2*	*7.6*	*6.2*	*7.4*		
Philippines							
Hydropower	3.2	3.3	3.3	3.4	3.5	3.5	8.729 GW (2030)
Geothermal	1.9	2.0	2.0	2.0	1.9	1.8	3.467 GW (2030)
Total	**5.1**	**5.3**	**5.3**	**5.4**	**5.4**	**5.3**	*Electricity output: 40% (2020), capacity 15.236 GW (2030)*
% share total installed capacity	*32.7*	*33.5*	*33.8*	*33.1*	*31.2*		
% share total power output generated	*33.7*	*36.0*	*33.9*	*26.3*	*28.5*		

Singapore

	(s)	(s)	(s)	(s)	(s)	(s)	(s)	
Biomass power	(s)	(s)	(s)	(s)	(s)	(s)	(s)	–
Total	**(s)**	**(s)**	**(s)**	**(s)**	**(s)**	**(s)**	**(s)**	**–**
% share total installed capacity	*0.2*	*0.2*	*0.2*	*0.2*	*0.2*	*0.2*	*0.2*	
% share total power output generated	*0.3*	*0.2*	*0.3*	*0.3*	*0.3*	*0.3*	*0.3*	

Thailand

	(s)	(s)	(s)	(s)	(s)	(s)	(s)	
Hydropower	3.5	3.5	3.5	3.5	3.5	3.5	3.5	0.324 GW additional small-hydro (2021)
Wind	0.0	0.0	0.0	0.0	0.0	0.1	0.2	
Solar PV	0.0	0.0	0.0	0.0	0.4	0.4	0.7	2 GW (2021)
Biomass power	0.0	0.7	0.8	0.8	0.8	0.8	0.8	3.63 GW (2021)
Total	**3.5**	**4.2**	**4.3**	**4.3**	**4.8**	**5.2**		***Primary energy: 25% (2021), 9.201 GW electricity (2021)***
% share total installed capacity	*9.2*	*14.1*	*10.4*	*8.9*	*8.9*			
% share total power output generated	*5.8*	*6.9*	*6.6*	*5.6*	*8.3*			

Vietnam

	(s)	(s)	(s)	(s)	(s)	(s)	
Hydropower	4.2	4.5	4.6	5.5	10.0	14.0	17.4 GW (2020); pumped storage 1.8 GW (2020), 5.7 GW (2030)
Total	**4.2**	**4.5**	**4.6**	**5.5**	**10.0**	**14.0**	***Primary energy: 5% (2020), 8% (2025), 11% (2050); Non-hydro electricity: 4.5% (2020), 6.0% (2030), 0.241 GW p/year average new (2006-2015), 160 MW p/year (2016-2025), 4.05 GW new (2025)***
% share total installed capacity	*36.2*	*35.2*	*31.7*	*36.2*	*45.2*		
% share total power output generated	*38.6*	*33.8*	*35.5*	*29.1*	*43.6*		

(continued)

Table 3.7 Southeast Asia's renewable energy profile (*continued*)

	Installed capacity, GW						Sector targets
	2004	**2006**	**2008**	**2010**	**2012**	**2013**	

Southeast Asia*

Hydropower	18.9	19.8	20.7	22.0	27.0	31.0	–
Geothermal	2.7	2.9	3.0	3.2	3.3	3.2	–
Solar PV	0.0	0.0	0.0	0.0	0.4	0.8	–
Biomass power	0.0	0.7	0.8	0.8	0.8	0.8	–
Total	**21.6**	**23.4**	**24.5**	**26.0**	**31.5**	**35.8**	–
% share total installed capacity	*16.7*	*18.5*	*17.3*	*16.9*	*19.3*		
% share total power output generated	*15.8*	*15.9*	*16.7*	*15.2*	*18.4*		

Sources: GWEC (2014), EPIA (2014), EIA (2014), IEA (2014), BP (2013), REN21 (2014).

Notes: All data are for power generation. Hydropower capacity figures do not include pumped storage. Biomass includes waste incineration. * Includes Brunei, Cambodia, Laos and Myanmar in addition to other six listed.

3.7.2 Southeast Asia's renewable energy policy agencies and stakeholders

Generally speaking, RE policy-making in Southeast Asia is strongly state-centric, in that private sector and civil society stakeholders exert relatively limited influence over policy formation. There are thus similarities here with China. Government and state-related organisations (e.g. SOEs) tend to dominate the sub-region's political economy of renewable energy development. The main state agencies responsible for RE policy in Southeast Asia and other stakeholders are the following:

- *Indonesia*: The Ministry of Energy and Mineral Resources is responsible for general policy and the National Development Board for strategic energy planning. Perusahaan Listrik Negara (PLN) is the state-owned electricity power company that dominates the generation market. The Indonesian Renewable Energy Society was formed in 1999 as a relatively loose network of government, businesses, scientists, and civil society representatives but with very limited political influence. The most prominent sector group is the Indonesia Geothermal Association, established in 1991, comprising around 500 members.
- *Malaysia*: The Energy Division of the Economic Planning Unit (in the Prime Minister's office) is primarily responsible for energy policy and strategic planning. The Ministry of Energy, Green Technology and Water, together with the Energy Commission, regulates the electricity industry and market and is heavily involved in RE policy-making. The Malaysian Industrial Development Authority promotes renewables industry manufacturing. The Business Council for Sustainable Development in Malaysia engages the business community in environment-related matters, such as renewables.
- *Philippines*: The National Renewable Energy Board of the Department of Energy is in charge of RE policy, and the Energy Regulatory Commission regulates the electricity market. The Renewable Energy Association of the Philippines represents RE sectoral associations, such as the National Geothermal Association of the Philippines, the Philippine Solar Energy Society, Association of Wind Energy Producers in the Philippines, and Wind Energy Association of the Philippines.
- *Singapore*: The Ministry of Trade and Industry formulates energy policy and leads the inter-ministerial Energy Policy Group. The Energy Market Authority (EMA) regulates the electricity industry, and the Economic Development Board (EDB) facilitates partnerships between foreign firms and research agencies on developing new RE technologies. Meanwhile, the Ministry of the Environment and Water Resources, the NEA, Clean Energy Programme Office, Energy Innovation Programme Office, and National Climate Change Committee oversee renewable energy, climate change, and energy-efficiency related policies. The Sustainable

Energy Association of Singapore is the city-state's only notable non-government actor representing green energy-related business interests.

- *Thailand*: The main policy-making agencies are the Energy Policy and Planning Office and the Department of Alternative Energy Development and Efficiency, both located in the Ministry of Energy. In addition, the following bodies are involved in RE policy: the Electricity Generation Authority of Thailand, the Provincial Electricity Authority, the Ministry of Science and Technology, the National Science and Technology Development Agency, the Ministry of Industry, Ministry of Agriculture and Land Use, Office of Natural Resources and Environmental Policy and Planning, the Board of Investment, and the Thailand Greenhouse Gas Organisation. The Renewable Energy Institute of Thailand Foundation performs mainly public information and debate functions and is primarily focused on bioenergy issues. In addition, the Thai Photovoltaic Industries Association serves the interests of the country's fast expanding solar PV industry.
- *Vietnam*: The Energy Department of the Ministry of Industry and Trade is responsible for energy policy and strategy-making, while Electricity of Vietnam is the state-owned power company, which monopolises the electricity industry and market. The quite recently established Wind Energy Association Binh Thuan is Vietnam's most important renewables sectoral organisation.

In addition, there are a number of largely state-owned power generation companies across Southeast Asia (e.g. Indonesia's PLN) and a myriad of mainly SMEs as well as generally larger foreign ones producing renewable energy equipment.

3.7.3. The evolution of Southeast Asia's renewable energy policy

Southeast Asia's earliest RE policies had a strong sector-specific focus. Government initiatives on developing hydropower and geothermal date back many decades, but these were primarily project-based in nature and focused on specific installations. More defined renewable energy policies for nationwide development began to emerge in the late 1970s. In 1978, the Philippines government introduced its Act to Promote the Exploration and Development of Geothermal Resources. A year later in 1979, Singapore established a biomass power generation scheme with mandatory obligation measures for power companies to utilise municipal solid waste feedstock. Malaysia's 1981 Four Fuel Diversification Strategy especially targeted hydropower (small and large) for further development, while relative late-starter Indonesia first introduced a multi-sector RE programme in 1990 that afforded special priority to geothermal and small hydro. By that time, the other late-starters, Thailand and Vietnam, began developing their own RE policies and a more multi-sector approach was apparent in the sub-region from the early 2000s onwards.

Most Southeast Asian countries have, like China, adopted an installed capacity-oriented approach to renewable energy policy. Singapore is the exception here, adopting a more technology-oriented approach due to two main factors: first, because the city-state has significant spatial constraints on installation development, and second, because it is by far Southeast Asia's most developed economy with strong techno-innovatory capacity (see Box 3.2). In some parts of the sub-region, economies of scope have been partly instrumental in the development of certain RE sectors. For example, Singapore and Malaysia's technological expertise and value-chain capabilities in semiconductors and electronics have provided a strong foundation for their solar PV manufacturing industries. Thailand's strengths in agro-processing, agriculture, and oil/gas confer it with advantages in bioenergy production. In each country case, these are the sectors that have been afforded policy priority for development. Most Southeast Asian governments used a mixture of direct financial support (e.g. subsidies, grants) and market-based mechanisms (e.g. tax incentives) to foster early renewable energy development (IEA 2014). Only Singapore from the ASEAN-6 group studied has eschewed the use of FiTs and other forms of subsidising renewables consumption. The government stated this is because of ideological reasons (i.e. only commercial criteria should apply to energy market decisions) but practical factors also apply. Solar PV is the nation's main renewable energy option and, as discussed in Chapter 5, this sector is fast approaching grid parity in many countries; therefore, it is increasingly less dependent on public funding support.[16]

Over time, the general legislative and policy frameworks for renewables in Southeast Asia has strengthened, improving the regulatory and financial environment in which RE sectors could flourish. At first, these frameworks were subsumed into broader energy-related legislation, such as Malaysia's aforementioned 1981 Four Fuel Diversification Strategy and Thailand's 1992 Energy Conservation Programme, which comprises two sub-programmes to promote renewables development in rural areas. Other policy reforms and new legislation introduced from the 1980s onwards across the sub-region helped create more competition in the electricity market, allowing renewable energy producers more scope to supply national grids (Pacudan 2005a, 2005b). Examples include Indonesia's Energy Law (2007) and Electricity Law (2009), Malaysia's Electricity Supply Acts (1990, 2001), the Philippines' Electric Power Industry Reform Act (2001), Singapore's Energy Market Authority Act (2001), and Thailand's Energy Business Act (2007).

More RE-specific frameworks were introduced from the early 2000s onwards. Vietnam's Renewable Energy Action Plan of 2001 was both multi-sector and multi-scale in approach. In the same year, Malaysia rolled out its Small Renewable Energy Programme, which enabled small-scale RE power

Box 3.2 Singapore: new technology over new installation?

Singapore is often thought of as a distinctly different part of Southeast Asia in many ways. It is by far the sub-region's most advanced economy and richest society. Moreover, it is best thought of a city-state rather than a nation-state, and it should be compared to other global cities such as Hong Kong, Tokyo, Seoul, and Shanghai in terms of understanding its development, position in the world, and specific challenges it faces. Singapore is an almost completely urbanised sovereign territory with extremely limited natural resources of its own but with very high levels of techno-industrial development. It is Southeast Asia's principal high-tech hub for manufacturing, research, and development.

Therefore, it is not surprising that the government has taken a strong technology-oriented approach to renewable energy policy. Its 1991 Joint Research with Universities on Environment Related Research scheme was the first in Southeast Asia to provide R&D support on renewables (Table 3.1). Most of Singapore's RE policy initiatives since have aimed at integrating renewable energy technology and equipment into the city-state's buildings and urban-industrial fabric, such as the Solar Capability Scheme on solar PV applications. Yet, for a state that has a good reputation for strategic policies on developing emerging industries, Singapore has been surprisingly reluctant at prioritising renewable energy sectors, such as solar PV, for power generation. However, it has become an important hub for manufacturing and R&D activity. For example, Norwegian firm REC operates the world's largest integrated PV wafer, cell, and module manufacturing plant in Singapore, which began production in 2010. The company also moved its R&D operations to the city-state. Incentives and other forms of support provided by the EDB were crucial in attracting this strategic industry investment from Europe.

While Singapore's accession to the Kyoto Protocol of the United Nations Framework Convention on Climate Change in 2006 did lead to the government affording more attention to renewables and other clean energy technologies thereafter, Singapore still remains the only East Asian country not to set any specific targets on installed RE development. Although the Ministry of Trade and Industry-led Economic Strategies Committee had recommended in the late 2000s that Singapore should aim for renewables to contribute to 5 percent of electricity generation by 2030, this was reportedly resisted by the Singapore Power Grid and EMA, who argued that the grid would become far more vulnerable if such a 'high' share of electricity depended on intermittent solar PV.[17]

Catalysing R&D and new technology demonstration or 'technology primer' projects constitute the main core of Singapore's RE policy, which also extends to other aspects of energy policy generally, such as smart grids.[18] Furthermore, the city-state has positioned itself as a test-bed market for certain new green energy technologies (Jacobs and Sovacool 2012), such as electric vehicles, for which the government has worked in partnership with Renault-Nissan, Daimler, Mitsubishi, and other automotive companies. In addition to REC's aforementioned R&D operations on solar PV, many other major foreign renewable energy firms, such as Vestas, Gamesa, Panasonic, Trina Solar, Siemens, and Yingli also have new technology development centres in the city-state (EDB 2012). These firms are supported by the efforts of the Solar Energy Research Institute of Singapore, established in 2008. Biomass waste generation has been the main area of Singapore's limited installation capacity. Indeed, the government continues to stress the limited renewable energy options available to the city-state on power generation due to various spatial and resource constraints (Singapore National Climate Change Secretariat 2012). This narrow scope for multi-sector development has led to renewables being deeply embedded within Singapore's 'clean energy' policy, which has historically placed greater emphasis on energy efficiency and lower carbon-intensive energy usage–what the government generally refers to as the objective of energy sustainability.[19] There are notable similarities to the approaches of developed Northeast Asia (i.e. Japan, South Korea, Taiwan) to green energy development, as examined earlier.

Although one could argue that Singapore could do considerably more toward expanding the use of renewables for its own power generation, new development on RE technologies being carried out in the city-state (e.g. floating solar PV, building-integrated wind turbines) could make important contributions to the global development of renewables. As a regional financial services hub, Singapore is also playing a kind of facilitating role in foreign RE project development, albeit still rather limited at this point.[20]

producers to supply the electricity grid for the first time. Indonesia's 2003 Geothermal Law meanwhile created a stronger legal basis for further developing its largest renewables sector, which included greater deferring of policy autonomy to local governments, such as in offering financial incentives to private investors and the operation of special business permits. This was followed in 2004 by its Green Energy Policy, a more general multi-sector policy framework aiming to improve infrastructure development, public awareness, international co-operation, and R&D activity in renewables. A few years later, the Philippines passed its 2008 Renewable Energy Act, which brought a raft of new legislation and financial incentives, and included FiT and RPS schemes. Compared to Northeast Asia, Southeast Asian countries were relatively late in introducing RPS and FiT legislation. Thailand was the first from the region on both accounts, in 2004 and 2007 respectively. They also lagged behind Northeast Asia on setting targets for future renewable energy sector development; Malaysia and Vietnam were the first to do so in 2001.

It is interesting to note, however, that multi-sector strategic planning on renewables began earlier in Southeast Asia than in Northeast Asia, with the only exception being South Korea. Malaysia and Vietnam were again at the sub-region's vanguard, both introducing their first plans in 2001 (Table 3.1). Other ASEAN countries followed suit in the early 2000s, in the sub-region's first phase of RE strategies. These were later superseded or followed by a second phase of more coherent, ambitious, and substantive renewable energy strategy-making in the late 2000s and early 2010s. A summary of this subject follows on a country-by-country basis:

- *Indonesia*: The National Energy Blueprint (2005–2025) introduced in 2005 was a general energy policy strategy that included substantive plans on renewables development. This was soon followed by the 2006 National Biofuel Roadmap (see Chapter 8). The government revised its targets significantly upward and strategic RE planning was further augmented by supporting measures and short-term targets introduced in the Mid-Term Development Plan 2010–2014, which had a stronger focus on lower carbon development generally than previous macro-plans.
- *Malaysia*: The Small Renewable Energy Programme (2001) was integrated into the Eighth Malaysia Plan (2001–2005) and devised to also complement the Five Fuel Policy introduced in the same year. This was followed by the Renewable Energy Development section of the Ninth Malaysia Plan (2006–2010), and then the 2010 National Renewable Energy Policy and Action Plan/Renewable Energy Act.
- *Philippines*: Its Renewable Energy Policy Framework 2003–2013 was a plan for attaining various development targets by 2013, in effect seeking to double the nation's RE installed capacity. It additionally aimed to make the Philippines the region's main solar PV manufacturing hub and largest wind energy producer. Five years later, the 2009 Renewable Energy Act provided the core legislative basis for a new National Renewable Energy Programme 2011–2030.
- *Singapore*: The city-state's strategic planning on renewables is integrated into the 2007 Comprehensive Blueprint for Clean Energy.
- *Thailand*: Its Strategic Plan for Renewable Energy Development (2004) was amongst Southeast Asia's earliest most comprehensive strategic development plans, with a strong multi-sector focus and multi-policy components. This was followed by the Renewable Energy Development

Plan 2008–2022, which was soon superseded by the more ambitious Alternative Energy Development Plan 2012–2021.

* *Vietnam*: The country's Renewable Energy Action Plan (2001) was part of its broader 2001–2010 Master Plan of Power Development. Extending electrification to remote areas through RE deployment was a core objective. Renewables were an integral component of Vietnam's National Energy Development Strategy 2008–2020. Similar to Thailand, this was soon superseded by the National Power Development Plan 2011–2030, which set even higher goals for renewable energy development.

Both the substance and ambitions of Southeast Asia's strategy-making on renewables development and new developmentalism generally strengthened from the late 2000s onwards. These strategies also dovetailed into new lower carbon development plans (e.g. Vietnam's Green Growth Strategy, Cambodia's National Strategic Plan on Green Development 2013–2030, Malaysia's Green Technology Strategy) or broader new developmentalist macro-plans (Table 2.1, p. 44). Aside from installed capacity targets, they also afforded greater priority to renewables as future strategic industries. Second-phase strategic plans also laid the platform for subsequent important policy measures to follow, such as revised FiT schemes aiming to further diffuse renewable energy technologies.

3.7.4 Southeast Asia's renewable energy development: main future challenges

A major challenge facing Southeast Asia, especially on an individual country basis, concerns the diversification of renewables development on power generation. Most ASEAN countries have come to depend heavily on hydropower plus another reasonably scaled-up sector, such as geothermal, biomass, or solar PV (Table 3.7), and for many where this limited diversification has occurred it has been only very recently. Deeper regional or even bilateral cooperation on RE electricity generation may offer some kind of solution, based on a more robust ASEAN Power Grid programme or cross-border trade arrangements, such as that which has operated for some years between Laos and Thailand, where the former has exported hydropower electricity to the latter. The greater exchange of renewables power generation across national borders within Southeast Asia would allow for greater economies of scale to be captured and accordingly attract higher levels of investment. However, this would depend on establishing stronger technical compatibilities among different national grids, as well as whether regional strategies on renewable energy development are compatible with the aims of national-level strategies.

Another challenge relates to state capacity and development capacity issues. Although RE policy in Southeast Asia has undoubtedly strengthened–not least because of a greater demonstrated commitment to lower carbon development–there remain serious questions concerning whether the sub-region's governments are able to fully and effectively implement their policies and strategic plans on renewables. As discussed in Chapter 2, and with the exception of Singapore, Southeast Asia's developing nations lack the same level of state capacity (e.g. technocratic and institutional aspects) enjoyed on the whole by their Northeast Asian counterparts, as well as generally weaker development capacity (e.g. infrastructural, commercial-industrial, and human capital) that presents additional challenges regarding the engagement of business in all aspects of RE development. In addition, in many Southeast Asian countries (most notably the largest of all, Indonesia), tensions exist between national government and relatively autonomous local governments (provinces, cities) over policy implementation generally–a challenge pertinent to China, as earlier discussed.

The main competitive challenge facing renewables in the sub-region arguably comes from the substantial subsidies enjoyed by fossil fuels that in 2012 amounted to around US$51 billion, with this figure having doubled over five years (IEA 2013b). This relates to electricity generation and transportation,[21] and subsidies are notably high in Indonesia, Malaysia, and Thailand. If these were removed, it would allow RE sectors to compete on a more level playing field with oil, gas, and coal, and in addition help mitigate carbon emissions and promote energy efficiency. However, it will be politically difficult to do so given that governments have hitherto used fossil fuel subsidies as a populist measure to garner support. Subsidised fossil fuel consumption has also long been viewed as an important socio-economic policy of providing low prices for a necessity good in a low-income society. Switching fossil fuel subsidy funds to investing in the expansion of renewables instead is now being viewed as a more serious option than before due to the stronger commercial viability of RE technologies and more committed efforts on low carbon development generally in the sub-region (IEA 2013b). By the early 2010s, no Southeast Asian nation had nuclear power generation capacity, but all ASEAN-6 nations were conducting feasibility studies or had plans (the most advanced being Vietnam) in this sector despite the 2011 Fukushima disaster (Singapore International Energy Week 2012, Desker 2013). This also presents a future potential competitive challenge to renewable energy in the sub-region.

3.8. Conclusion

This chapter has examined the general evolution of East Asia's renewable energy policies, setting these in their broad development contexts, identifying key policy stakeholders, and highlighting the main challenges lying ahead for the future development of renewables. It was shown that strategic planning has become an increasingly prominent feature of East Asia's RE policy-making, and that this has become more substantial and ambitious in most parts of the region since the late 2000s especially. The chapter then offered more detailed evidence of East Asia's new developmentalism, yet this needs to be critiqued.

In most parts of East Asia–especially in its developing country areas–renewables had initially been viewed as a solution to providing off-grid electricity to remote communities. Its more highly developed economies also placed an early emphasis on R&D support measures, which is consistent with international norms. Initial RE policies thus focused on these two issues for some time, with additional ad hoc measures aimed at promoting a wider deployment of renewable energy installations and technologies. By the early 2000s, the region's RE policy architecture became more structured around medium- to long-term strategic plans, which often were superseded before their term period by more comprehensive plans with upward revised targets on renewables development. We can broadly identify a first phase of RE strategic planning that occurred across the region during the early to mid-2000s, and then a second phase from the late 2000s onwards that generally coincided with the green growth fiscal stimulus policies introduced in response to the 2008–2009 financial crisis (Chapter 2). Thus, this second-phase strategic planning on renewables was founded on a strengthening of state capacity and more ambitious intent toward lower carbon development.

Certain patterns in this new developmentalism could be observed over East Asia. China and Vietnam's renewable energy policy and strategic planning has always been deeply embedded in its macro-development plan frameworks, such as China's FYPs. The Chinese government in particular has brought growing direct financial support to bear on the development of most RE sectors. Its strategic plans have also incorporated a growing number of market-based policy support mechanisms that correspond with the country's wider economic reform process. In first-phase strategic planning, most governments in the region set short- to medium-term targets on renewables power generation and other fields. This especially applied to Japan and South Korea. However, much longer-term strategic development targets (e.g. up to 2030 or 2050) were set in second-phase planning; this indicated a firmer commitment to RE sectors as future strategic industries and their contribution to meeting future energy security challenges. East Asian governments were not only also clearly inter-linked renewable energy strategies with the country's climate change strategy and development plans and policy measures under the emergent greater influence of ecological modernisation.

Despite the greater strategic intents that East Asian governments have recently demonstrated for promoting renewables, in Northeast Asia (excluding Taiwan) new developmentalism has too been strongly focused on nuclear power. China and South Korea have the world's most ambitious plans on the future development of their respective nuclear sectors, and notwithstanding the aftermath of the 2011 Fukushima disaster Japan is also likely to revert at some point to nuclear as a foundational element of its long-term energy security strategy. Indeed, the prioritisation of nuclear in formulating low carbon energy strategies in Japan and South Korea in the past largely explains the relatively slow development of renewables power generation in both countries. More generally, in East Asia's most developed economies–Japan, South Korea, Taiwan, and Singapore–renewables development has been overtly technology-oriented and also more preoccupied with fostering strategic industry capacity for production and export, rather than expanding installed capacity for generating electricity. Even though Japan, South Korea, and Taiwan have upwardly revised their future targets on installed capacity, in proportionate terms these are still not that ambitious when compared to China and many Southeast Asian countries.

In terms of power generation, it is evident in many parts of East Asia that strategic plans for increasing the deployment of renewables from the early 2000s onwards have had mixed results. As data presented in this chapter show, from 2004 to 2013, China more than tripled its installed renewables capacity in absolute terms, from 106.0 GW to 378.9 GW. In contrast, Japan could only manage an additional 16.5 GW increase to its total over those years, with increases of 2.7 GW for South Korea, 1.2 GW for Taiwan, and 14.2 GW for the whole of Southeast Asia. Even in China's case, these impressive figures are set in the context of a very fast expanding power generation sector generally, where renewables' share of total capacity increased just a few percentage points (24.7 percent to 30.5 percent) over the same period. This does, then, call somewhat into question just how effective East Asia's new developmentalism has hitherto been to date for catalysing the expanded deployment of renewables, especially outside China. One could argue that it is still too early to make an overall assessment of how effective much of East Asia's second-phase RE strategic planning will prove to be, given that we are still at the beginning of the process, and it will could take some time before we begin to see its longer-term effects and impacts.

This chapter also noted the growing public support for RE policies and low carbon development, generally in East Asia. Business and civil society are becoming more active and engaged stakeholders in renewable energy development per se, particularly in the region's more open democratic societies, where there is wider contesting influence in shaping energy policy. At the same time, energy SOEs remain an important extension of government RE policy across many parts of East Asia, most significantly in China (Box 3.1). These enterprises continue to dominate many aspects of renewable energy industry value chains, especially in power generation. Here, largely uncontested national (e.g. KEPCO in South Korea, TaiPower in Taiwan, PLN in Indonesia) or regional (*Denjiren* companies in Japan, the big five power companies in China) SOE monopolies have limited the scope for new competition from independent private-sector firms entering the market. These SOEs have in many ways been gatekeepers of renewables power generation, being also the same firms that dominate conventional fuel sectors.

Overall, RE policy-making and power generation in East Asia still remains a largely state-centric process. In Japan specifically, energy-related government agencies have maintained a strong strategy-making capacity but have been hampered by constant change in political leadership in the state's executive arm. In South Korea, it remained unclear for a while if GGS-framed renewable energy policies would take a new strategic direction after new President Park took office in 2013. For obvious reasons, continuity in strategic planning on renewables is much less of an issue in mono-regime states, such as China and Vietnam. Although all states have to be flexible in adapting their strategies on renewable energy development in light of new trends and intervening variables arising, it is important to remain committed to strategy's core elements over their duration for them to maintain credibility and stakeholder confidence.

Notes

1 This typically involves paying a generation tariff for all electricity generated and then paying an additional export tariff to the producer's excess electricity generated supplied back to the grid.
2 The World Health Organisation (2012) estimated that air pollution contributed to 1.3 million premature deaths in China in 2011, which was the fourth-leading mortality risk factor behind diet, high-blood pressure, and smoking.
3 http://www.nytimes.com/2013/03/30/world/asia/cost-of-environmental-degradation-in-china-is-growing.html (accessed May 29, 2014).
4 In addition, the Inner Mongolia provincial government operates its Western Inner Mongolia grid system.
5 Supported by the World Bank and Global Environment Facility.
6 The 12th FYP set the intermediate target of achieving a 17 percent reduction by 2015.
7 http://www.reuters.com/article/2011/05/25/idUS38212+25-May-2011+BW 20110525 (accessed May 28, 2014).
8 METI's predecessor organisation form was the Ministry of International Trade and Industry.
9 The Japan Association for Water Energy Recovery, Japanese Wind Power Association, Geo-Heat Promotion Association of Japan, Solar System Development Association, Japan Geothermal Developers' Council, Geothermal Research Society of Japan, Research Committee on Climate Change of the Architectural Institute of Japan, and the Japan Wood Pellet Association.
10 Under its Comprehensive Review of Japanese Energy Policy, the government also set a target of renewable energy accounting for 10 percent of primary energy supply by 2020, which had not been further revised by the mid-2010s.
11 Formed in 2013 and superseding the Ministry of Knowledge Economy, which itself was created in 2008.
12 The government has a controlling 51 percent share of KEPCO's ownership.
13 Other power generation companies include Korea District Heat Corporation, Korea Water Resources Corporation, Posco Energy, GS Power, and MPC Yulchon.
14 http://www.koreaherald.com/view.php?ud=20131110000342 (accessed May 25, 2014).
15 Research interview with Singapore government officials (EMA, EDB), January 2013.
16 Moreover, where the Singapore government does offer subsidies on new capital investment on renewables (through the EDB), there must be a research goal

or new innovation outcome delivered in partnership with a research institute (research interview with government official, Singapore, January 2013).

17 Research interview with Energy Studies Institute analyst, Singapore, January 2013.

18 This is managed through the National Innovation Challenge programme of Singapore's National Research Foundation and the EDB's Spring Singapore scheme (research interviews with Singapore government officials, January 2013).

19 Research interviews with Singapore government officials from various agencies, January 2013.

20 Research interview with Singapore Economic Development Board officials, January 2013.

21 Oil is the principal subsidised fuel, and thus subsidies mainly concern transportation.

4 Wind energy

4.1. Introduction

Wind energy is currently the largest non-hydro renewables sector, and it has been at the forefront of strategic planning on renewables in most parts of East Asia. The theoretical global energy resource potential of wind energy is second only to solar, and it is far greater than all fossil fuels combined. Since the mid-2000s, worldwide installed wind capacity has grown at annual average rates in excess of 20 percent, as has wind turbine production (GWEC 2014, windpower.net 2014). Ambitious government and business plans for future expansion strongly indicate that wind energy will be an important strategic industry for years and even decades to come. Nowhere is this more clearly evident than in East Asia.

4.2. Wind energy: an overview

4.2.1. General evolution of the sector

For many centuries, human societies have harnessed wind energy for various utility purposes. Windmills have been used to pump water and basic agricultural processing in many parts of world.[1] The first electricity-generating wind turbines were developed in Europe and the United States from the late 19th century onward; these had very low-level kilowatt capacities.[2] Experimental, larger-capacity wind turbines were developed around the mid-20th century, assisted by technological developments in aeroplane manufacturing (e.g. propellers, mono-plane wings), with many turbines being grid-connected (Fleming and Probert 1984, Gipe 1991, Kaldellis and Zafirakis 2011, Musgrove 2010, Righter 1996).

Modern wind turbines convert the kinetic energy of moving air into mechanical and then electrical energy. The rotational energy of a moving turbine is converted into rated power by its generator, which then passes through a transformer before supplying power to an electricity grid or other system. The power output of wind turbines rises exponentially with wind speed, and areas that have constant and high wind densities are naturally prime locations for developing wind farms (e.g. offshore and high altitudes) to ensure the best potential and most predictable power generation. The threshold minimum 'cut-in' speed that modern large turbines require to produce electricity is approximately 3–4 metres per second, and the maximum 'cut-out' speeds of most turbines are in the 20–25 metres per second range; higher speeds would cause structural and component damage. Due to wind speed intermittencies and the non-dispatchability problem (see Chapter 1), wind energy is essentially a supplement to constant and predictable forms of energy generation. However, its potential optimum share in the total energy mix will increase with grid expansion as more wind resources are harvested across larger inter-connected areas.

An international wind energy industry began to emerge by the late 1970s, with development initially concentrated in North America and Europe. Offshore wind farms were first developed by Denmark in the early 1990s, accounting for just 2.2 percent of global installed wind capacity by 2013 (GWEC 2014). Although it is much more expensive to develop than onshore wind, offshore wind offers higher capacity loads due to stronger and more constant sea winds. The largest wind turbines in development are for offshore farms. They are less prone to siting conflicts (visual and noise related) and land-based transportation constraints in the wind farm construction process, as offshore wind turbines are invariably manufactured in nearby seaports. Although operating and maintenance (O&M) costs are normally higher than for onshore, offshore farms offer much greater scope for capturing scale economies. Offshore wind farms are situated in relatively shallow water areas (normally up to 20 metres deep) and typically up to 20 km from shore. Both distance parameters have increased over time and will continue to do so with improvements in engineering technology, thus extending offshore wind into higher energy-yield zones. Many of the world's largest proposed new wind farms are offshore, and the growing involvement of oil, gas, and large civil engineering firms in the offshore sub-sector has intensified competition. For Northeast Asia's densely populated economies of Japan, South Korea, and Taiwan in particular, offshore wind is likely to become a strategically important renewables option.

4.2.2. Growth and expansion

Wind energy has experienced around an 18-fold overall increase in global installed power generation capacity from 2000 (18.0 GW) to 2013 (318.1 GW), with an average annual growth rate of 21 percent over the 2008–2013 period, and it has added more new capacity to the renewables total than any other sector except large-hydro (GWEC 2014, REN21 2014, WWEA 2013a). Together, the world's wind farms have an electricity generation capacity of well over twice that of solar energy and over three-and-a-half times that of biomass. Wind energy currently contributes approximately 3 percent of total worldwide power generation capacity. This is a small share, but it has more than doubled in the last 5 years. The number of countries with grid-connected wind farms continue to grow, rising to 102 by 2013 (Windpower. net 2014).

The United States initially led the way in global wind energy generation, by 1990 accounting for 79.9 percent of the global total; Europe accounted for 19.2 percent and the whole Asia and Oceania region just 0.9 percent. However, a decade later, the positions of the United States and Europe became reversed, with the European Union being responsible for 73.3 percent of global installed capacity by 2000; the United States had 14.2 percent at this point, while East Asia held just a 2.7 percent share (Table 4.1). Germany and Spain have long been Europe's leaders on this front, partly explained by having large wind turbine manufacturers, such as Enercon, Siemens, RE Power, Nordex, Gamesa and Ecotecnica (Table 4.2). For some time, Danish company Vestas has been the world's largest turbine producer and remains an important technological leader.

As the 2000s progressed, East Asia–and in particular, China–began to make an increasing impact on the industry globally. In 2000, the country's installed capacity was just 352 MW, around the same as the United Kingdom. From 2004 to 2009, China's capacity level doubled annually; by 2010, it had overtaken the United States as the world's leading nation on installed capacity. By 2013, China's capacity level was 91.4 GW, 28.7 percent of the global total, remaining well ahead of the United States (61.1 GW) and the leading European countries of Germany (34.3 GW) and Spain (23.0 GW). The contribution of another emerging power, India, should also be noted–now ranked fifth in the world with 20.2 GW. The European Union's share of total global wind energy capacity had fallen to 36.9 percent by 2013, but it still remains the world's largest producer with 117.3 GW capacity. In terms of installed capacity, East Asia's wind energy sector is very much dominated by China, with its share of the region's total rising to 95.6 percent by 2013; this completely overshadowed neighbouring economies such as Japan, Taiwan, and South Korea, whose collective capacity level was less than Portugal's that year. China drove a 22-fold increase in East Asia's wind energy capacity from 2006 to 2013, from 4.3 GW to 95.6 GW. Consequently, the region's share of the global total has risen significantly over this period, from 5.8 percent to 30.0 percent (Table 4.1).

4.3. Techno-innovation and production

4.3.1 General trends

Wind energy has matured into a mass production industry. Techno-innovatory advances and intensifying competition have helped to drive production efficiency. In 1980, the typical cost range of wind-generated electricity was US$0.60 to US$0.70 p/kWh, but by 2011 it had fallen to as low as US$0.05 p/kWh (IEA 2004, REN21 2012). Taking an illustrative

Table 4.1 Wind energy development, East Asia and global, 2000–2013

Installed capacity (MW)

	2000	2006	2007	2008	2009	2010	2011	2012	2013
China	352	2,599	5,912	12,210	25,810	44,733	62,364	75,324	91,412
Japan	142	1,309	1,528	1,880	2,083	2,304	2,501	2,614	2,661
Taiwan	3	188	280	358	436	519	564	571	614
South Korea	0	176	192	278	348	379	407	483	561
Thailand	0	0	0	0	0	0	8	112	223
Mongolia	0	0	0	2	2	2	2	1	50
Philippines	0	25	25	25	33	33	33	33	33
Vietnam	0	0	0	1	9	30	31	31	31
Indonesia	0	1	1	1	1	1	1	1	1
East Asia total	**497**	**4,298**	**7,938**	**14,755**	**28,722**	**48,001**	**65,911**	**79,170**	**95,586**
% share of world total	*2.7*	*5.8*	*8.5*	*12.2*	*18.0*	*24.4*	*27.7*	*28.0*	*30.0*
Germany	6,095	20,622	22,247	23,903	25,777	27,191	29,060	31,308	34,250
Spain	2,535	11,630	15,145	16,740	19,149	20,623	21,674	22,796	22,959
United Kingdom	409	1,963	2,389	3,288	4,051	5,204	6,540	8,445	10,531
France	68	1,567	2,455	3,404	4,492	5,970	6,684	7,564	8,254
Italy	427	2,123	2,726	3,736	4,850	5,797	6,737	8,144	8,552
Denmark	2,417	3,136	3,125	3,160	3,465	3,749	3,871	4,162	4,772
Sweden	241	571	831	1,067	1,560	2,163	2,907	3,745	4,470
Portugal	83	1,716	2,130	2,862	3,535	3,706	4,083	4,525	4,724
Other EU	510	3,144	3,722	4,328	5,811	7,402	10,063	12,961	18,777
European Union	**13,225**	**48,031**	**56,517**	**64,713**	**74,919**	**84,074**	**93,947**	**106,041**	**117,289**
% share of world total	*73.3*	*64.8*	*60.2*	*53.5*	*46.9*	*42.8*	*39.5*	*37.5*	*36.9*
United States	**2,564**	**11,575**	**16,823**	**25,237**	**35,159**	**40,180**	**46,919**	**60,007**	**61,091**
% share of world total	*14.2*	*15.6*	*17.9*	*20.9*	*22.0*	*20.4*	*19.7*	*21.2*	*19.2*
India	1,267	6,270	7,850	9,587	11,807	13,066	16,084	18,421	20,150
Rest of world	490	3,948	4,799	6,613	9,161	11,310	14,817	19,067	23,989
World	**18,040**	**74,122**	**93,927**	**120,903**	**159,766**	**196,630**	**237,669**	**282,587**	**318,105**

Sources: GWEC (2014), windpower.net (2014).

Table 4.2 World's top-10 wind turbine producers, 2005 and 2013

2005				2013			
Rank	*Company*	*Country*	*Capacity (MW)*	*Rank*	*Company*	*Country*	*Capacity (MW)*
1	Vestas	Denmark	3,200	1	Vestas	Denmark	4,910
2	Enercon	Germany	2,700	2	Goldwind	China	4,120
3	Gamesa	Spain	1,900	3	Enercon	Germany	3,670
4	GE Wind	United States	1,300	4	Siemens	Germany	2,770
5	Siemens	Germany	1,100	5	GE Wind	United States	2,470
6	Suzlon Group	India	900	6	Gamesa	Spain	2,060
7	RE Power	Germany	900	7	Suzlon Group	India	1,990
8	Goldwind	China	700	8	Guodian United Power	China	1,500
9	Nordex	Germany	500	9	Mingyang	China	1,310
10	Ecotecnica	Spain	300	10	Sinovel	China	1,240

Sources: BNEF (2011), Navigant Research (2014).

example from China, operating generation tariffs in large-scale wind farms in Hebei and Xinjiang provinces were 3,850 yuan per MW in 2011, compared to 6,200 yuan per MW capacity just a few years earlier in 2008. In many countries, wind energy can now compete quite effectively on price against fossil fuels; at its most efficient, it is on cost parity with grid-connected hydropower. The wind sector also may be considered as a mainstream advanced technology industry, attracting fast-growing research and development (R&D) funding and venture capital investment since the early 2000s. As discussed in Chapter 1, the wind energy industry has cluster linkages to various other high-tech sectors, such as nanotechnology, aerospace, energy storage electronics, meteorology, software, and marine engineering. Developed country firms still retain prominent technology advantages in wind energy, although this is likely to be challenged over time by companies from China and other emerging Asian nations (IEA 2009b, Lee et al. 2009).

Leadership in wind energy manufacturing has shifted increasingly from Europe and North America to Asia, and especially to East Asia. In 2005, only one East Asian firm (Goldwind) was ranked in the world's top-10 wind turbine producers. However, as Table 4.2 shows, by 2013 there were four, all from China–Goldwind, Guodian United Power, Mingyang, and Sinovel. Chinese turbine producers were by this time not only producing around half the world's wind turbines but had also become major exporters and were improving techno-innovation capabilities (CNREC 2013b, Zhou et al. 2012).

More generally, the majority of turbine production takes place in just six countries–China, India, Denmark, Germany, Spain, and the United States–and producers from these countries still dominate their regional markets. Japanese, Taiwanese, and Korean firms are also expanding their wind turbine production operations as the industry expands nationally and globally.

4.3.2 Wind energy: more specific technological developments

Early grid-connected wind turbines had rated power capacities around 20–30 kW, but they now range from around 500 kW to 8 MW, the most common of which lie in the 1.5 to 3 MW range. Over time, the length of turbine blades have increased from an average of around 10 metres in the 1980s, to 30–40 metres by the late 1990s, to up to around 70–80 metres by the early 2010s. By 2013, there were 23 manufacturers worldwide that were either producing or developing wind turbines of 5 MW or greater; nine of these were from East Asia: DeWind (Daewoo), Dongfang, Guodian United Power, Hyundai, Mingyang, Samsung, Shandong Swiss Electric, Sinovel, and STX (Windpower.net 2014, GWEC 2014). Vestas' 8 MW model was the world's largest turbine in production at this time, and three firms were developing 10 MW turbines: Chinese firm Shandong Swiss Electric, Norwegian company Sway, and US-owned Windtec, whose machine will have rotor blades over 90 metres long and a diameter of 190 metres. In the Shanghai metropolitan area, just one such 10 MW turbine could potentially meet the electricity demand of around 20,000 households; in West Europe, it could supply between 5,000 to 7,000 households (WWEA 2012b). Meanwhile, the very small wind turbine (i.e. normally sub-0.1 MW capacity rating) industry and market has steadily grown due to deepening worldwide demand among households, communities, organisations, and construction firms. By the end of 2011, an estimated 730,000 such units were installed, with China accounting for close to 70 percent of these (WWEA 2013b). Small wind turbines are providing electricity in many remote areas of developing countries, assisted by the declining costs of grid-connected inverter technology.

Wind farms can be developed with relatively short lead times, typically 3–5 years–which compares very favourably with most other energy sectors (Roland Berger 2010). The average size of wind farms has also consistently increased over time; the largest have increased from around 190 MW capacity during the early 2000s to around 1 GW by the early 2010s, and they are expected to rise to 1.5 GW in the 2016–2020 period (REN21 2014, Bloomberg New Energy Finance 2011). In addition, there is a growing trend for community wind farm projects in Europe, North America, and parts of East Asia, most notably in Japan. Improvements in material technologies and component supply chain systems are extending wind turbine lifespans, which are currently around 20–25 years. As wind turbine technology has improved and reduced both production and O&M costs, wind farms in lower-resource areas have become more cost competitive, thus extending the geographic scope for wind energy development. Over their lifetimes, initial investment costs of onshore plant construction are approximately 75–80 percent of total costs, while the range for offshore is 50–60 percent as O&M costs are comparatively higher. Technological leader firms are developing cheaper and lighter composite materials, sensors related to extreme elements, and advanced blade coatings; these all mainly relate to developing more robust offshore turbines (Lee et al. 2009). Overall, wind energy is subject to high rates of techno-innovation where efficiency rating levels are improving on a constant basis (Lee et al. 2009, GWEC 2012).

4.3.3. Trade and foreign direct investment trends

Although international trade and foreign direct investment (FDI) have become a more important feature of the wind energy industry, there is still a tendency toward localisation. This can be explained by the inherent nature of the production processes involved and recent developments in industry structure. Raw material costs comprise a large percentage of total costs (approximately 60–90 percent) for wind turbine manufacture. This is likely to remain the case if average sizes of turbines continue to increase, making their basic material elements proportionately high in the value chain relative to other parts, such as gearboxes and generators. Labour represents a small percentage of total costs, thus limiting the production cost benefits of relocating turbine manufacture to low-wage countries in East Asia. Furthermore, the aforementioned increasing size of turbine blades and also towers (now up to around 200 metres plus tall and over 300 tonnes in weight) means that transportation costs can be very high if moved over considerable distances, hence compelling producers to source locally.

Nevertheless, being a multi-component product (a standard utility-scale wind turbine has up to 8,000 parts, mainly located in the nacelle where all the generating components are housed), there is scope for sourcing equipment internationally. The parts that are produced in house by major assembler producers and the parts that are outsourced to external supplier firms varies across different value-chain segments, depending on the production economy, industry structure, and technological capacity factors (see Table 1.4, p. 23). As a consequence of localisation, international trade intensity in the sector has generally decreased since the mid-2000s, and China's growing production self-sufficiency combined with its position as the world's number-one market and producer has a major contributing factor. In 2004, China imported 82 percent of all its wind energy equipment installed; by 2010, this share had dropped to just around 10 percent.[3] Increasing vertical integration in the industry as turbine producers seek more quality control on component manufacture is another factor driving localisation. Overall, domestic sourcing in the world's major wind energy markets has been on the rise. The industry is also more oligopolistic compared to solar photovoltaics (PVs), structured around a relatively small number of large firms that have become regional hub producers.

Foreign direct investment in the wind energy industry has been largely motivated by market-driven factors. For some time, FDI was mostly concentrated in the United States and Europe. Most recently, China has emerged as the third most important destination for overseas wind turbine production; in East Asia, other still relatively small but fast-growing markets also could attract higher levels of FDI in forthcoming years. Future sector trade could be largely driven by FDI, as foreign investing firms import components from their established overseas-based suppliers. Some of these may follow major turbine assemblers into the foreign host nations, further increasing FDI levels. Larger wind turbine producers and innovators are also globalising their R&D operations, an example being Vestas' techno-innovation centre established in Singapore in 2008.

4.4. Obstacles to expansion

Political support for wind energy remains relatively weak in the major East Asian economies of Japan, South Korea, and Taiwan. Generally speaking, there remains a significant policy gap between future capacity targets and planned policy measures (including financial support mechanisms) aimed at realising those targets, and moreover a lack of stability and reliability in market frameworks in the wind energy sector (WWEA 2011). Wind farm planning processes continue to be slow, cumbersome, or even over-burdened in certain cases, such as in countries like China, where there has been strong growth in site development applications.

Compared to other renewable energy (RE) sectors, wind energy faces relatively few technological obstacles. High rates of techno-innovatory advancement are expected to continue. Turbines of 20 MW capacity could be designed and constructed in the near future (EWEA 2011). Spatial constraints and public opposition to onshore wind has led certain countries to push for offshore wind, which presents certain technical challenges. Resource-related obstacles will become more apparent as the wind industry becomes further scaled up. Chapter 1 noted, for example, that motors in advanced wind turbines use a highly magnetic rare earth element, neodymium, which is almost exclusively produced at present in China. The GWEC (2012) has noted human skills shortages that exist in some parts of the sector, this being particularly relevant in China and elsewhere in East Asia (Li 2010, Li and Ma 2009, Li *et al* 2007, Liu and Kokko 2010). Skilled labour shortages contributed to rising wind farm investment costs in the late 2000s, which additionally rose due to increased raw material costs, such as steel, copper, aluminium, carbon fibre, and cement (Wiser and Bolinger 2010). The rising cost of steel was a notable cause of rising prices for wind energy between 2005 and 2009 (REN21 2012).

Continued disputes over intellectual property right infringements between developed country companies and Chinese firms in particular may hamper future efforts at fostering techno-innovatory collaboration on wind energy development. In addition, the growing number of bilateral disputes over wind turbine trade at the World Trade Organization (WTO) has soured inter-firm relations. China has again been the main target of US and EU complaints in cases taken to the WTO Dispute Settlement Mechanism, with Chinese firms being accused of trade dumping practices.

In many countries, there has been growing public opposition to wind farm development, mainly due to visual and noise-related problems (IPCC 2011, WWEA 2012a). Wildlife groups have also complained of bird and bat fatalities through collisions with wind turbines, although evidence would suggest that other tall constructions are far more culpable: the rate for communication towers is 50 times greater than wind turbines, for vehicles is 850 times greater, and for buildings/windows is 5,820 times greater (GWEC 2008a). Rotating blade shadow flicker can a further annoyance, affecting both people and wildlife. Such complaints may be expected to rise given the increasing average size of wind farms and their growing numbers across landscapes. However, many of these hazards and annoyances are avoided in offshore wind.

4.5. China

4.5.1. Introduction

As previously noted, China has become on most accounts the world's leading nation in wind energy development. Over recent years, the sector has contributed around a fifth of annual net additions to the country's power generation capacity, roughly the same as hydropower (Table 3.3, p. 82). China also hosts many of the world's largest wind farms, and the government's 12th Five Year Plan (FYP) and national energy strategies foresee wind energy's share of electricity generation increasing steadily over time. It is viewed as an emerging new 'clean industry,' as well as a means to spur rural economic development. However, as we later discuss, there exist significant challenges ahead for further scaling up the Chinese wind energy sector.

4.5.2. Policy development

The Chinese government's prioritisation of wind energy development over most other RE sectors is due in part to the existing competitive advantages the country possessed in related engineering industries. During the 1990s, various state support measures were introduced to promote indigenous turbine production and technology development, based significantly on joint venture arrangements with European firms such as Vestas and Gamesa. Goldwind (started as the state-owned Xinjiang Wind Energy Company and becoming

Goldwind Science and Technology Company in 1998, then a private company in 2001) and Sinovel (formed in 2006 as a state-owned enterprise [SOE], then becoming private in 2011) and other such Chinese companies consequently emerged out of this process. The sector's development was further spurred by a 1995 government power purchasing arrangement arising from the China Electric Power Act that obliged the national grid operator to procure wind-generated electricity at prices and profit levels aimed at sustaining the growth of wind farm developers (mostly SOE power companies, such as Guodian Corporation; see Box 3.1, p. 84) and turbine producers (GWEC 2007a). However, this measure did not secure formal legal status, and the lack of sanctions for grid companies failing to comply meant that most refused to pay the higher wind energy prices (Lema and Ruby 2007).

In response, a new Ride the Wind programme and parallel policy mechanisms were introduced in 1997 to provide loan finance to help foster indigenous industry development, install larger wind turbines in the initial four designated wind resource provinces (Inner Mongolia, Xinjiang, Zhejiang, and Hebei) over the 9th and 10th FYP periods, and fulfil a local content requirement on foreign investor producers to source 20 percent minimum of components and materials from domestic suppliers (Li 2010, Xia and Song 2009). The programme also encouraged further foreign technology transfers, with German company Nordex being the first to participate in the scheme with local firm Xian Aero Engine Company (IRENA 2012b). Goldwind soon followed by signing technology licensing agreements with German firms Jacobs (600 kW turbines) and RE power (750 kW turbines) in the late 1990s, and Sinovel signed a 1.5 MW technology license with Furlander in 2004 (Ru et al. 2012). Windey, Mingyang, Shanghai Electric and XEMC-Darwind also signed similar licence agreements with European and Japanese firms during the same period[4] (GWEC 2007b, Li and Zhu 1999).

Four key policy initiatives of the early to mid-2000s helped lay a firmer foundation for wind energy development in China. First, the 2000 National Debt Wind Power Programme offered financial support for wind farms constructed using locally made turbines and other equipment. Second, in 2002, the government restructured the monopoly State Power Corporation into five separate power-generation companies and two grid companies; ultimate responsibility for RE policy was thereafter conferred to the National Development and Reform Commission (NDRC) National Energy Bureau, established in 2003 (Chapter 3). However, there was now a multiplication of state actors to coordinate across the wind energy sector, which became exacerbated by the proliferation of private and state-owned wind turbine industry producers in subsequent years. Third, the 2003 Wind Power Concession Programme introduced China's first feed-in tariff (FiT) scheme and targets to create up 20 wind farms with between 100 to 200 MW capacity through competitive bidding, as well as a national 20 GW target by 2020. A 50 percent local content rule on turbine production was applied to the concession projects, revised upward in 2004 to 70 percent but later abolished in 2009. By the end of 2006, there were 15 wind concession projects with a total capacity of 2,550 MW (Xia and Song 2009). Fourth, the 2006 Renewable Energy Law (REL) legally obliged the two newly formed state grid companies to purchase at least 5 percent of their electricity from wind energy produced from the five new power generation companies (GWEC 2007a, 2007b).

With gradually ramped up public financial support and a more certain and coherent policy-legislative environment, China's wind energy sector began to grow rapidly. Over 2005–2009, national installed capacity levels doubled annually, rising from 1,266 MW in 2005 to 25,810 MW by 2009 (Table 4.1). By the end of the 11th FYP plan (2006–2010), capacity reached 44.7 GW, far exceeding the plan target of 5 GW by 2010. In 2009, the NDRC introduced revised 20-year contracted FiT rates for the four designated wind resource provinces, over and above which competitive market bidding operators submitted 'tender tariff' rates (Zhao et al. 2012a). These were significantly favourable rates compared to other RE sectors, further spurring the industry's growth (GWEC 2010a, 2010b).

The 12th FYP set a target of 100 GW by 2015 (including 5 GW of offshore), including 70 GW from the Wind Base Programme launched in 2008, comprising mega-scale projects in Inner Mongolia, Xinjiang, Gansu, Hebei, Jilin, and Jiangsu (GWEC 2010b, NDRC 2011). The first Wind Base project started construction in Jiuquan, Gansu province for on a cluster of eight large wind farms and a US$2.9 billion grid development plan (Li 2010). A longer term target of 150 GW grid-connected capacity by 2020 was initially set by the 12th FYP, with the great majority of this (138 MW) to come from the Wind Base provinces (GWEC 2011, Kang et al. 2012). The 2020 target was later revised upward to 200 GW (including 30 GW offshore), and a special technology-oriented development plan for wind energy was introduced in 2012. This was essentially a strategic industry policy aimed at fostering capacity in high-tech large turbines in three key areas:

- 3 to 5 MW class direct-drive permanent magnet synchronous turbine generators and associated component design and manufacture
- 7 MW wind turbine design, manufacture, installation, operation, and maintenance
- Design capability for 10 MW offshore turbines and components for future production and installation.

This marked a clear progression from the 600 kW turbines targeted for development under the 9th FYP and developing 2 to 3 MW technology machines under the 11th FYP (Wang et al. 2012). Furthermore, the 12th FYP placed much greater emphasis on wind energy techno-innovation generally (Zhang et al. 2013). In addition, a new Wind Energy Resource Assessment Centre was established to better map wind resource availability and to create an improved database on intellectual property rights issues (CNREC 2013b,

IEA 2013a). It should be noted, however, that the Chinese government has supported wind energy R&D activity through various schemes over the years, most importantly through the National High Technology Research and Development Programme, Major State Basic Research Development Programme, National Key Technology R&D Programme, and Special Fund for Development of Renewable Energy (Zhou et al. 2012).

China's provincial and city governments have played a particularly important role in developing this renewable energy sector. They exercise significant decision-making autonomy within the national policy framework (e.g. approving projects below 50 MW capacity) as well as exercising their own local policies. Inner Mongolia, with currently over a quarter of national installed capacity, was China's first province to devise its own distinct provincial-level policy, in 2006 issuing rules on wind farm planning, facilitating new meteorological surveys on wind resources and feasibility studies on possible new projects, and specifying budget resources, administrative processes, and timetables for project development. Local institutionalisation of wind energy policy has been replicated in other parts of China, albeit often with a different emphasis. For example, whereas Inner Mongolia has concentrated on optimising installed capacity, Jiangsu Province has focused on promoting domestic industry production. Guangdong, Jilin, Gansu, and Xinjiang provinces have also formulated their own wind energy policies, although not to the same well-defined degree (Liu and Kokko 2010).

4.5.3. National market and industry development

Meteorological surveys first conducted in the late 1970s found that China's best wind resources lay mostly to the north and west (Liao et al. 2010, Xiao and Hang 2009, Yang et al. 2011). The first wind resources zoning map was published in 1981, and initial state-funded demonstration projects were developed from the early 1980s, with the first becoming operational in 1986 at Rongcheng, Shandong province (IRENA 2012b). China's first grid-connected wind farm was constructed at Daban City, Xinjiang Uygur Autonomous Region in 1988 with funding from the Danish government (Brennand 2001, Liu and Kokko 2010). Foreign aid support followed from France, Germany, and Spain for demonstration projects and small-scale wind farms based on imported turbines and other equipment (Han et al. 2009, Lema and Ruby 2007, Li 2010, Xia and Song 2009, Zhang 2012, Zhao et al. 2009). Up to 2004, wholly foreign-owned firms or Sino-foreign joint venture enterprises outnumbered domestic firms in turbine manufacturing in China (Wang et al. 2012).

Thus, in the early stages of China wind energy development, foreign firms (especially European) played a very significant role. However, gradually strengthening state support for developing indigenous production and technological capabilities led to the inexorable rise of domestic Chinese wind turbine and component manufacturers from the mid-2000s, leading to the country becoming a net exporter of wind energy equipment by the end of the decade (GWEC 2009, 2010a). By 2013, Chinese firms were producing around 80 percent of all installed wind turbines in the country, compared to 56 percent in 2007 and 25 percent in 2004 (GWEC 2008b, 2014). The rise of Chinese wind turbine producers has intensified competition globally and also nationally within China. By the early 2010s, China was home to around 70 domestic wind turbine manufacturers (up from just 6 in 2004 and 40 in 2007), and the top national market rankings of the very largest producers have changed regularly. Key wind industry production clusters are located in northern China, at Baotou in Inner Mongolia (Sinovel, Goldwind), Jiuquan in Gansu province (Sinovel, Goldwind), and at Tianjin city (Vestas, Gamesa, Suzlon), with these being geographically close to the country's largest 'wind bases' (Zhao et al. 2012b). The country's wind energy producers are represented by the China Wind Energy Association, which has grown in power and influence as the industry has grown. China has also developed the world's largest production base of very small turbines, with output rising from just a few thousand a year in the early 1980s to over 150,000 units annually by the early 2010s. China's mass production of low priced, small-scale turbines–exported to more than 100 countries by 2011–has provided remote and rural communities around the world with better opportunities to generate off-grid electricity (WWEA 2012b).

Approximately 90 percent of investment in China's wind farm development derives from the country's five state-owned power generation companies (see Box 3.1, p. 84), along with Shenhua Group (the nation's largest coal company) and the China Guangdong Nuclear Power Holding Corporation (Li 2013, Liu and Kokko 2010). Some of China's largest turbine producers are also SOEs (most notably Shanghai Electric Group), and most have their roots in the power generation, aerospace, astronautics, and heavy engineering industries, such as Sinovel Wind Corporation, Goldwind Science and Technology Corporation, and Dongfang Electric Group. China's turbine producers work closely with state agencies generally. For example, Sinovel's offshore R&D centre has received financial support from the NDRC and the National Energy Administration (NEA). A few turbine producers (e.g. Goldwind, Huayi Electric Apparatus Group, XEMC-Darwind) have also vertically integrated downstream into wind farm development and operation (Li 2013).

Since the mid-2000s, China's larger wind turbine producers have increasingly progressed from technology licensing agreements to engaging in joint and independent research and development activity, with the support of the aforementioned government R&D programmes and strategic plans. For example, Goldwind forged a closer technology partnership with German firm Vensys over the 2000s, buying a 70 percent owning stake in the company in 2008 and thereafter building a global R&D network based on three research centres (in Beijing, Xinjiang, and Germany), thus enabling Goldwind to establish stronger indigenous design capabilities. Meanwhile, Sinovel's joint R&D venture with American company Windtec in the late 2000s on 3 MW turbine development helped to provide the technology platform on which the

Chinese firm was later producing 6 MW turbines by 2011 (Ru et al. 2012). The number of Chinese firms submitting wind energy technology patents in China since the late 2000s has gradually risen, whilst those from foreign companies has fallen. According to Ru et al. (2012), all of the country's major turbine producers had developed significant indigenous R&D capabilities by the early 2010s. Despite these advances, the techno-innovatory capability gap between China's wind turbine manufacturers and their most advanced foreign counterparts remains considerable (Zhou et al. 2012).

4.5.4. Main challenges ahead

4.5.4.1. Grid infrastructure and connectivity problems

Because most of China's best wind resources and large-scale wind farms lie in remote areas far from the nation's main 'load centre' cities in the coastal and southern provinces, grid transmission connectivity distances are great and required investments in new grid infrastructure are enormous (Li et al. 2012). Zhang and Li (2012) summarised the challenges of integrating large-scale wind energy in China as follows: the uncoordinated development of capacity and power grids, lack of appropriate technical standards for integration, insufficient clarity over corporate responsibility for grid connection, and poor economic incentive structures for grid companies to use wind energy. Consequently, a significant level of China's installed wind energy is not actually delivering electricity to grid.

Investment in China's grid infrastructure generally has not kept pace with rapidly expanding power generation capacity, resulting in many power stations operating at low generation load levels. This affects wind farms in the same way as coal-fired power stations. Although Japan and most other Organisation for Economic Co-operation and Development (OECD) nations have invested comparatively more in grid infrastructure than power generation since the late 1970s, the converse is true for China (Li et al. 2012). It is a predicament that applies equally to the grid's geographic coverage and transmission capacity: for example, the Northeast China grid network is based on a 500 kV, system which has proved rather limited in the wind base zones (Zhao et al. 2012c). According to Luo et al. (2012), the main root of the problem stems from the restructuring of the energy sector in 2002 and its consequent sub-division of interests amongst power plant operators, electricity generation firms, grid companies, national government, and local government. This created coordination difficulties on grid planning and construction, as well as various actors blaming others for not sufficiently investing in new grid infrastructure. Luo et al. also argued that a key reason for the almost uncontrollable expansion of wind farm development in China is due to the aforementioned 50 MW capacity approval rule. To evade coming under relatively stricter NDRC approval at this threshold capacity level, a large number of wind farm projects rated at 49.5 MW have emerged, which only required comparatively looser local government approval, further weakening the national government's grip on the sector's development overall.

As Table 4.3 indicates, China had 180 wind farms rated at 25–49.5 MW by 2013. The approval process for wind farms (especially smaller ones) has also hitherto been much quicker than for constructing new transmission lines. Wind farm developers are willing to wait some months or even years for grid connectivity, which has contributed to the capacity–generation gap. The lack of centralised control over wind farm development up to the early 2010s has more generally led to a situation where installed capacity levels have shot considerably well beyond targets set by the NDRC and other economic plan coordinating agencies in Beijing.

Table 4.3 Largest wind farms in East Asia

Wind farm	Country	Capacity (MW)	Details	Plant status
Jiuquan, Gansu	China	12,710	5,160 MW by 2010	Under construction (completion date unknown)
Da'an	China	1,000	2,500 GWh p/a, 250 turbines	Under construction (completion date unknown)
Shangdu	China	504	1,259 GWh p/a	Under construction (completion date unknown)
Taipusiqi	China	450	1,123 GWh p/a	Under construction (completion date unknown)
Tongyu	China	407	1,018 GWh p/a	Under construction (completion date unknown)
Xiangyang	China	401	1,001 GWh p/a	Operational 2011
Danjinghe	China	400	1,000 GWh p/a, 207 turbines	Operational 2009
Zhangwu	China	350	875 GWh p/a	Operational 2010
Dabancheng	China	340	850 GWh p/a	Operational 2010
Huitengliang	China	300	750 GWh p/a, 200 turbines	Operational 2010
Kaiyuan	China	300	750 GWh p/a, 200 turbines	Operational 2011
Bayanzhuoer Wulanyiligeng	China	300	750 GWh p/a, 200 turbines	Operational 2009
Tongliao Beiqinghe	China	300	750 GWh p/a	Operational 2011
Korat II (West Huay Bong)	Thailand	270	90 turbines	Under construction (2016)
Zhenlai	China	209	522 GWh p/a	Operational 2010
Korat I (West Huay Bong)	Thailand	207	90 turbines	Operational 2013

(continued)

Table 4.3 Largest wind farms in East Asia (*continued*)

Wind farm	Country	Capacity (MW)	Details	Plant status
Hongsongwa	China	205	513 GWh p/a, 66 turbines	Operational 2005
Chenjiang Xiangshui	China	201	502 GWh p/a, 134 turbines	Operational 2012
Beidaqiao 4	China	201	502 GWh p/a, 134 turbines	Operational 2010
Sunjiaying	China	201	502 GWh p/a, 268 turbines	Operational 2010
China has another 21 wind farms rated in the 150 to 200.5 MW range either in operation or under construction				
Mutsu	Japan	148	370 GWh p/a	Operational 2008
Wind 1	Taiwan	101	254 GWh p/a	Operational 2009
China has 29 wind farms rated in the 100 to 149.5 MW range either in operation or under construction				
Bien Dong	Vietnam	99	310 GWh p/a, 66 turbines	Operational 2013
Gangwon	South Korea	98	245 GWh p/a, 49 turbines	Operational 2005
Pagudpud	Philippines	81	27 turbines	Under construction (2014)
Shin Izumo	Japan	78	195 GWh p/a, 26 turbines	Operational 2009
Changbin I	Taiwan	76	189 GWh p/a, 33 turbines	Operational 2007
Yeongdeok	South Korea	68	169 GWh p/a, 41 turbines	Operational 2004
Nunobiki	Japan	66	165 GWh p/a, 33 turbines	Operational 2007
Tomamae	Japan	64	160 GWh p/a, 56 turbines	Operational 1999
Yeong Yang	South Korea	61	153 GWh p/a, 41 turbines	Operational 2008
Jeju	South Korea	60	150 GWh p/a, 20 turbines	Under construction (2014)
Khao Kor	Thailand	60	150 GWh p/a, 26 turbines	Operational 2013
Soya Misaki	Japan	57	142 GWh p/a, 57 turbines	Operational 2005
Noheji	Japan	51	125 GWh p/a, 25 turbines	Operational 2008
Nagashima	Japan	50	126 GWh p/a, 21 turbibes	Operational 2008
Miaoli	Taiwan	50	125 GWh p/a, 25 turbines	Operational 2006
China has 47 wind farms rated in the 50 to 99.5 MW range, either in operation or under construction				
Takine Ojiroi	Japan	46	115 GWh p/a, 23 turbines	Operational 2010
Taichung I	Taiwan	46	115 GWh p/a, 20 turbines	Operational 2009

Table 4.3 Largest wind farms in East Asia (*continued*)

Wind farm	Country	Capacity (MW)	Details	Plant status
Ikata	Japan	43	106 GWh p/a, 15 turbines	Operational 2003
Kamaishi	Japan	43	107 GWh p/a, 43 turbines	Operational 2004
Guanyin	Taiwan	43	109 GWh p/a, 19 turbines	Operational 2010
Taegisan	South Korea	40	100 GWh p/a, 20 turbines	Operational 2008
Yeongdeok	South Korea	40	99 GWh p/a, 24 turbines	Operational 2004
Hsinchu	Taiwan	38	95 GWh p/a, 17 turbines	Operational 2009
Houlong	Taiwan	34	85 GWh p/a, 16 turbines	Operational 2012
Irozaki	Japan	34	85 GWh p/a	Operational 2010
Iwaya	Japan	33	83 GWh p/a, 16 turbines	Operational 2001
Samdal	South Korea	33	82 GWh p/a, 11 turbines	Operational 2009
Azuchioshima	Japan	32	80 GWh p/a, 16 turbines	Operational 2007
Mutu Ogawara	Japan	32	78 GWh p/a	Operational 2001
Nishime	Japan	32	79 GWh p/a, 17 turbines	Operational 2012
Kunimiyama	Japan	30	75 GWh p/a, 15 turbines	Operational 2009
Taoyuan	Taiwan	30	75 GWh p/a, 20 turbines	Operational 2005
Binh Thuan	Vietnam	30	76 GWh p/a, 20 turbines	Operational 2010
Hiyama	Japan	28	70 GWh p/a	Operational 2010
Tokiwa	Japan	27	66 GWh p/a, 11 turbines	Operational 2010
Taichung II	Taiwan	26	66 GWh p/a, 12 turbines	Operational 2006
Hachiryu	Japan	26	63 GWh p/a, 17 turbines	Operational 2006
Minami Kyushu	Japan	26	65 GWh p/a, 20 turbines	Operational 2009
Bangui Bay I	Philippines	25	62 GWh p/a, 10 turbines	Operational 2005

China has 180 wind farms rated in the 25 to 49.5 MW range either in operation or under construction

Wind farm	Country	Capacity (MW)	Details	Plant status
Bangui Bay II	Philippines	8	20 GWh p/a, 5 turbines	Operational 2009
Thweppana	Thailand	8	20 GWh p/a, 3 turbines	Operational 2013
Oelbubuk	Indonesia	1	12 GWh p/a, 12 turbines	Operational 2004

Sources: windpower.net (2014) database, various media.

Notes: 'Operational' status relates to when the wind farm first started to generate power in the initial phase of development.

A key part of the problem concerns policy incentives and strategic planning geared toward installation rather than power generation output, leading to 'excess' installed capacity problems. Zhang et al. (2013) contended that this approach was primarily driven by strategic industry policy motives, in that setting ambitious targets on installed wind energy was conceived as an important driver for expanding China's turbine manufacturing base. Poor coordination between government agencies (urban, environment, industrial policy) on wind farm development was also clearly evident (Luo et al. 2012).

Given that wind energy is a fast emerging and potentially very profitable industry in China, wind farm developers have been able to attract considerable investments. The rapid growth of China's wind energy sector during the 2005–2010 period was largely due to an initial spurt of firms entering the market seeking first-mover advantages during a time when the government had created a more robust and predictable policy environment, especially after the introduction of the REL in 2006 (Xia and Song 2009). The ensuing speed at which wind farms were being constructed put pressure on existing planning procedures, leading to significant time lags in operational status being awarded (Martinot and Li 2010).

The lack of appropriate grid connectivity technical standards or codes specifically for wind farms, and poor compliance by operators where they do exist, has also resulted in an underuse of wind energy generated and bad management of the grid transmission system during high wind-speed periods, leading to power generation cut-outs in parts of the grid (Kang et al. 2012). More generally, there was a rising trend of turbine breakdown and disconnection incidents from the mid-2000s to early 2010s. For example, Zhang et al. (2013) reported that, during the first 8 months of 2011, there were 193 separate turbine disconnection incidents; in February of that year, a combined 840 MW power loss from short-circuit problems affected 598 turbines. The State Electricity Regulatory Commission meanwhile blamed the concomitant rise in turbine breakdowns over the period on sub-standard components and connectivity equipment, and thus ultimately weak or weakly applied technical standards and codes in manufacturing and grid connection technology (State Electricity Regulatory Commission 2011). In 2006, the State Grid Corporation of China introduced the Technical Rules for Connecting Wind Farms to Power System regulation in an attempt to improve operator conformity to grid connectivity standards. However, the rules were largely ignored because they were not legally binding (GWEC 2010b). In February 2011, the China Electricity Council (CEC) published its draft *Operational Standards for Wind Power Dispatch* document, designed to be China's first industrial standard for wind power integration, yet this code was also legally non-enforceable. Thus, by the early 2010s, there were still no effective national technical standards for grid connection of wind power (Zhao et al. 2012c).

There also have been weak incentives for new grid infrastructure investment. In China's electricity generation industry, a planned market applies for base power needs and an open trade market exists for incremental power needs between producers and consumers. This affects how electricity is traded on an inter-provincial basis, and thus how wind energy generated in remote areas is sold to large load centres some distances away; only incrementally produced wind energy is effectively tradable across the regions. Moreover, because there has been no national standard price formula for inter-provincial power transmissions, prices are determined by bilateral negotiations between provincial authorities, which have created significant transaction costs. By the early 2010s, there was also still no mechanism to compensate grid companies for the inevitable power losses arising from long-distance transmissions, thus offering the grid companies little incentive for facilitating inter-provincial wind energy trade (Zhao et al. 2012b, 2012c).

As a consequence of these problems, it was estimated that between a quarter to a third of China's installed wind energy capacity was not providing power to the country's grid systems. According to Yang et al. (2012), from 2007 to 2010, the average capacity factor of wind energy in China was just 16.3 percent, which is outside the usual 20–40 percent range. However, a GWEC (2011) report noted that the capacity-generation gap was being overstated, primarily owing to the method used by the CEC to calculate operational capacity. According to the GWEC (2011: 32–33), this is because the CEC 'only counts entire wind farms that have been connected to the grid with a Power Purchase Agreement, have undergone a testing procedure and have been accepted; and for which the national grid operator … has started to pay electricity bills.' Wind farm operators may still be providing energy to the grid that is not recorded under this 'full commissioning and approval' calculation method.

The 12th FYP seeks to address problems regarding China's grid infrastructure and regulatory gaps on wind energy. In 2012, the NEA introduced the Wind Farm Development and Management Interim Rules and Regulation, which stipulates that wind farms must acquire complete approval from the authorities before commencing operation, as well as other measures that aim to strengthen quality control over installations. A new Safety Management of Wind Farms procedure has been brought into force, and the government is now encouraging development of wind farms closer to main load centres, albeit in low wind speed areas. Furthermore, a new grid code and 17 other technical standards have been introduced, as well as operators' use of low-voltage ride-through technology (GWEC 2012). National government has also taken more direct control of wind farm development within the 12th FYP strategic plans, wrestling power away from local government. Investments totalling RMB3.8 trillion (US$580 billion) for 'strong and smart grid' development up to 2020 has additionally been implemented, which should in time alleviate many of these grid connectivity problems (GWEC 2013, Li 2013). However, for smart grids to perform optimally, they should be allowed to operate on certain decentralisation principles and be reactive to consumer demand, which will require the Chinese power generation sector to become more open and competitive rather than staying monopolised by a few SOEs (Solidiance 2013).

4.5.4.2. Offshore wind in China

By 2013, China was globally ranked fifth in installed offshore wind, with an accumulated 428.6 MW capacity (GWEC 2014). Its coastal provinces of Hainan, Guangxi, Guangdong, Fujian, Zhejiang, Shanghai, Jiangsu, Shandong, and Liaoning have significant offshore wind resources, with the best being located in Fujian province in the Taiwan Straits, followed by the north in Zhejiang (Chen 2011, Hong and Möller 2012, Zhang et al. 2011). As part of its 2007 Medium- and Long-Term Renewable Energy Development Plan, the Chinese government proposed the construction of pilot 100 MW offshore wind projects by 2010 and the development of 1 GW offshore wind farms by 2020 (NDRC 2007a). In addition, R&D capacity in the sub-sector would be promoted. China's first offshore wind farm (a 102 MW installation near Shanghai) became operational in 2010, providing electricity for approximately 200,000 households (GWEC 2010c, WWEA 2012b). The 12th FYP later set national offshore wind targets of 5 GW by 2015 and 30 GW by 2020. The nation with the next most ambitious target by the latter date is the current world leader in this sub-sector, the United Kingdom, with 25 GW.

At the same time, a number of coastal provinces announced their medium- and long-terms plans for offshore wind development. Despite having the nation's best resources, Fujian province has the lowest plan targets, just 300 MW by 2015 and 1.1 GW by 2020. Jiangsu has the most ambitious plans, consistent with its very active wind energy policy generally, with targets of 4.6 GW and 9.5 GW, respectively, for the same years. The provincial government has planned four 1 GW scale 'Offshore Three Gorges' projects at Rudong, Dongtai, Dafeng, and Qidong (Hong and Möller 2012, Zhang et al. 2011). Shandong has the next most ambitious plans, with 3 GW (2015) and 7 GW (2020) targets set. Aside from the usual technical challenges facing the sub-sector, the key challenges for offshore wind in China are similar to those for onshore: strengthening indigenous techno-innovation capacity, improving grid infrastructure and regulatory environments, and developing more effective incentives for business investment in the sector (Chen 2011, Lui et al. 2013).

4.5.4.3. Clean development mechanism

The Clean development mechanism (CDM) scheme of the Kyoto Protocol (see Box 1.2, p. 32) has been a vital instrument to financing wind farm development in China (Pechak et al. 2011). In the first year of the scheme's operation, only 52 wind energy projects had been approved by Beijing in 2006 (GWEC 2007a). By September 2013, of the 2,609 CDM-registered wind energy projects worldwide, 1,517 of these (58.1 percent of the total; see Table 1.5, p. 34) were located in China. Existing data structures only allow for very broad approximations of what exact contributions CDM project investments have made to delivering clean energy in developing countries. In addition, a number of projects may relate to the same wind farm through various and often incremental stages of its development, while other projects may have involved single turbine demonstration installations. Notwithstanding the CDM's importance in helping finance wind farms in China, including incentives provided to investors, its share of total funding in the nation's wind sector remains relatively small, especially when compared to levels of Chinese state spending (GWEC 2010b). Furthermore, government rules stipulate that only companies that are majority Chinese-owned can claim CDM benefits, thus compelling foreign investors in wind farm projects to defer majority control to a Chinese partner if they are to enjoy CDM revenues (Yang et al. 2010). By the late 2000s, with China dominating CDM projects globally, the scheme's examining board started to impose stricter assessments of project applications in the country. Consequently, by February 2010, a total of 15 projects were declined on the additionality clause, and the CDM Examining Board also proposed an additionality audit on existing China-based projects (GWEC 2010b). Despite these factors, China's wind energy sector is likely to continue attracting a large share of CDM financing over forthcoming years, as long as the scheme itself remains in operation.

4.6. Japan

4.6.1. Introduction

Despite Japan having the second largest installed wind energy capacity in East Asia and being amongst the region's most advanced and largest wind turbine producers, the development of the country's wind market and industry has been disappointing by OECD standards and most international comparisons. It has also been overshadowed by solar energy–a sector where Japan's government, companies, and research organisations have devoted far more attention and resources than any other in the renewables spectrum (Moe 2012). Government support for wind energy R&D has, for instance, been a fraction of that devoted to solar PV, and the vast majority of research and development in Japan's wind sector is undertaken by private enterprises (Valentine 2011). Japanese firms such as Mitsubishi and Hitachi have considerable innovatory capacity in wind energy technology, which is arguably applied more in foreign wind farm projects than in Japan itself. However, in the early 2010s the government announced ambitious new plans to support Japan's fledgling offshore wind industry, with state assistance to help to develop and commercialise innovative new floating turbine and installation technologies. The nation's best onshore wind resources are located in the remote, sparsely populated northern regions that have limited grid connections to the extremely large and populated load centres of Tokyo, Osaka, and Nagoya (Esteban et al. 2010, Tsuchiya 2012).

4.6.2. Policy development

For over two decades, the Sunshine Project, a multi-sector programme launched in 1974 (Chapter 3), provided the only Japanese government support for wind energy development. This was more or less limited to funding a handful of demonstration installations from the early 1980s onwards, the first being a singular 300 kW turbine at Okinoerabu Island (Harborne and Hendry 2009, Ushiyama 1999a). After the Sunshine Project programme review of 1990, the government sought to expand their wind energy partnerships with Japan's ten regional power companies (*Denjiren*) and emerging turbine manufacturers. It was not until the mid-1990s that the government introduced wind energy–specific policy measures. The 1996 New Renewable Energy Target initiative set an installed capacity objective of reaching 3 GW by 2010, which subsequently the nation failed to attain (Table 4.1). In 1998, a new subsidy scheme for R&D and wind farm development was introduced, helping to initially spur national installed capacity (Maruyama et al. 2007). By the mid-2000s, state support for wind energy R&D was terminated and funds were redirected to improve grid performance and power quality (GWEC 2008b, Harborne and Hendry 2009).

Prior to the introduction of renewable portfolio standards (RPS) legislation in 2003, *Denjiren* power companies negotiated contracts to independently purchase electricity from wind farm operators over 15- to 17-year periods. As national wind energy capacity levels increased, these companies imposed 'introduction limitation quotas' on wind energy to limit their exposure to financial risk (because of still relatively high generation costs at that time) and technical risks relating mainly to intermittent power supply. Any wind energy generated above these quota levels was competitively tendered. This practice, at the time unique in the world, was first initiated by two *Denjiren* power companies (Hokkaido and Tohoku) with high wind resources and then emulated by others, essentially giving these firms control over how much wind-generated electricity in aggregate was produced in Japan each year. The *Denjiren* companies were generally reluctant to add wind energy to their generation portfolios, putting a notable constraint on any government aspirations to develop the sector (Chapter 3). In 2003, the power company quotas amounted to just 330 MW, while the bids tendered totalled 2,400 MW, indicating a significant mismatch between what wind farm developers wished to supply to the market and what the power companies actually wanted to supply to the electricity grid. The 2003 RPS law changed this, obligating the *Denjiren* to generate minimum quantities of electricity from renewables set by national government via the Agency for Natural Resources and Energy (Inoue and Miyazaki 2008, Maruyama et al. 2007).

However, the pace of growth in Japan's installed wind energy capacity slowed in the immediate years after the 2003 RPS, suggesting it had little impact on the sector. No new policy measures relevant to wind energy were introduced for the rest of the decade, and the decision in the aftermath of the global financial crisis to curtail subsidies for wind energy development in 2009 was a further setback. Until the government's FiT scheme was extended to include wind and other renewables (not just solar) in July 2012, there were in effect no policy incentives in place supporting wind energy over this three-year period (Ushiyama 2012). It was no surprise, then, that the sector's installed capacity growth rate slowed during this time (Table 4.1).

While new wind FiT rates were considered generous by international comparison[5] and relative to other RE sectors in the same scheme, a loophole reportedly exists that allowed the *Denjiren* to refuse wind farms connection to the grid (Ushiyama 2012). There additionally remain strict regulations on using land in forest reserves, farmland, and nature reserves, and environmental laws protecting against bird strikes. Previously, a 2007 building code that classified wind turbines 60 metres high or taller as buildings had effectively paralyzed the Japanese wind market for around a year due to compliance to the very complicated and time-consuming planning procedures involved. Although this process was later streamlined, in October 2012 a new law required more stringent environmental impact assessments of wind farms over 10 MW capacity, in effect applying to more or less every proposed new project. These assessments were expected to take around three to five years to complete and add an anticipated extra JP¥100 million (US$1.3 million) cost to each project investment plan.[6]

A more general concern, particularly expressed by Japan's wind energy industry firms, has been the lack of a long-term strategic target (aside from the aforementioned missed 2010 one) and an integrated policy plan for the sector's installed capacity development (GWEC 2012). Even the Japanese government's 2010 New Growth Strategy macro-plan for decarbonising the economy (Chapter 3) did not include a wind energy industry strategy, aside from a new zoning policy on locating wind farms. The country's predominantly technology-oriented approach to renewable energy development provides a general explanation for this, and thereby the relative neglect of installed capacity. As we later discuss, in 2012 the government announced its plans to fast-track development of offshore wind over forthcoming years. It also allocated JP¥43.4 billion (US$560 million) of state support for wind energy development for the 2013 fiscal year, with JP¥25.0 billion of this on grid extension investment and JP¥15.1 billion on offshore wind development. Just JP¥2.0 billion was earmarked for advanced wind turbine R&D (GWEC 2013). This would seem to indicate a possible shift in Japanese government policy in this sector toward a more installed capacity approach in the country's new post-Fukushima energy policy.

4.6.3. National market and industry development

In 1997, the country's capacity level stood at 18 MW, rising steadily to 2,661 MW by 2013 (2.8 percent of the East Asian total) and contributing 0.5 percent of national electricity generation output (GWEC 2014). A high number of its

224 wind farms operating by this time were very small, comprising just between one to four turbines (Windpower.net 2014). The average size of Japan's wind farms in 2013 was about 12 MW, compared to China's average of 138 MW. The comparatively sluggish growth of Japan's wind energy market and industry is reflected in the country's profiling in recent GWEC annual wind energy reports. Those published in 2006, 2007, and 2008 conveyed more or less the same information on Japan each year, and the country was even omitted altogether in the 2009 and 2010 editions. Japan's largest wind farm is the Mutsu installation (148 MW); there were six plants operating in the 50 to 99.5 MW range and 13 in the 25 to 49.5 MW range by 2013 (Table 4.3).

There are currently four Japanese wind turbine producers: Japan Steel Works (JSW), Hitachi (Subaru wind division),[7] Mitsubishi Heavy Industry, and Komai Tekko, which specialises in making smaller turbines. In terms of the domestic market for wind turbine installations, JSW dominated with a 48 percent share in 2011, followed by the German firms REpower (17 percent) and Enercon (11 percent), and then Fuji (10 percent), Mitsubishi (7 percent), and General Electric (7 percent). Mitsubishi were developing a 7 MW offshore turbine by 2012, the largest by far of any Japanese firm and in line with the largest devices in development being constructed in the industry worldwide. In other parts of the value chain, Japan's main producers of wind turbine bearings are NSK, JTEKT and NTN, all of which have long experience as suppliers in various large industries, such as automotive. Key producers of wind turbine generators are Hitachi, TMEIC, Meidensha, and Yasukawa Electric (GWEC 2012, Ushiyama 2012). Japanese firms are working at the industry's technological frontier. For example, the Japan Wind Development Corporation and battery maker NGK Insulators are jointly developing sodium-sulphur battery accumulators, a new power-storage technology integrated into wind farm installations designed to help solve power intermittency problems.[8]

4.6.4. Main challenges ahead

4.6.4.1. Wind resource and spatial constraints

Japan faces similar challenges to China regarding the exploitation of the nation's best wind resources, which are located in remote, sparsely populated areas far from main load centres. Around 70 percent of the country is mountainous, presenting certain technical and logistical challenges when developing wind farms in non-urbanised areas. Because many wind energy developers have to build on hilly terrain, installation costs are around twice the levels as the United States.[9] After the 2011 Fukushima disaster, there has been growing political support for rationalising and expanding Japan's grid infrastructure to improve electricity trading among regions (GWEC 2012). However, even with political backing, it could take up to a decade before the nation's grid infrastructure is sufficiently extended into remote northern (Hokkaido and Tohoku) and southern (Kyushu) areas, where Japan's best wind resources are located. The government's aforementioned 2013 JP¥25.0 billion plan for new grid extension investment is perhaps the first important step in this direction.

4.6.4.2. Dealing with the forces of nature

Japan's geographic location and topography make it particularly susceptible to extreme weather conditions and natural disasters, as so devastatingly shown by the March 2011 earthquake and tsunami. Interestingly, almost every one of the 190 wind turbines situated in the epicentre Tohoku region survived the disaster with no operational damage sustained. Even the Subaru 2 MW turbines at the Kamisu semi-offshore wind farm, which received a 5-metre tidal-wave hit from the tsunami, survived and were back in service 3 days later, after the grid system came back online (GWEC 2012). This was a testament to the integrity of Japanese wind turbine manufacture and installation engineering. The country is also in a typhoon zone and is buffered by high wind turbulence generally and violent thunderstorms. Extreme weather periods in 2004 and 2007 caused significant damage to wind farms, leading to tough safety 'J-class' standards being developed for turbines (GWEC 2007a, Moe 2012). Installations located in mountainous areas, and hence higher altitudes, have been especially prone to lightning strikes. According to data from the mid-2000s, half the wind farms located on certain Sea of Japan coastal regions were at some point damaged by lightning (Urashima 2007). Lightning protection technologies and the relocation to lower altitudes offer solutions to this problem.

4.6.4.3. Offshore wind in Japan

As a large island and a technologically advanced economy, Japan is a natural candidate for offshore wind development. However, the country is surrounded mostly by deep water, which presents technical engineering challenges to constructing offshore wind farms.[10] Dealing with objections from the politically powerful fishing lobby is another challenge (Moe 2012). In 2009, the New Energy Development and Industrial Technology Organization started to map Japan's offshore wind potential and attempted to locate the best demonstration sites (WWEA 2012b). The following year, five experimental low-MW range projects were approved by the government in collaboration with Fuji Heavy Industries, Mitsubishi Heavy Industry, and Japan Steel Works (Ushiyama 2012). By 2013, Japan had 49.6 MW of installed offshore wind capacity and was globally ranked ninth. Up to 254 MW of new offshore wind capacity was in development at this time. Japan's Ministry of Environment estimates the nation's theoretical offshore wind potential was 1,573 GW and current realistic potential is 1,000 GW (GWEC 2013, 2014). The government has substantially increased funding support for offshore wind R&D, from just JP¥200 million (US$2.6 million) in 2008 to JP¥5.2 billion (US$68 million) in 2012,[11] and the vast majority of the earlier noted JP¥15.1 billion (US$196 million) state budget allocated for offshore wind was directed to demonstration test installations. Although they are more expensive to develop than conventional offshore turbine installations, floating variants offer Japan a long-term solution to its deep seawater predicament. Fixed lines are up to 200 metres long and are anchored to the seabed, providing for their stability on the sea surface.[12] Floating offshore wind turbine technology, out of necessity, is therefore likely to be an area where Japan will develop world-leading expertise. For example, 'wind-lens' technology (based on a specially designed inward curving ring attached around the turbine blade perimeter) developed by the International Institute for Carbon-Neutral Energy Research at Kyushu University presents new installation options for floating offshore and moreover is claimed to potentially improve turbine efficiency by two to three times, as well as operate at significantly quieter levels than conventional devices.[13]

4.7. South Korea

4.7.1. Introduction

Despite the stated ambitions of the South Korean government's Green Growth Strategy (GGS) launched by the Lee Myun-bak administration in 2008 to invest heavily in renewable energy, the country's wind energy sector has shown limited signs of dynamism and expansion since then. Growth has been hampered by complex planning regulations and procedures, as well as local public opposition to onshore wind farms (GWEC 2012). This, together with being a very densely populated and mountainous territory with plenty of coastline, has led to a recent push for offshore wind. In further similarity to Japan and Taiwan, South Korea is moving toward a more balanced installed capacity-oriented and technology-oriented approach to its wind energy policy.

4.7.2. Policy development

South Korea's first wind energy policy measures were embedded in the 2001 Basic Plan for New and Renewable Energy Development, Utilisation, and Promotion, in which the sector received prioritised R&D support for establishing initial demonstration projects. Part of the plan entailed the introduction of East Asia's first RPS scheme to the electricity retail market, but with the relatively unambitious aspiration of renewables achieving a minimum 1 percent of electricity generation—and this applied only to the Korea Electric Power Corporation (KEPCO), the state-owned firm dominating the power generation industry and grid system (Chapter 3). Targets for raising wind energy capacity to 1,137 MW by 2010 and 2,250 MW by 2012 were set in the 2001 plan as well. Also in 2001, the government introduced the region's first

FiT scheme for wind over a guaranteed 15-year period, up to a 250 MW capacity limit per project. The 2001 plan was soon revised by a Second Basic Plan launched in 2003, which increased RPS targets to renewables supplying 5 percent of the country's total primary energy by 2011, offered new low interest loans and grants (25 percent capital costs for wind projects), and provided new R&D funding support for wind energy technology development.

In 2008, the government launched its new One Million Green Homes Programme, which offered new subsidies for installing small wind turbines in residential areas. Meanwhile, its FiT scheme was proving rather ineffective at promoting wind energy and other RE sector development, and it was replaced by a new RPS programme in January 2012 based on minimum 20-year contracts. By this time, South Korea had abjectly failed to reach the 1,137 MW wind energy target by 2010, only achieving 379 MW of installed capacity by that year. Under the 2012 RPS scheme, the largest power utilities were required to either generate or purchase wind energy and other RE-generated power at set minimum shares, beginning at 2 percent by 2012 and rising incrementally to 10 percent by 2022. Given that renewable energy's share of the nation's electricity output increased from 1.7 percent to only 2.1 percent over the 2004–2011 period, this will require a significant step-change up in capacity expansion in wind and other RE sectors. As part of the GGS and its associated long-term energy strategy, the government set revised targets for wind to reach 15.7 GW by 2022 and 23.0 GW installed capacity and 50,000 GWh of actual power generated by 2030, by then covering an estimated 10 percent of future expected electricity demand. Even taking into account the long-range nature of this target, it is very ambitious, assuming an average of 1 GW plus of new capacity to be added each year to the grid. Up to 2013, it had taken South Korea over a decade just to accumulate around 0.5 GW of installed wind energy (Table 4.1).

4.7.3. National market and industry development

For a country of its economic development level, South Korea was a relative latecomer to the wind energy industry. The country's first grid-connected installation became operational at Haengwon, Jeju Island (two 300 kW turbines) in 1998.[14] In 2000, national installed capacity had reached 6 MW, rising to 192 MW (12 wind farms, 120 wind turbines) by 2007 and 561 MW by 2013, which represented only 0.2 percent of the country's electricity generation (GWEC 2008b, 2014). By this time, there were 46 wind farms with an average capacity size of just 13 MW, fractionally higher than Japan's. Three-quarters of these farms operate with a single-figure number of turbines in the sub-10 MW range (Windpower.net 2014). The country's largest wind farm is the 98 MW installation at Gangwon, which became operational in 2005. There were three other operational or under construction installations in the 50 to 99.5 MW range and another three in the 25 to 49.5 MW range by 2013 (Table 4.3). The second largest wind farm, Yeongdeok (68 MW), was initiated through the CDM scheme of the Kyoto Protocol (GWEC 2008b).

After the government put in place policies to support R&D and showcase demonstration technologies in wind energy from the early 2000s onwards, it did not take long for South Korea's large diversified *chaebol* companies with relevant engineering competences to develop industrial capacity in the sector. The countries eight main wind turbine producers are DeWind (Daewoo), Doosan, Hanjin, Hyosung, Hyundai, Samsung, STX, and Unison, represented overall by the Korean Wind Energy Industry Association. All have made substantial investments in the industry, applying well-established expertise as well as economies of scope advantages from heavy and marine engineering, aeronautics, information communication technology, and material science. In 2006, many of these firms were already producing 3 MW rated turbines, then at a cutting edge of the sector's new technology development (GWEC 2007a). By 2014, Samsung was developing 7 MW offshore turbines, and Hyundai was producing 5.5 MW models. There are now also a growing number of Korean companies involved in wind energy component manufacture and export. This corporate strategy is aligned with the state's own industrial strategy: The government has plans to create wind energy manufacturing hubs in the country in order to achieve its GGS target of capturing a 10 percent global market share of green technology industries by 2020.

4.7.4. Main challenges ahead

4.7.4.1. Overcoming public opposition

There are acute spatial constraints to developing wind energy in South Korea given the dense population and mountainous terrain. Like in Japan and Britain, opposition from the South Korean public to onshore wind farms has hindered the sector's development. The scarcity of publically accessible countryside has led to well-organised local protests against wind farm development to the extent that only a quarter of installation proposals are formally passed through South Korea's tough planning permission regulations. However, there is evidence that public opposition in the country is waning in light of the 2011 Fukushima disaster in nearby Japan and the effects of climate change (GWEC 2012).

4.7.4.2. Offshore wind in South Korea

It is conceivable that South Korea could become a world leader in offshore wind in the coming years, based on an increasingly favourable policy environment, private sector investments, strengthened techno-industrial capacities of Korean firms, and state strategic planning (GWEC 2012, Oh et al. 2012). In 2010, the Ministry of Knowledge and Economy released a strategy roadmap for offshore wind development based on a three-stage plan, with the target of establishing 2 GW of offshore capacity by 2020—more than four times South Korea's total wind capacity in 2012. By that year, the country had only installed 5 MW of offshore wind capacity. In the strategy roadmap's first phase, a 100 MW demonstration offshore farm located off the southwest coast will be completed by 2014 using 3 to 7 MW turbines from eight domestic manufacturers. The second phase entails a further 900 MW of capacity to be in operation by 2016; in the third and final stage, an additional 1,500 MW will be constructed through a competitive tendering process over 2017–2019. The combined 2.5 GW offshore capacity will come under KEPCO's supervision. Raising offshore capacity to this level will cost an estimated US$8.2 billion of investments (GWEC 2013). In addition, local governments and power companies have their own plans to develop a possible 4.5–6 GW of offshore wind capacity and a new production cluster. For example, the Korean Southern Power company plans to install 500 MW capacity in four southwest coast locations. Nearby, the Shinan local government in south Cheolla province is developing a large-scale wind energy industry cluster for offshore and onshore wind over the next 20 years (WWEA 2012b).

4.8. Taiwan

4.8.1. Introduction

The island of Taiwan has abundant onshore and offshore wind resources. It has only recently begun to realise its wind energy potential. The government started to promote wind farm development from 2000 onwards, but prototype technology development programmes began two decades prior to this. Taiwanese firms have considerable techno-innovatory capabilities on which to found a robust, expanding wind energy industry. Like its Northeast Asian neighbours, Taiwan is looking to develop an offshore sub-sector in the future, especially on its west coast and offshore islands, where wind speeds are particularly strong (Huang and Wu 2009, Hwang 2010).

4.8.2. Policy development

Initial endeavours to develop Taiwan's wind energy sector date back to the 1980 Wind Turbine Prototype Development Programme led by the Industrial Development Bureau (IDB) to develop test device technologies up to 150 kW capacity. This was the first wind-specific policy legislation introduced in East Asia. The 1980 programme, however, was abandoned in 1990. Although Taiwan's wind energy sector benefitted from generic investment tax credits and R&D policy support in the latter half of the 1990s, it was not until 2000 that the foundation for sectoral development was laid (Lee and Shih 2011, Lin et al. 2009). In this year, the government initiated a new 5-year programme for subsidising wind energy demonstration projects for up to 50 percent of total installation costs while simultaneously providing technical support to domestic firms, for example, through research facilities such as the Energy and Environment Laboratories of the Industrial Technology Research Institute

(Chen et al. 2010a). Three demonstration wind farms were consequently created, the first being the Mai-Liao installation in Yunlin County, which was completed in November 2000 and operated by Formosa Heavy Industries. Others followed on Penghu island in 2001 (operated by TaiPower) and at Chupei, Hsinchu County (operated by Tien Long) in 2002. The wind energy demonstration programme was terminated after 4 years in 2004.

Due to the long delay in passing Taiwan's Renewable Energy Development Act (REDA; see Chapter 3), the state-owned enterprise TaiPower announced in November 2003 that it would procure wind energy and other RE power from independent power producers via 10-year guaranteed power purchase agreements, later extended to 15 years (Lin et al. 2009). This provided incentives required for wind farm projects to take off on the island's west coast (Chiang 2004) and was eventually replaced by the new FiT scheme brought in by the REDA in 2009. In the meantime, it was clear from 2005 onwards that the government took an increasingly strong industrial policy approach to wind energy. This corresponded to simultaneous efforts to strengthen Taiwan's RE policy, stemming from the 2005 Second National Energy Conference, the 2007 Strategic Review Board (SRB) meeting of the Bureau of Energy (BOE), the 2008 Framework for Taiwan's Sustainable Energy Policy, and the 2009 Green Energy Industry Sunrise Plan. In these government strategic initiatives, wind energy was viewed as an emerging green energy industry where Taiwanese firms could exploit existing, relevant techno-innovatory capabilities to become internationally competitive players in this field (Liou 2010, 2011). The key policy developments on promoting wind energy were as follows:

- *Renewable Energy Equipment Industry Developing Plan, 2005–2008*: Introduced by the IDB, this plan provided R&D support and subsidies to develop key components, including small and large wind turbine generators. The plan also included international cooperation measures with twin aims to foster technology transfers from foreign firms selling or producing wind turbines in Taiwan, thus helping to improve domestic industrial capacity, as well as to raise Taiwanese business engagement with the sector's international production networks.
- *Key Components for MW-class Wind Turbines Development Plan, 2006*: Managed by the BOE and running parallel to the previous policy, this plan was based on NT$1.3 billion (US$44 million) of R&D investment. It aimed to integrate existing technological competences in Taiwan's electrical, engineering, and materials sectors to establish broader foundations for a wind energy industry.
- *National Energy Technology Plan, 2010*: This plan included a National Science Council programme to advance R&D in small and mid-scale wind turbine components, introduce a new national testing and authentication platform for these components, and further develop Taiwan's capacity for offshore wind power turbine engineering and marine technology.

The common underlying basis of these policies was developing a broad base of wind energy component manufacturing because Taiwan lacked a MW-class turbine producer. The government's plan was to establish a multiple-tier supplier value chain to create conditions for whole-assembly turbine makers to emerge, as well as to enable Taiwan's dense concentration of small- and medium-sized enterprises in related sectors to expand operations in this new strategic industry. Despite the growing funds diverted to RE sectors in Taiwan during this period, wind energy received less than 5 percent of the total, with around 40 percent going to bioenergy and more than 50 percent to solar (Liou 2011). Meanwhile, the government's initial target for wind energy (1,500 MW by 2020, set in the 2003 plan) was revised upward in the 2007 SRB meeting: a shorter-term target of 980 MW by 2010 (just over half this level was actually achieved by that year), then 1,480 MW by 2015 and 3,000 MW by 2025 (BOE 2007). These targets were in turn superseded by new, longer-term targets set in the 2011 New Energy Policy of 4,200 MW by 2030, where onshore wind would contribute 1,200 MW and offshore would contribute 3,000 MW.

4.8.3. National market and industry development

Taiwan's total installed wind energy capacity only broke through the 100 MW barrier in 2005. By 2014, there were 22 grid-connected wind farms in Taiwan, which were all onshore; the largest was the 100.8 MW 'Wind 1' installation, which became operational in 2009. In addition, Taiwan had two wind farms in the 50 to 99.5 MW range by 2012, and six in the 25 to 49.5 MW range (Table 4.3). Total installed capacity had reached 614 MW by 2013, and Taiwan's average wind farm size was 28.2 MW, which is well over twice that of Japan's and South Korea's averages. TaiPower owns and operates most of the island's wind farms. In 2010, TECO became Taiwan's first whole-assembly manufacturer of MW-class turbines, making 2 MW machines with a view to scaling up to mass production in forthcoming years. TECO and Formosa Heavy Industries have produced large turbine generators for some time, while Atech Composites mass produces turbine rotor blades. In addition, an increasing number of Taiwanese manufacturers of very small-scale turbines have emerged, and here Taiwan's industry has achieved almost 100 percent technological independence and development capacity (Liou 2011).

Taiwan had no offshore wind farms by the early 2010s. In 2010, the government helped to form the Taiwan Offshore Wind Power Alliance, comprising 18 companies from the energy, engineering, and manufacturing sectors. Future plans include a collaboration with Formosa Wind Power and Fuhai Wind Farm to install a number of large offshore wind turbines on the western coast by 2015, with the state covering up to 50 percent of construction costs (GWEC 2014). By early 2014, another dozen or so other offshore projects were at the design stage in Taiwan.[15]

As a consequence of the IDB's aforementioned Renewable Energy Equipment Industry Developing Plan (2005–2008), international collaborations have cultivated a series of technological transfers and learning in Taiwan's wind energy industry. For example, Vestas established production facilities on the island in 2008 and signed a technology licence agreement with TECO on developing engineering expertise for offshore wind turbine generators, as well as an agreement to source a third of its components from local firms. This included a contract for substantial sourcing of carbon fibre materials from Formosa Heavy Industries and pylon materials from China Steel. The development of these and other relationships with Vestas additionally helped to further internationalise Taiwanese company production networks in the industry. To consolidate corporate network relationships and nurture common international competitive positions, a number of business associations in Taiwan's wind energy industry were formed, namely the Taiwan Wind Turbine Industry Alliance (2005), Wind Power Generator Equipment R&D Alliance (2005), Taiwan Wind Energy Association (2006), and the Taiwan Small and Medium Wind Turbine Association (2009).

4.8.4. Main challenges ahead

The majority of Taiwan's population lives along the western seaboard, where the island's best wind resources are located. While this has advantages in terms of load-centre proximity, it has raised problems in terms of situating wind farms far enough away from residential areas. Given the dense population of Taiwan generally, there are stringent regulations on land use and competitive demands on scarce land resources (Huang and Wu 2009), thus presenting challenges similar to those facing Japan and South Korea concerning future onshore wind energy development.

Valentine (2010) reported that there have been some significant differences of opinion over Taiwan's wind energy potential, which may undermine support for the sector's future development. In late 2000s, TaiPower estimated that the island's technical potential was 4,600 MW for onshore wind and 9,000 MW for offshore wind—thus, 13,600 MW in total. However, due to various mitigating factors (e.g. land-use restrictions, economic constraints, competition for land development), TaiPower's predictions for realisable potential were only 1,000 MW onshore and 1,200 MW offshore. This was in contrast with private firm Infravest's much higher realisable potential figures of 3,000 MW and 5,000 MW, respectively, based on more optimistic assumptions regarding future industry cost reductions, improving government incentive measures, smarter site location techniques, and diminishing local objections to wind farms. Interestingly, much of Infravest's analysis was based on the German model of how a well a wind energy sector could develop given a conducive policy and regulatory environment, whereas TaiPower's more pessimistic predictions reflected the government's caution over public opposition to wind farm development. This included offshore development, where environmental groups have complained that the construction of wind farms at sea could damage the habitat of the endangered white-beaked dolphin.

4.9. Southeast Asia

4.9.1. Introduction

Wind energy has been a relatively low-priority renewables sector in Southeast Asia because only a few of its nations have significant wind resources. Singapore, Malaysia, Brunei, and Indonesia are located in the inter-tropical convergence zone (a low-pressure area around equatorial latitudes, also known as the 'doldrums'), where there are relatively calm winds in normal weather conditions. For example, average wind speeds in Singapore are around 2 metres per second. The sub-region's best wind potential lies to the north in areas of Vietnam, Cambodia, Myanmar, Laos, the Philippines, and Thailand. Of these, only the Philippines, Thailand, and Vietnam have started to substantively promote a wind energy sector, with a focus on installed capacity and no plans to foster industrial capacity development.

4.9.2. Policy development

Policies for supporting wind energy development in Southeast Asia remain generally weak. Very few sector-specific policy initiatives have been introduced and state support has invariably been embedded within general programme of support. In the Philippines, the 1997 New and Renewable Energy Programme (Ocean, Solar, and Wind Energy Law) carried the first policy instruments to promote wind energy development in the country, mainly aimed at small demonstration projects of sub-1 MW capacity. This had no apparent impact, however, on wind farm installation. Later, the 2003 Renewable Energy Policy Framework outlined a long-term aim of the nation becoming Southeast Asia's largest wind energy producer. The Philippines' first grid-connected wind farm became operational the following year at Bangui Bay (see below). In 2008, the government's Renewable Energy Act legislated for new FiT and RPS schemes for various RE sectors including wind, and most recently set a target of 2,378 MW capacity by 2030.

Measures to support wind energy in Thailand began in 1994 with the Small and Very Small Power Purchase Agreements programme, which was similar to the Philippines' with an initial focus on fostering small-scale deployment of wind turbines but with an emphasis on grid-connected projects. This was extended in 2002 to include a new very-small power producers (sub-1 MW) category, and a nationwide feasibility survey was conducted to identify the most suitable site areas, mainly located in northeastern Thailand (Phuangpornpitak and Tia 2011). In the country's 2004 Strategic Plan for Renewable Energy Development, wind energy was one of many sectors chosen for strategic support through the introduction of a RPS scheme, financial incentive programmes, and R&D funding for renewable energy technologies. Thailand later introduced a FiT scheme from 2007 onwards, which offered additional incentives for wind energy investors. The 2008 Renewable Energy Development Plan set an installed capacity target for wind of 800 MW by 2022, which was subsequently revised upward to 1.2 GW by 2021 by the superseding 2011 Alternative Energy Development Plan, with a new FiT scheme with relatively generous benefits for wind farm developers. This strategic planning has provided significant impetus for Thailand's wind energy sector (Table 4.1).

Meanwhile in Vietnam, the government's 2001 Renewable Energy Action Plan incorporated wind energy into its rural electrification strategy. In the later-introduced National Energy Development Strategy (2008–2020), wind was a priority RE sector to help achieve the target of delivering 5 percent of national electricity generation by 2020 from 'new energy.' Later, the National Power Development Plan (2011–2030) set the ambitious goal for wind of 1 GW by 2020 and 6.2 GW by 2030.

4.9.3. National markets and industry development

The Philippines' best wind resources are in its central region islands, the far north on Luzon island, and Palawan island to the west (Sibayan 2010). A demonstration 1.5 MW wind farm was constructed at Basco in 2002 by the state-owned National Power Corporation. Thereafter, the Bangui Bay wind farm complex was developed on the far north coast, comprising the two wind farms of Bangui Bay I and II; these were the only plants in operation by early 2014. A much larger 81 MW wind farm at Pagudpud, also in the Illocos Norte province, was under construction at this time.

By the early 2010s, Thailand was developing two wind farms at Korat, West Huay Bong in the northeastern province of Nakorn Rachasrima. Korat I (207 MW) was completed in 2013, whereas Korat II (270MW) is due for completion in 2016. These were developed largely through private-sector financing, led by Thai company Wind Energy Holding, with limited government support. Thailand also had a third, much smaller 7.5 MW Theppana wind farm (three 2.5 MW Goldwind turbines), located in Chaiyaphum province, which become operational in August 2013 (Table 4.3).

The best wind resources in Vietnam are located on its southern coast, mountainous central regions, and to a lesser extent in the far northern coast (Nguyen 2007, Petro Vietnam 2009). By 2013, the Binh Thuan wind farm was Vietnam's only operational installation; it had gradually expanded to 30 MW capacity over the years. Meanwhile, a 99 MW wind farm was constructed in the same year at Vinh Trach Dong in Bac Lieu province, and the government was reported to have plans to develop a 5,000 MW wind energy complex at this location with the assistance of a US$1 billion loan arranged between the Vietnam Development Bank and the US Eximbank.[16]

4.9.4. Main challenges ahead

Taking into account Southeast Asia's human and natural geography, one of the most promising future areas of its wind energy development concerns the deployment of off-grid small turbines in higher elevation remote areas, where wind resources are generally better than in lower elevation zones. As noted in Chapter 1, electrification rates (i.e. the share of national populations served by the electricity grid) in many parts of the sub-region are comparatively low. Thus, this particular wind energy sub-sector could offer good power generation options in Southeast Asia's more remote areas. Indeed, off-grid small wind turbines are already being used across the sub-region (Nguyen 2007, Phuangpornpitak and Tia 2011, Sibayan 2010), and this modality of wind energy use may prevail for some years yet over grid-connected large turbines. Even though it has relatively poor wind resources, Southeast Asia has only developed a tiny fraction of its wind energy potential. The above-mentioned plans to develop larger-scale installations in the Philippines, Thailand, and Vietnam in the coming years, as well as relatively ambitious installed capacity targets, may yet provide the much-needed momentum require for the wind energy sector to prosper in this part of East Asia.

It would seem unlikely that Southeast Asia will develop an industrial production hub for wind energy equipment in the near future. Those nations in the sub-region with relevant techno-innovatory capabilities (e.g. in high-end electronics, engineering, and material science) are Singapore and Malaysia. However, they lack wind resources and thus the incentive to develop an indigenous wind energy industry in any strategically significant sense. European firms, such as Siemens, have supplied most of the wind turbines operating in Southeast Asia to date. If the Philippines, Thailand, and Vietnam do develop a sufficient critical mass market at home in the future, they have an opportunity to engage in the industry's international production networks, manufacturing lower-end components initially and then possibly moving to whole turbine assembly at some future point.

4.10. Conclusion

Wind energy is highly significant for a number of reasons. It is the largest non-hydro renewables sector in power generation and has become an important emerging strategic industry, where East Asian firms and countries are strengthening their competitive position. Increasing levels of state and business investments have been channelled toward wind energy development, spurring impressive advances in techno-innovation. More broadly, in the context of 'long-wave' techno-economic development theory, wind energy is seen in the region generally as an important element of the emerging new green or eco-industry paradigm.

Notwithstanding notable spatial constraints that confront them, Japan, South Korea, and Taiwan have the economic and technological resources to develop a much larger wind energy sector than they have to date. All three have considerable potential for the more expensive and technically challenging offshore development. Most parts of Southeast Asia do not face the same spatial constraints but lack wind energy resources in the first instance, as well as the economic and technological resources to significantly expand the sector. China faces the least constraints in the region concerning the above-mentioned factors. More generally, China has made a significant impact on the global wind energy sector. Its strategic development approach to the sector charts back to the late 1990s, with the government initially promoting foreign technology transfers before fostering indigenous production and innovation capacity. The Chinese government has since cultivated the symbiotic development between industrial production and installed generation capacity. Increasing levels of state support were channelled toward the industry as China's international position in the industry strengthened over time, and wind energy became an early important motif of the nation's ecological modernisation by the late 2000s. The latest phase of this strategic industry policy is the Special Technology Development Plan for Wind Energy, introduced in 2012 as part of the 12th FYP, with an aim to make China a global player in high-tech large turbine manufacturing.

The record of East Asian governments meeting their strategic goals on wind energy generation has been mixed. China has constantly exceeded its targets on installed wind capacity, whereas its Northeast Asian neighbours (Japan, South Korea, and Taiwan) have thus far consistently failed to meet theirs to date. All three are, moreover, relative underachievers on installed wind capacity by international comparison with other economies at similar gross domestic product/income levels; South Korea is arguably the worst. A small number of Japanese firms, such as Mitsubishi, have been active players in turbine production for some time, but their market position has been relatively static. From a 'new energy' perspective, wind energy has not been afforded as much of a strategic industrial priority as solar or nuclear for some decades (Chapter 3). More recently, however, the Japanese government has been looking to offshore wind as a promising strategic option in its energy policy. Against the background of the South Korean state's GGS, firms such as Samsung have made increasing investment commitments to the industry and have expanded the nation's turbine production capacity. However, questions remain regarding whether a growing dependence on foreign investor firm suppliers will be required to meet the government's current wind-energy development targets. Additionally, GGS targets on exporting RE products and the strong export-orientation of the nation's wind energy equipment manufacturers is indicative of a possible intensification of strategic industrial competition among East Asian nations. Relatively stronger state support in Taiwan has created more favourable conditions for business investment.

Power-generation SOEs have played a significant role in East Asia's wind energy development. In many cases, wind farm developers are the same state-owned companies operating coal-fired, gas-fired, and nuclear power stations. Some may also dominate the grid (e.g. TaiPower, KEPCO) by buying wind energy–generated electricity and thereafter distributing it. Other SOE wind farm developers are national oil companies, such as PetroVietnam and China's CNOOC. Private business in East Asia is only really substantially engaged in equipment manufacturing. Thus, in power generation, exercises of state capacity in this sector mainly concern the government's provision of policy support and strategic direction to other state agencies.

Finally, although wind energy development in East Asia has been a high-profile example of the region's new developmentalism—spurred by growing state capacity support and ecological modernisation thinking on energy policy—the pattern and quality of state capacity intervention in the sector has been somewhat patchy. China's experience most clearly reveals the need for 'smarter' state capacity and national-level governance, and this especially relates to better coordinated policies on wind farm installations, grid connectivity, and infra-structure development. This chapter showed, for example, how Chinese government subsidies provided strong incentivisation for expanding installed capacity but insufficient incentives on actual power generation output. Poor intrastate agency coordination between national governments, local governments, grid companies, and power generation companies was largely to blame. This first sector-specific chapter has shown that the quality of state capacity is important; it not just a matter of its quantitative intensification. This is, to some extent, typical of fast-growing strategic industry sectors, where policy-makers often find themselves at the bottom of steep learning curves due to quick-changing market and industry circumstances. Very recent policy initiatives have shown that the Chinese government is trying to better understand and accordingly address the problems arising from the dynamic and rather chaotic period of wind energy development, as well as the general need for smarter governance and exercises of state capacity.

Notes

1   Vertical axis windmills are thought to have existed on the Persian-Afghan border around 200 BC, and horizontal axis windmills in the Netherlands and parts of the Mediterranean from around the 14th century onwards.
2   The first electricity-generating wind turbine was a 12 kW device constructed in Cleveland, Ohio, USA in 1888. In the early 20th century, a number of 25 kW devices were installed across Denmark.
3   http://www.ictsd.org/bridges-news/bridges/news (accessed May 29, 2014).
4   XEMC-Darwind was in fact a merger between China's XEMC and Dutch company Darwind.
5   http://www.sunwindenergy.com/news/japan-feed-tariff-scheme-confirmed (accessed May 27, 2014).
6   http://www.bloomberg.com/news/2012-03-29/floating-windmills-in-japan-help-wind-down-nuclear-power-energy.html (accessed May 27, 2014).
7   Hitachi took over the Subaru wind turbine division of Fuji Heavy Industries in March 2012.
8   http://www.greentechmedia.com/articles/read/japans-wind-power-problem-828 (accessed May 29, 2014).

9 http://www.rechargenews.com/regions/asia_pacific/article307337.ece   (accessed May 29, 2014).
10 In contrast, the current world leader in offshore wind, Britain, has many shallow sea-shelves around its coastline, making it less technologically challenging and costly to develop wind farms at sea.
11 http://www.bloomberg.com/news/2012-03-29/floating-windmills-in-japan-help-wind-down-nuclear-power-energy.html (accessed May 29, 2014).
12 http://www.bloomberg.com/news/2012-03-29/floating-windmills-in-japan-help-wind-down-nuclear-power-energy.html (accessed May 29, 2014).
13 http://www.bellona.org/articles/articles_2012/far_east_wind (accessed May 29, 2014).
14 http://www.kweia.or.kr (accessed May 29, 2014).
15 http://www.4coffshore.com (accessed May 29, 2014).
16 http://www.cleanbiz.asia/news/mekong-delta-500-mw-wind-plant-underway (accessed May 29, 2014).

5 Solar energy

5.1. Introduction

Of all the energy sectors thus far developed (renewables, fossil fuels, and nuclear), the technical power potential of solar energy is by far the greatest. Annual solar irradiation reaching the planet is enough to satisfy the current global energy demand around 10,000 times over (EPIA 2011b, IPCC 2011). The vast majority of East Asia's landmass is in the global Sunbelt zone, thus making it a region with huge solar energy potential. For example, Hang et al. (2007) estimated that captured solar radiation falling on just 1 percent of China's 'wasteland,' located in its northern and western regions (26,300 km^2), could potentially generate 1,300 GW of electricity, which is well in excess of the country's present electricity consumption levels (see Chapter 2). The solar energy sector comprises three main sub-sectors: solar photovoltaics (PVs), concentrating solar power (CSP), and solar water heaters (SWHs). This chapter is structured around the study of these three different technologies.

5.2. Solar energy: an overview

5.2.1. General evolution of the sector

5.2.1.1. Introduction

Over the course of human history, various kinds of solar energy technologies have been devised by civilisations around the world. From the mid-19th century to the early 20th century, solar energy was used to generate steam by capturing the sun's heat to run engines and irrigation pumps (Smith 1995). As we later note, the modern solar energy industry, based on photovoltaics, emerged in the 1950s. In addition to having the greatest power potential of all renewables, the solar energy sector also has one of the widest ranges of technology applications, being able to deliver electricity generation, heating, cooling, and fuels for various purposes (IPCC 2011).

5.2.1.2. Solar photovoltaics

Photovoltaic cells are by far the most commonly used form of solar energy technology, converting light into electric current through photoelectric reaction effects. More specifically, each cell is composed of layers of semi-conducting material (e.g. silicon) that create an electric field when exposed to sunlight, the intensity of which determines the amount of electricity generated. Solar PV cells work even in cloudy or other low-light conditions, and their generation efficiency ratings have progressively improved over time. Cells are connected together into larger units, generically known as modules, the most common of which are the solar PV 'panels' that are most frequently seen on rooftops. These are enclosed between a protective glass or other transparent cover and a weatherproof backing usually made from thin polymer (IEA 2009c, EPIA 2011b). The energy required to manufacture a solar PV installation usually takes between 0.5 and 3 years to recover in terms of energy generated, depending on its location. Solar PV is arguably the most versatile renewable energy technology, having multiple applications and varying more than any other energy technology in terms of power generation devices. It can be integrated into very small electronic machines, such as calculators and watches, or comprise solar park installations that supply electricity for many thousands of households.

Solar PV technology first emerged in the 1950s and has developed over three generations. First-generation PV is based on polysilicon (also referred to as crystalline silicon) that is mined, melted, and then cooled to form silicon ingot blocks. Wafer-thin slices then are cut and transformed through various treatment processes to create solar PV cells, and thereafter modules or panels. This generation type still dominated the market in the mid-2010s, with around an 80 percent share (EPIA 2014). Second-generation PV is based on thin-film technology, whereby extremely thin layers of photosensitive semiconducting material are deposited on low-cost substrate backing such as glass, stainless steel, or plastic. This has opened up new possibilities for integrating solar PV into the fabric of buildings and various end-consumer products. Third-generation PV involves new, largely experimental technologies that are still mostly at the demonstration stage, including the development of organic PV and thermo-PV cells that have the ability to function to relatively high output efficiency levels, even in dim light.

Other examples are concentrator PVs that use a special lens to concentrate sunlight onto PV cells and thus are best deployed in very sunny locations; they have close connections with concentrating solar power technologies, discussed later. In addition, research is being conducted on the development and application of quantum dots and nanotechnology particles in PV cells, especially in concentrator photovoltaics (EPIA 2011b, Wu and Mathews 2012). Solar PV is furthermore a widely used distributed energy technology that has enabled millions of households, businesses, and other organisations to become independent energy 'prosumers' (see Chapter 1).

5.2.1.3. Concentrating solar power

CSP is a solar thermal technology that uses 'concentrator' mirrors to convert solar rays into high-temperature heat, which is in turn converted into mechanical energy using heat engines (e.g. steam turbine) and other generator devices that then produce electricity. It also has the potential for producing other energy carriers, such as solar fuels. Whereas solar PV can generate electricity even in cloudy conditions, CSP requires clear skies and direct sunlight to operate, and therefore installations tend to be situated in sunny arid zones, such as deserts. Captured solar energy used in CSP plants is measured as direct normal irradiance (DNI). Estimates suggest that the earth's deserts can produce enough CSP energy in 6 hours to equal the world's total annual energy consumption (European Policy Centre 2009). Optimal DNI levels are usually located in 15°–40° latitude zones in the north and south. The equatorial zone in between is normally too cloudy and wet during the summer months, and higher latitudes are too cloudy and cold. Much of Southeast Asia lies in the equatorial band (15° north/south), so it is not ideally suited to CSP. Furthermore, the sub-region's relatively high humidity levels increase atmospheric scattering and limit direct normal radiation (Lidula et al. 2006). China, Japan, Korea, and Taiwan lie in the optimal latitude zone, but only China has the arid spaces required to develop utility-scale CSP plants, especially in the country's western and northern provinces (Chien and Lior 2011).

There are currently four types of CSP technology: parabolic trough, Fresnel mirror, power tower, and solar dish collector (Wolff et al. 2008). These technologies tap into the beam element of solar irradiation. Different types of concentrators bring the solar rays to a 'point focus' or 'line focus' onto heat-collection elements (e.g. absorbers and receivers), where the energy-generation process occurs by the use of special conducting fluids and other materials. CSP installations comprise arrays of concentrators that track the sun's movements during daylight hours and concentrate the solar irradiation of a magnitude of 70 to 100 times greater than the collected sunlight, similar in principle to how magnifying glasses can concentrate sunlight to small focal points. CSP's main advantage over solar PV lies in its ability to store and transmit power 24 hours per day through the use of heat storage technology (most commonly molten salts), whereas solar PV's generation of power is limited to sunlight periods only. Thus, CSP can dispatch electricity to the grid when required, including after sunset to match evening peak demand periods. By providing backup sources of power, it can play a vitally important role in helping to integrate larger amounts of intermittent renewable sources, such as wind and solar PV, into the grid. CSP technologies furthermore have potential for supplying process heat for industry; cogeneration of heating, cooling, and power; and more specialised uses, such as water desalination (IEA 2009d).

5.2.1.4. Solar water heaters

Basic roof-mounted SWHs were first developed during the 1900s in the United States, and the modern development of the technology began in the 1960s. SWH collectors or panels are normally rooftop-installed devices that look quite similar to their solar PV module counterparts and generate hot water for domestic or industrial use. Open-loop or direct SWH collectors are made up of a series of parallel tubes containing potable water, which is heated through solar radiation and then fed through directly as the hot water supply into the system. In the closed-loop or indirect method, a heat-transfer fluid passes through the collector tubes and then to the heat exchanger, where water is heated. The latter method is often used in areas where water freezing and scaling are problems. Passive systems rely on the circulation of water in the system through the natural thermosiphon process, whereas active systems use pumps. Solar water heaters can be found in most countries around the world, even in relatively cool temperate zones.

5.2.1.5. Other sub-sectors

Thus far, we have examined various forms of active solar energy technologies that entail conversion or storing processes. By contrast, passive solar concerns simply collect the energy without converting the sun's heat or light into other forms. For example, passive solar heating involves the manipulation of solar irradiance on buildings to create comfortable conditions inside them through the use of glazing, other transparent materials, and control devices to manage heat gain and loss, thus negating the use of electric air conditioning or pump appliances. Another example is solar cooling for buildings, which is achieved through the use of solar heat to create thermodynamic refrigeration absorption or adsorption cycles. Solar energy for lighting is based on engineering and architectural design rather than being a conversion-based technology, and demonstration-stage solar fuels use solar energy to deliver hydrogen- or hydrocarbon-based energy carriers through artificial photosynthesis processes—the most important energy output being the production of hydrogen.

5.2.2. Growth and expansion

The first solar PV cell was produced in 1954 in the United States by Bell Laboratories, and the use of solar PV cells in orbiting satellites from the late 1950s helped to initially develop the modern solar energy industry (Green 2005, Timilsina et al. 2012). The sector saw steady growth up to the mid-1980s, but then cheap oil led to a period of only moderate growth for the next decade

(Bradford 2006). Leadership in this sector had switched from the United States to Japan and Europe by the mid-1990s. Solar PV is now the third most important renewable energy (RE) sector in terms of installed electricity generation capacity, after hydropower and wind energy. Since the early 2000s, it has also been the fastest growing mainstream power generation sector (EPIA 2012a, 2012b). Over 2008–2013, installed worldwide PV capacity has recorded a 55 percent annual average growth rate. In absolute terms, total global capacity increased from 1.5 GW in 2000 to 7.0 GW in 2006 to 138.9 GW by 2013 (Table 5.1).

Europe has remained the dominant force in solar PV electricity generation, with this position gradually strengthening from the mid-2000s onwards. Its share of the global total has risen from 47.4 percent in 2006 to 58.7 percent by 2013. Meanwhile, East Asia's share gradually declined for most of this period, from 26.2 percent in 2004 to 11.9 percent in 2011, mainly owing to Japan letting slip a once world-leading position. In 2000, Japan accounted for around 40 percent of global installed capacity and 35 percent of solar PV cell production, but it did not keep pace with its main European rivals during the years that followed (EPIA 2011a). The United States, which like Japan was a past pioneer in this field, has too fallen back in this field over time. However, East Asia has experienced resurgence in installed PV capacity due to activist policy support in a number of leading countries from the region, most notably China and Japan. East Asia's capacity level has consequently risen sharply from 9.1 GW in 2011 to 35.1 GW by 2013 (Table 5.1). China has emerged as not only as the region's East Asia's principal producer of PV equipment but also its leading installer, with its capacity level having increased from 0.9 GW in 2010 to 18.8 GW by 2013. Japan's capacity level rose from 3.6 GW to 13.6 GW over the same years, while South Korea, Thailand, and Taiwan also have registered strong recent growth. East Asia's continued rising investment in installed PV capacity will most likely make it the global leading region in the near future, overtaking Europe.

The first concentrating solar power plants were developed in the 1980s, first in the United States and then in Europe. Early commercial plants were established in California from 1984 to 1991, assisted by state government support measures, but falling fossil fuel prices led to a withdrawal of this support in the early 1990s. Of the three main solar sub-sectors, CSP remains the smallest in terms of energy generated worldwide. There are, however, plans to develop large-scale CSP plants–the most ambitious of which are mega-scale projects that have been proposed for the Sahara ('Desertec') and Gobi ('Gobitec') deserts. These would require enormously expensive super-grids–initially €40 billion for Desertec, according to the European Policy Centre (2009)–to distribute the energy over many thousands of kilometres to their main markets (the European Union for Desertec and Northeast Asia for Gobitec). Worldwide CSP installed capacity remains very small by energy sector comparison but is growing very quickly, from just 0.4 GW to 3.4 GW by 2013, with the vast majority of this capacity located in the United

Table 5.1 Solar PV development, East Asia and global, 2006–2013

	Installed capacity (MW)							
	2006	**2007**	**2008**	**2009**	**2010**	**2011**	**2012**	**2013**
China	80	100	145	373	893	3,093	8,300	18,800
Japan	1,708	1,917	2,147	2,627	3,618	4,914	6,914	13,600
South Korea	36	81	358	524	656	812	1,064	1,467
Thailand	0	0	0	0	28	149	359	704
Taiwan	2	4	9	19	32	102	206	392
Indonesia	1	3	6	8	17	20	20	20
Malaysia	5	7	9	11	11	13	36	57
Singapore	0	0	0	2	4	6	10	15
East Asia total	**1,832**	**2,112**	**2,674**	**3,564**	**5,259**	**9,109**	**16,909**	**35,055**
% share of world total	*26.2*	*22.2*	*17.0*	*15.5*	*13.1*	*11.9*	*16.5*	*25.2*
Germany	2,899	4,170	5,979	9,785	17,193	26,678	32,411	35,715
Italy	47	117	456	1,173	3,494	12,754	16,361	17,928
Spain	148	690	3,398	3,415	3,784	4,191	5,166	5,340
Other Europe	213	280	721	1,984	5,306	8,093	15,162	22,505
Europe	**3,307**	**5,257**	**10,554**	**16,357**	**29,777**	**51,716**	**69,100**	**81,488**
% share of world total	*47.4*	*55.7*	*66.9*	*70.5*	*74.4*	*74.2*	*67.7*	*58.7*
United States	**624**	**831**	**1,173**	**1,650**	**2,528**	**4,383**	**7,777**	**12,100**
% share of world total	*8.9*	*8.8*	*7.4*	*7.1*	*6.3*	*6.2*	*7.6*	*8.7*
Other	**1,223**	**1,300**	**1,441**	**1,704**	**2,556**	**5,281**	**8,390**	**10,213**
% share of world total	*17.5*	*13.3*	*8.7*	*6.9*	*6.2*	*7.7*	*8.2*	*7.4*
World	**6,980**	**9,443**	**15,772**	**23,210**	**40,019**	**69,684**	**102,156**	**138,856**

Sources: EPIA (2011a, 2012a, 2013, 2014), REN21 (2014).

States and Spain. As we later discuss, China has plans to scale up its CSP sub-sector.

By 2012, there was 258.0 GW_{th} of installed solar water heating capacity. China dominates this market, with 180.4GW_{th} (69.9 percent) of this total. The next ranked East Asian country was Japan (ninth), with just 3.1 GW_{th} (REN21 2014).

5.3. Techno-innovation and production

5.3.1. General trends

Solar energy is arguably the most technologically dynamic renewables sector. The rate of techno-innovatory advancement has been very high compared to most energy fields. This has been particularly evident in solar PV, as the deeper commercialisation of the sub-sector has brought increasing levels of investment into research and development efforts to drive efficiency gains in power outputs and production costs.

The solar PV vertical production chain can be divided into four main parts: the manufacture of high-quality polysilicon, then silicon wafer manu-facture, followed by the production of solar PV cells, which are then used as the main components of PV modules (Zhao et al. 2011). In 1982, the cost of solar PV modules stood at US$27.00 per Watt (p/W) produced, but by the early 2000s had fallen to around US$5.00 (Timilsina et al. 2012). The price of PV modules remained generally flat from 2004 to 2008 at around US$3.50 to US$4.00 p/W. Although significant technology and production efficiency improvements were achieved during this time, a shortage of polysilicon and sharp rise in its market price[1] constrained production and thus placed coun-teracting upward pressure on module price levels. The ensuing rise in profit margins, though, brought in new investments and expanded capacity in polysilicon production and downstream manufacturing. By December 2009, the average market price for PV modules had halved in a year, to US$2.00 p/W.[2] Price levels fell below the US$1.00 p/W mark for the first time in late 2011, which for many observers was viewed as grid parity (Aanesan et al. 2012, Branker et al. 2011, Laird 2011, Yang 2010). According to the IEA (2012c), China's major solar PV manufacturers were producing at below or around the US$0.90 p/W mark by this year.

Concomitant reductions in solar PV installation, maintenance, and financing costs have made the sector more competitive (World Economic Forum 2011). A complete solar PV installation involves balance-of-system (BoS) costs that, in addition to the price of solar PV modules or panels, includes those for installing switches, support racks, wiring, and inverters. From 2008 to 2012, BoS costs fell by an annual average of 16 percent, compared to 40 percent per year for PV modules during the same period. Costs to residential customers installing solar PV systems (i.e. module and BoS costs) had fallen over time from US$100 p/W in 1975 to US$8 by end of 2007, and then to just US$3 by 2012 (module costs US$1, BoS costs US$2). Estimates by Aanesan et al. (2012) indicate that the price level for residential consumers may fall to US$1 by 2020. From early 2009 to early 2012, the average levelised cost of energy (see Chapter 3) of first-generation solar PV had fallen by roughly half, from US$0.32 p/kWh to US$0.17 p/kWh. For second-generation thin-film PV, these costs decreased from US$0.23 p/kWh to US$0.16 p/kWh, respectively, with costs ranging across countries from as low as US$0.05 p/kWh to US$0.25 p/kWh (BNEF 2012). A similar trend is apparent regarding off-grid power, where solar PV can compete on increasingly equal terms with diesel and other oil-based generators (BNEF 2011, IRENA 2012c). Modest estimates predict that techno-innovatory improvements could lead to a further 50 percent reduction in solar PV generation costs from 2011 to 2020 (EPIA 2011c). Meanwhile, PV cell efficiency ratings are increasing by around 2 percent per year (EPIA 2011b).

Although solar PV has become increasingly competitive, the industry has recently suffered from over-capacity problems. High-profile firms have gone bankrupt or been taken over by others (e.g. Hanwha Solar One's purchase of Germany's Q-Cells), while some have significantly scaled back production. Chinese firms have especially had a significant global impact by driving down solar PV module costs and prices with the assistance of state policy support, leading to high-profile trade disputes with the United States and European Union, as later examined. Whereas in the early 2000s China accounted for a small fraction of global PV equipment manufacturing, it now produces almost two-thirds of the world's PV cells and modules. Other East Asian states, such as South Korea and Taiwan, have also become more significant producers, based largely on extending technological competences from their broad-based semiconductor industry to solar PV sector development. Intranational knowledge flows on first-generation PV technology have increased over time in both economies, but this is not so evident in China, whose firms still depend on foreign technology transfer. Public research agencies have, at varying times, played crucial roles in techno-innovatory capacity building in all three economies. South Korea's large *chaebol* conglomerates, such as Samsung and LG, however, have possessed sufficient independent capabilities to develop their own technology with a lower level of assistance from state research institutes and universities, whereas in China they have been crucial. Taiwan's position lies between these two; the dense clusters of small- and medium-sized enterprises (SME) that form the basis of its economy are engaged in smaller-scale technological improvements in comparison to their much larger Korean counterparts (Wu and Mathews 2012).

5.3.2. Trade and foreign direct investment trends

The solar PV industry has become more internationalised compared to its sister wind energy sector due to the different nature of production processes, value chain dynamics, and industry structure. Solar PV modules and their component parts are far less bulky and therefore more easily internationally transportable. The industry structure is also far less concentrated, with a much larger number of firms competing across value-chain segments (Table 1.4). Despite deepening vertical integration in the solar PV industry (i.e. firms extending their operations up and down the value chain), its international trade intensity is much higher than in the wind energy industry, which has become more localised (see Chapter 4). International production networks are particularly well developed amongst the major solar PV firms, with cluster technology interests in related sectors such as electronics and information and communications technology (ICT; e.g. Sharp, Hitachi). China has become by far the world's largest producer-exporter of solar PV equipment, valued at US$35.8 billion in 2011, up from just US$2.9 billion in 2007 (IEA 2012c).

However, foreign direct investment (FDI) in the solar PV industry been hitherto quite limited and confined primarily to 'greenfield' investment in new production, research, and design facilities. Initially, this mostly concerned European and US investments in East Asia, with the earliest important example being American firm SunPower's 2003 venture in the Philippines. This was soon followed by others in the mid-2000s: Aleo Solar in China, and Q-Cells and First Solar in Malaysia. Later, First Solar moved its major manufacturing capacity to Malaysia, and then in 2009 signed an agreement with the local government authorities of the Chinese city of Ordos in Inner Mongolia to develop the world's largest utility-scale PV project (Kirkegaard et al. 2010). Norwegian firm REC produces the majority of its production in Singapore and has established a major research and development (R&D) centre in the city-state. Chinese and other East Asian firms are also themselves beginning to invest in new production facilities overseas. This new emerging trend could further internationalise the solar PV industry to potentially a more substantial level.

5.4. Obstacles to expansion

Dynamic technological change and market growth in the solar PV industry has made it difficult for policy-makers to predict future trends and set policies accordingly. For example, many governments around the world have significantly underestimated the popular response to their feed-in tariff (FiT) schemes and have had to scale these back for budgetary constraint reasons. As module and BoS costs continue to fall over time and solar PV systems become more affordable, the main challenge for policy-makers will increasingly shift to develop the enabling infrastructure and market relationships to expand installed capacity, as well as foster a stronger 'prosumer' culture. There is the main socio-technical challenge facing future solar PV development.

As noted earlier, solar PV is an intermittent energy source; thus, it serves mainly as a complement to constant power generation sources, such as hydropower. Although molten salts storage technology helps CSP circumvent this constraint, parallel developments in general storage and battery technology will be crucial to the future prospects of both sub-sectors. Increasing solar PV module longevity and efficiency rates is another important technological challenge. Module efficiency degrades over time, depending on the quality build of the product, losing around a fifth to a quarter of its original power rating after 20–25 years; thus, modules are often replaced after this period. Current efficiency ratings (i.e. conversion of captured irradiation energy into generated power) for second-generation thin-film PV are also still quite low, at just 4–12 percent, compared to first-generation PV ratings, which typically are around 20–25 percent. The energy efficiency ratings of heat transfer fluids used in CSP are similarly low.

Solar energy technologies are dependent on a variety of natural resource derivatives, most importantly at present polysilicon for first-generation PV. More advanced thin-film PV depends on cadmium and tellurium, which are byproducts of zinc and copper processing, respectively. Solar water heaters also require the use of various mineral elements. As with all industries that depend on finite resources, a lack of investment in material recycling could prove to be a major obstacle to longer-term development.

Solar PV's flexibility of application—especially its building-integrated and building-attached options—means that it does not face the same geospatial constraints that wind energy must contend with, although large solar park and wind farm developers do of course share similar land use and locational challenges. The same applies to CSP plants, which are generally large-scale installations and require much land, which must also meet a number of specific topological and other requirements. This includes ground with less than a 3 percent gradient (less than 1 percent is considered ideal) due to solar ray angle factors, stable soil, low average wind speeds to avoid dust issues, and access to a large water supply, mainly for cooling power cycle equipment (Chien and Lior 2011, Cohen 2005). On average, 3–3.5 litres of water are needed to produce a kWh of CSP-generated electricity (Jones 2011). Although this still compares favourably to conventional energies (14.2–28.4 litres p/kWh for fossil fuel power stations and 31.0–74.9 litres p/kWh for nuclear), CSP plants are situated in arid areas where water is scarce and often expensive to source.

Although many energy companies see solar energy as an opportunity for growth and development, some are concerned that the falling costs of solar PV (in particular, to grid parity levels) will have negative impacts on their profits and future prospects generally. The unique scope of solar PV for enabling all elements of society to become electricity prosumers through building-attached photovoltaic (BAPV) and building-integrated photovoltaic (BIPV) could lead to significantly diminishing dependencies on energy companies to supply their own generated power. Depending on their paths of development and utilisation, the SWH and CSP technologies could have the same effect of narrowing the market base for corporate expansion. This could lead to some power-generating companies possibly obstructing the expansion of solar energy sector. Alternatively, they may use their scale advantages to develop large solar park installations that can offer even cheaper electricity to consumers than may be generated from their much smaller rooftop systems. It will depend fundamentally on whether these companies and their shareholders have a conservative or forward-thinking mindset.

Solar energy has many socio-technical and environmental advantages over other RE sectors–most importantly over wind energy, its closest competitor in the renewables field. Solar energy has relatively compact and unobtrusive applications that generate power and heat noiselessly with no mechanical movement involved, except for very slow-moving mirrors in CSP plants. Furthermore, second- and third-generation solar PV allow possibilities for the visual and material blending of applications into existing buildings more effectively in the future. Most of the problems arising under this category concern the toxic outputs produced in the solar energy equipment manufacturing process, where explosive gases, lead-based materials, and corrosive liquids are often used in significant quantities (Gottesfeld and Cherry 2011, IPCC 2011).

5.5. China

5.5.1. Introduction

China possesses the most substantial solar radiation resources in East Asia. The country's average annual solar radiation is around 1.5 MWh per m^2. Over recent years, China has become the main global production centre for the solar energy sector, manufacturing around two-thirds of the world's solar PV equipment and around three-quarters of SWHs. Furthermore, solar PV power-generation capacity has dramatically taken off in China since the early 2010s, and the country has installed more SWHs than the rest of the world combined. The Chinese government are in addition investing in exploiting the country's significant CSP potential. The burgeoning growth of the solar energy industry can be mainly attributed to active state support, where new policy initiatives are being very regularly introduced, and also dynamic Chinese business entrepreneurism.

5.5.2. Policy development

Experiments on developing basic solar PV technology in China started in 1958, and test applications were first introduced in the 1970s. This laid the basis for a more substantial R&D policy on solar energy spearheaded by the Ministry of Science and Technology (MOST) schemes of the 1980s: the 1981 Key Technologies R&D Programme and 1986 National High-tech R&D Programme. In the mid-1980s, two PV cell production lines were established, initially using reject material from the country's then fledgling micro-electronics industry (Yang et al. 2003). Later, the 1997 National Basic Research Programme and

1999 Fund for Technology Based Firms created added further support to R&D activity, helping fund a series of solar energy demonstration projects in partnership with the China Academy of Sciences and selected universities (Grau et al. 2012, Huo and Zhang 2012).

Solar PV's versatility on offering a number of off-grid options on power generation has made it a long-standing core element of China's rural electrification programmes. The technology has been an important source of electricity generation for schools, hospitals, and various utility services (e.g. freshwater supply pumps) in many remote areas of China (Zhao, R. et al. 2011). It was a core element in the government's landmark Brightness Programme (1996–2000), the country's first comprehensive RE policy scheme that provided new energy facilities in over a thousand rural townships and villages. However, at the programme's end, China's total installed solar PV capacity stood at only 8 MW. This was superseded by the Township Electrification Programme (2001–2005) at the start of the 10th Five-Year Plan (FYP), leading to the installation of over 700 PV systems with a total combined capacity of 51 MW. This was later superseded by the 2006 Village Electrification Programme at the onset of the 11th FYP, based on US$5 billion public investment to extend electricity to 3.5 million rural households, with solar PV being a key deliverable technology (Huo and Zhang 2012, IEA 2014).

Solar energy received general RE policy support from multi-sector initiatives and legislation over the following years, such as from the 2006 Renewable Energy Law (REL) and the 2007 Medium and Long-Term Development Plan for Renewable Energy. Article 17 of the 2006 REL obligated the government to create an ambitious programme of integrating solar water heaters into existing buildings and new constructions, starting in high solar irradiation areas of the country. In the same year, the Ministry of Housing and Urban-Rural Development (MOHURD) and the Ministry of Finance (MOF) jointly issued various support measures to this end at the national government level. Article 40 of the 2007 Energy Conservation Law, the National Development and Reform Commission's 2007 National Implementation Programme for Promoting Solar Thermal Applications, and Article 4 of the 2008 Energy-Conservation Ordinance further mandated the utilisation of solar energy technology in building construction. In 2007, local authorities in Gansu, Jiangsu, and Shenzhen introduced laws requiring buildings with fewer than 12 floors to install SWHs. Others city and provincial governments followed their lead, setting off a nationwide policy trend at this level. By 2011, there were 34 city and provincial municipalities that had enacted SWH obligations (Hu et al. 2012, Timilsina et al. 2012, Xie et al. 2012).

Meanwhile, the 2007 Medium and Long-Term Development Plan for Renewable Energy set targets for solar PV development: 300 MW capacity by 2010 (893 MW was actually reached) and 1.8 GW by 2020 (already well surpassed by 2011). A new FiT scheme arising from the plan and launched in 2008 also helped paved the way for China's first large-scale PV (LS-PV) solar park projects. In June 2010, the National Energy Administration (NEA) announced a call for tenders for 280 MW of solar park projects by province-based allocations: Inner Mongolia (60 MW), Xinjiang (60 MW), Gansu (60 MW), Qinghai (50 MW), Ningxia (30 MW), and Shanxi (20 MW). A year before, the Solar Roofs Programme, led by MOF and MOHURD, offered subsidies for capital investments in BAPV and BIPV applications as well as new small solar parks in remote areas.

Of even greater importance was the Golden Sun Programme, the main policy platform for China's solar energy sector into the early 2010s (EPIA 2011b, 2012a; Grau et al. 2012; Huo and Zhang 2012; Zhao, R. et al. 2011; Zhao, Z.Y. et al. 2011). Jointly launched by MOST, MOHURD, MOF, and NEA in July 2009, this programme provided US$15 billion of public invest-ment (including US$2.9 billion of investment subsidies) for developing PV installations of a minimum 300 kW generation capacity. The scheme's target was for a 500 MW increase in national capacity within 3 years, laying the foundation for a huge spike in installed capacity development in the country (Table 5.1). Revised FiT schemes introduced by national and local govern-ments combined with a dynamic entrepreneurial response to these policies from both firms and individuals put China's solar PV sector on a whole new growth trajectory (IEA 2012c). In this fast-changing market and industry environment, policy-makers have had to almost constantly adjust their positions. In November 2012, the government revised its subsidy support rates for solar PV installation, providing greater incentives to investors—although only for a limited period, to June 2013; thereafter, subsidies were reduced.[3] In the same context, solar energy development targets have been regularly revised due to the sector's highly dynamic nature. During 2011 and 2012, China's state authorities were reported to have made three upward revisions to these targets.[4] Initially at its outset in March 2011, the 12th FYP had set a target of installed capacity expansion to 5 GW by 2015. However, after data revealed in early 2012 that over 3 GW had already been achieved by the end of the previous year, the target figure was over quadrupled to 21 GW. Then, in September 2012, after monthly data showed a pattern of continued rapid growth, targets were once more significantly revised, now to 40 GW for solar energy installed capacity by 2015 (which included 1 GW for the CSP sub-sector); these targets rise to 50 GW and 3 GW, respectively, by 2020. In the SWH sub-sector, the Chinese government have maintained their target of 300 million m^2 by 2020 (Hu et al. 2012). The 12th FYP also set a number of other technical targets, including the following by 2015:

- First-generation PV efficiency conversion rating to reach at least 20 percent; second-generation thin film to reach at least 10 percent and achieve basic levels of commercialisation
- Strengthening of the indigenous design capacity and system-integrated equipment for 100+ MW LS-PV installations
- Reducing power generation costs to the 0.5 Yuan p/kWh level (equivalent to around US$0.06 p/kWh)

- Reducing polysilicon production costs by 30 percent and achieving domestic supply capacity of polysilicon of at least 50 percent
- Establishing a national technical standard system and product testing platform
- Developing 24 new demonstration projects and 28 innovation platforms, and establishing more than 50 standards
- Implementing a National 973 Programme for R&D on next-generation superefficient PV cells with 40 percent conversion ratings.

As well as playing an active role in promoting installed solar energy capacity, China's local governments have also engaged on the industrial development side, especially those cities and provinces where major producers are based. In Jiangsu, home to the world's two largest solar PV cell producers, Suntech and JA Solar, the provincial government released its plan in 2006 to establish five international manufacturers and set a PV cell production target of 1.2 GW by 2010 (Zhao et al. 2011). Suntech and JA Solar each alone exceeded this target by that year. In 2008, Jiangsu's Development and Reform Commission issued a PV installation plan and accompanying FiT scheme with a new additional on-grid capacity target of 400 MW by 2011 (Huo and Zhang 2012). The aim was that much of this new capacity would be supplied by the province's own fast expanding companies.

As Huo and Zhang (2012) observed, the Chinese state has used different types of policies to support solar PV development, depending on the nature of the market, business actors involved, and the level of applications. For example, capital subsidies have been mainly used for promoting off-grid PV in remote areas because low income levels in rural communities mean they cannot easily cover initial investments costs. A combination of capital subsidy and net metering has been used for grid-connected BIPV and BAPV to help relieve the financial pressure on small-scale investors. For solar park projects, FiT schemes have been the primary incentive scheme because the market actors are usually companies with sufficient capital resources to make the initial investment required for the project's take-off stage. Public bidding has been generally used for solar park development, which has encouraged market competition in the sector's supply chain.

5.5.3. Domestic market and industry development

5.5.3.1. Solar PV

The development of China's solar PV market and industry corresponds closely with the substantiation of government policy and strategy toward the sector; the aforementioned 1996 Brightness Programme helped to consolidate the early stages of this development. By 1998, annual PV module production was a mere 2.3 MW capacity equivalent, with a fifth of this being exported. Meanwhile, installed PV capacity had increased from 1.8 MW in 1990 to

19.0 MW by 2000; the vast majority of this was imported and used for off-grid applications in remote parts of the country (Tan et al. 2012, Zhao 2001). At this time, China still lacked indigenous capabilities for PV cell manufacture. The aforementioned policy support measures brought in under the 11th FYP (2006–2010) and 2007 Renewable Energy Plan was to change this. For example, the newly created Institute of Solar Energy and Tianjin Institute of Power Sources collaborated with business on value-chain development and R&D activity to help establish a fast-expanding foundation for industrial-scale PV equipment manufacturing (Wang and Zhai 2012, Yang et al. 2003). In 2000, China's share of global PV production was just 2 percent, rising to 5 percent by 2004 and 15 percent by 2006. By 2008, the annual national output of PV modules had risen to 2.6 GW (up from 2.3 MW a decade earlier), then to 8.0 GW by 2010 and 23.0 GW by 2012. Correspondingly, China's share of global production increased to 63 percent by 2012 (EPIA 2011a; Huo and Zhang 2012; IEA 2012c, 2013a; REN21 2013). By 2013, six of the world top-ten PV module producers were Chinese (Table 5.2).

Most of China's solar PV equipment production (typically around 90 percent) has been exported over the last decade. However, a gradually increasing share of indigenous output is being used for the domestic market due to fast-rising national installed capacity levels. The Chinese PV industry is also very competitive with high rates of business turnover. The IEA (2012c) reported in 2011 that more than half of SMEs in the sub-sector went bankrupt, while small and large firms alike have scaled back production considerably due to over-capacity problems. Chinese dynamic entrepreneurship has led to production supply capacity even outpacing dramatic recent increases in demand. Industry consolidation around the nation's largest firms

Table 5.2 World's top-10 solar PV module producers, 2006 and 2013

2006			**2013**		
Rank	*Company*	*Country*	*Rank*	*Company*	*Country*
1	Sharp	Japan	1	Yingli Green Energy	China
2	Q-Cells	Germany	2	Trina Solar	China
3	Suntech	China	3	Sharp	Japan
4	Motech	Taiwan	4	Canadian Solar	China
5	Solar World	Germany	5	Jinko Solar	China
6	China Sunenergy	China	6	ReneSola	China
7	Kyocera	Japan	7	First Solar	United States
8	Isofoton	Spain	8	Hanwha Solar One	South Korea
9	Schott	Germany	9	Kyocera	Japan
10	Sanyo	Japan	10	JA Solar	China

Sources: BNEF (2011); PV-tech.org (http://www.pv-tech.org/news/first_solar_and_suntech_led_2011s_module_manufacturer_rankings_says_lux_res; accessed May 28, 2014), PV-magazine.com (http://www.pv-magazine.com/news/details/beitrag/pv-module-industry-forecast-to-see-18-growth-in-2013_100011097/#axzz2ZxiI86Kt; accessed May 28, 2014).

has deepened as these firms have become more vertically integrated along the solar PV value chain and have exploited their expanding production scale advantages. By 2012, China had between 20 and 30 manufacturers with substantial production capability in polysilicon, along with more than 60 module producers also making PV wafers and cells (IEA 2013a).

By 2010, of China's total installed PV capacity of 893 MW in this sector, only 411 MW was grid-connected. At this time, most of the country's solar PV applications were still small-scale installations for residential or commercial building use in mainly remote areas, with relatively few utility-scale solar parks in operation (Tan et al. 2012). The situation drastically changed, though, as the sharp increases in national installed capacity during the early 2010s were due almost entirely due to new on-grid applications. In 2011, national capacity jumped to 3,284 MW total, with on-grid projects accounting for 99.4 percent of the new added capacity (IEA 2012c). Solar park installations made up 80 percent of this increase, and BIPV and BAPV together were 19.2 percent. This ultimately stemmed from the 2007 Renewable Energy Plan's policy support measures for LS-PV development (Zhang et al. 2012). By 2013, China had constructed 12 solar PV parks, each with at least 30 MW capacity, the largest of which was the 200 MW Huanghe Golmud plant (Table 5.3). A 2 GW mega-project is also under development at Ordos City in Inner Mongolia, which is due for completion by 2019. Most of these installations are built and operated by the country's state-owned power company (see Box 3.1, p. 84). In sum, China is emerging as a world leader in large solar park development, which is largely a state-directed and state entrepreneurial process.

5.5.3.2. Concentrating solar power

By the late 2000s, China had constructed only two demonstration CSP plants: a small 10 kW installation funded by MOST during the 10th FYP and a 1 MW plant under the 11th FYP (Wang, Z. 2010). The 12th FYP (2011–2015) heralded more ambitious plans, setting CSP capacity targets of 1 GW by 2015 and 3 GW by 2020. New projects took off during the early 2010s, the largest of which was the long-term development of a 2 GW plant at Shaanxi province, a 1 GW plant in Qinghai province, and a 300 MW plant in Xinjiang province. The first phase (10 MW installed) of a 50 MW plant in Delingha was in place by 2013 (REN21 2014). Additional 100 MW projects also were under development in Sichuan, Golmud, Ningxia, Xian, Guangdong, and Gansu.[5]

Annual summits on CSP development involving business, government, and scientific stakeholders were inaugurated in 2011. According to the Chinese Academy of Science, national CSP capacity could reach 3 GW as early as 2015 based on current project development plans. Certainly, there is huge CSP energy potential in China, which possesses more than 2.6 million km^2 of desert (27 percent of its territory), mainly located in central, northern, and western provinces. Based on estimates of CSP plants in the United States,

Table 5.3 Largest solar parks in East Asia

Solar park project	Country	Capacity (MW)	Output	Plant status
Ordos City, Inner Mongolia	China	2,000	3,875 GW.h p/a	Planned for completion by 2019 (by four stages)
Huanghe Golmud, Qinghai	China	200	317 GW.h p/a	Operational 2012
Gansu, Jiayuguan	China	100		Operational 2013
Ningxia, Qingyang	China	100	150 GW.h p/a	Operational 2013
Xitieshan, Qinghai	China	100	164 GW.h p/a	Operational 2011
Lopburi	Thailand	84	106 GW.h p/a	Operational 2011
Oita City	Japan	82		Operational 2014
Datong, Shanxi	China	80		Operational 2013
Kagoshima, Nanatsujima	Japan	70		Operational 2013
Beixiao (Shilin), Yunnan	China	66	77 GW.h p/a	Operational 2011
Hongsibao, Ningxia	China	50		Operational 2011
Huaneng Geermu, Qinghai	China	50	99 GW.h p/a	Operational 2012
Weidi	China	50	69 GW.h p/a	Operational 2013
Aichi	Japan	50	Approx 80 ha	Operational 2014
Ayutthaya (Bangchak)	Thailand	44		Operational 2012
Jiayuguan, Gansu	China	40	67 GW.h p/a	Operational 2012
Dahe, Henan	China	40	68 GW.h p/a	Operational 2012
Yangbajing, Tibet	China	30	55 GW.h p/a	Operational 2012
ATN Solar	Philippines	30		Operational 2013
SinAn	South Korea	24	35 GW.h p/a	Operational 2008
Gemas	Malaysia	10		Operational 2013

Sources: Various media sources.

approximately 1 km² of land with high solar radiation is required for around 50 MW of capacity, not including thermal storage and hybridization processes (Chien and Lior 2011). Therefore, if 20,000 km² of China's deserts were used for CSP installations (just 0.7 percent of its total desert space) a 1,000 GW power generation capacity would be achievable from the sub-sector.

5.5.3.3. Solar water heaters

China is both the world's largest manufacturer and user of SWH equipment. Its installed capacity increased multi-fold over the 2000s, from 6.1 million m² in 2000 to around 317 million m² by 2013 (Table 3.3, p. 82; Xie et al. 2012). The government's target of reaching 800 million m² by 2020 will spur future industry development. Chinese SWH manufacturers produced 71.6 million m² of collector area in 2013 (REN21 2014). Hu et al. (2012) argued that the expansion of the SWH industry has been based on the early scientific research foundations laid down in the 1970s and 1980s, technology transfers in the 1990s (especially on all-glass vacuum tubes), and the commercialisation of the sub-sector in the 2000s. By the end of this decade, there were more than 5,000 enterprises in China's SWH industry, of which about 1,800 were assembly manufacturers and the remainder component suppliers.

5.5.4. Main challenges ahead

5.5.4.1. Technical standards and techno-innovation

A key problem in China's burgeoning solar PV industry has been weaknesses or gaps in technical standards legislation, causing safety risks and constraints in most aspects of the sub-sector. The fast rate of techno-innovatory progress has meant that government ministries and grid companies have had to often play catch up with most recent technological breakthroughs (Huo and Zhang 2012). This is also an issue for China's SWH industry (REN21 2011). As noted earlier, though, establishing a national technical standard system and product testing platform by 2015 is a key sector target of the 12th FYP, which if realised should help address these issues. Furthermore, notwithstanding China's long history of solar energy R&D, its budgets for new scientific research and development remain relatively modest by international comparison. Chinese firms may have substantially improved the technological competitiveness and quality-build of their products over recent years, but they are still largely technology followers rather than leaders (Solidiance 2013). This is most notably evident in second-generation and third-generation solar PV. In addition, Huo and Zhang (2012) observed that there was a lack of coordination and information sharing amongst the country's centres of solar energy research, leading to duplications of effort. However, as a consequence of the 12th FYP's emphasis on strengthening China's indigenous innovatory capabilities, we may expect Chinese enterprises to more effectively compete in new solar energy technologies in the near future.

5.5.4.2. Coordinating industry development policies

It was noted how local governments have embarked on active solar energy industry development policies in recent years. The devolvement of economic policy-making on the city and provincial levels has meant that a myriad of policies have emerged. Given the huge size and regionalised nature of the Chinese market, local policy diversity across the country is not necessarily a problem, but issues can arise when local governments do not coordinate manufacturing promotion policies with others related to facilitating the expansion of installed solar energy capacity, such as grid connectivity, labour skills, transportation, finance, and R&D. On grid connection, local governments need to coordinate their policies and strategic plans with those of national government, as well as grid companies, to ensure that infrastructure development keeps pace with installed capacity growth in a sequential and systematic manner (Huo and Zhang 2012). Otherwise, solar park developers in China in particular may face similar grid-connectivity problems encountered by their counterparts in the wind energy sector.

5.5.4.3. Foreign protectionism

As first discussed in Chapter 1, China's growing domination of solar PV manufacturing over recent years has led to the United States and European Union taking the Chinese government to the World Trade Organization over alleged unfair trading practices. American solar PV producers especially have complained that the capital subsidies received by their Chinese counterparts have gradually priced Western firms out of the market. Although many American and European solar PV companies have indeed gone bust due largely to Chinese competition, the extent to which lower PV equipment prices in China can be attributed to subsidies remains a contentious point. More affordable Chinese solar products may be more attributable to captured cost efficiencies rather than state support. Whatever the case, these trade disputes may constrain access to developed country markets in the future if significant countervailing duties or quota limits are placed on Chinese imports. More internationalised PV producers, such as Suntech, JA Solar, and Yingli Green Energy, will look increasingly to emerging country markets if protectionism from the United States and Europe persists. Similar problems may arise with SWH should the sub-sector significantly expand outside China. A similar situation may arise in the CSP sub-sector should Chinese firms develop a dominate position here also.

5.6. Japan

5.6.1. Introduction

Mainland Japan lies between 30 and 45 degrees latitude north and has relatively high solar radiation levels, especially in the southern parts of the country (Tsuchiya 2012). Over the late 1990s and early 2000s, it was the world leader on installed solar PV capacity; however, since then, its growth rate has been slower than most competitor countries (IEA 2012d). Flat performance of the Japanese economy is partly to blame; however, there have been signs of a more robust expansion of the nation's solar energy industry. Its firms still remain technology leaders in the PV sector, and many Japanese firms are further internationalising their operations. In the government's post-Fukushima energy policy, there is moreover new impetus to realise Japan's solar energy potential, which hitherto has been almost exclusively confined to the solar PV field.

5.6.2. Policy development

Introduced as a response to the 1973–1974 oil crisis, the Japanese government's 1974 Sunshine Project was East Asia's first ever modern renewable energy policy initiative. While it aimed to foster the development of renewables generally through measures to promote new R&D activity and demonstration applications, solar PV was its core sector. Sharp, Sanyo, and Kyocera were among the major companies working with the government on the Sunshine Project and later went on to organise the Japan Photovoltaic Energy Association.

It was not until the early 1990s, though, that the government took the first policy initiatives to significantly scale up the industry. In 1992, it introduced a buy-up system for PV-generated surplus power aimed at the electric utility companies. Japan's 1994 Residential PV System Dissemination Programme was one of the world's earliest subsidy measures for installing residential PV systems. These two programmes were followed in 1996 by government set targets for installed solar PV of 4,820 MW by 2010, which it failed to realise (Table 5.1). Meanwhile, the 1997 Promotion for the Development and Dissemination of Solar PV Systems policy aimed to further stimulate R&D in the sub-sector, promote international joint ventures, and develop industrial solar energy applications (Avril et al. 2012). This cluster of policies during the early to mid-1990s had a notable impact on Japan's solar PV industry growth and techno-innovatory advances (Bradford 2006, Moe 2012). However, there was a lull in Japan's solar energy policy after the 1997–1998 East Asian financial crisis as the government grappled with other pressing economic issues.

Specific attention to solar energy returned again in 2004 with the launch of Japan's PV Road Map 2030 (NEDO 2004). This was the most comprehensive strategy on solar PV development then yet devised in East Asia, led by the country's NEDO (Aratani 2005). Its main elements were a revitalised R&D programme, a series of technical targets (e.g. cell efficiency ratings to reach 30 percent by 2030, generation costs to fall to grid parity for residential use by 2010), and financial assistance programmes to help realise these targets. In 2006, the Road Map started its 4-year R&D for Next-Generation PV Systems new development plan to further augment Japan's technological strengths in this sector (IEA 2007b). The revised PV Road Map 2030+ (Plus) was introduced in 2009, bringing technology development targets forward 5 years to 2025 while setting new technical targets extended through to 2050 (e.g. efficiency rating of 40 percent by that year) and new production targets for Japanese industry for the home and overseas markets (NEDO 2009). There were revised installed capacity targets (2–3 GW by 2020, 6–12 GW by 2030, 25–35 GW by 2050), but clearly the prime emphasis was on qualitative technological improvement rather than quantitative expansion per se. Furthermore, with a target for overseas production set at 300 GW and installed domestic capacity at only 35 GW by 2050, the overall aim of the programme was to devise new cutting-edge technology and mainly supply the international market. Japan began to export more solar PV modules than it produced for domestic consumption in 2003.

Around this time, a new subsidy programme for residential PV systems was introduced in 2008, superseding the 1994 scheme and aiming to have 70 percent of new homes installed with PV modules. In 2009, Japan also introduced its first national FiT scheme, this being specifically for solar PV and only for surplus power generated by residential and business units sold back to the grid, as well as a Green Heat Certificate scheme for solar hot water systems. Along with the PV Roadmap 2030+ programme, this constituted a new government push to revitalise the industry after the country had lost its leading position in global installed capacity in the early 2000s. In 2009, the government revised upwards its targets for installed capacity solar PV to 28 GW by 2020 and 53 GW by 2030. By 2011, 875 local governments and municipalities had reportedly initiated their own subsidy programmes for residential PV systems (IEA 2012e). The Japanese government's revised, multi-sector FiT scheme introduced in the aftermath of Fukushima in 2012 offered 20-year contract schedules and among the most generous tariffs for solar PV worldwide (BNEF 2012b). This helped spur a new resurgence in national PV industry production (IEA 2013b).

As well as promoting solar energy at home, the Japanese government overseas development assistance policy includes a number of 'green aid' measures for helping install PV applications in developing countries, especially in Asia. These are known as Smart Community PV Demonstration Projects, and important examples of these have been developed for the Delhi-Mumbai Industrial Corridor in India, Chongqing in China, and industrial district locations in Java, Indonesia as part of wider projects for clean energy development in dense, fast-growing urban zones where pollution is an especially acute problem (IEA 2012e). The Japan International Co-operation Agency plays a key coordinating role in these projects, as well as other overseas aid initiatives on off-grid PV system development in rural areas in developing regions around the world.

5.6.3. Domestic market and industry development

5.6.3.1. Solar PV

The Japanese PV industry began to emerge in the 1950s. Sharp Incorporated was an early world leader in manufacturing solar panels for lighthouses, satellites, and other applications (Moe 2012). Japan was one of the first countries to develop a substantive solar PV production and R&D base. Installed capacity stood at 19.0 MW in 1992, when national data began to be collected, rising to 636.8 MW by 2002 (IEA 2003a). At the start of this period, off-grid non-residential applications accounted for 80 percent of the total capacity, but a decade later this situation had reversed, with grid-connected PV taking an 89 percent share by 2002 based on a nationwide spread of distributed small-scale rooftop installations. This remains the dominant pattern of installed PV in Japan. Meanwhile, in 1998, the country overtook the United States to become the world's largest PV equipment producer, accounting for 40 percent of the industry's global output. In 2002, the country's total production levels stood at 258.0 MW, around ten times higher than China's at the time, and by 2006 Japanese companies were producing more than half the world's solar cells and modules (Renewable Energy and Energy Efficiency Partnership 2013).

As well as being a world-leading solar PV manufacturer, Japan remained one of largest generators of PV electricity through the 2000s, accounting for around a quarter of the world's total in the middle of the decade. Both installed capacity and manufacturing levels in Japan have gradually increased over time in absolute terms. Its installed capacity increased from 1,708 MW in 2006 to 13,600 MW by 2013 (Table 5.1). However, in 2005 it was overtaken by Germany as the world leader (IEA 2012d), and Japan no longer enjoys its long-held dominant position in East Asia's solar PV electricity generation. The same applies to manufacturing: although Japan produced 2.5 GW of cells and modules in 2011–about 10 times the level a decade earlier–this was now a fraction of Chinese production. Whereas in 2006 Japan had three companies in the world top-ten PV module manufacturers (with Sharp being the largest), in 2012 it had none thus ranked. However, a new resurgence in Japan's PV industry spurred by the aforementioned stronger policy support from government led to Sharp and Kyocera breaking back into the top-ten rankings in 2013 (Table 5.2). Japan's imports of solar PV equipment, mainly from China, began to rise by the early 2000s but at the same time Japan's exports to other countries remained higher than products made for the domestic market. Despite intensifying competition from their Chinese rivals, Japan maintains very strong indigenous capacity for new innovation and formidable competitive advantages in higher-tech parts of the industry. Thus, the share of second-generation thin-film PV products is much higher than China and other East Asian countries, at around a quarter of the total (IEA 2012e).

By the early 2010s, there were 13 companies in Japan listed as PV cell and/or module manufacturers (up from 10 in 2006), including Sharp, Kyocera, Sanyo/Panasonic, Mitsubishi, Hitachi, Fuji, and Canon. All of these firms are renowned manufacturers of electronic products and have used their economies-of-scope advantages to help develop a substantive national PV industry, in the same way as semiconductor firms in South Korea and Taiwan have (as noted previously). Sharp has remained the historic industry leader, and it is also the most vertically integrated (wafer to modules).[6] The company started developing solar cells in 1957, mass producing them in 1963, and by the 1970s was using them as a power source for calculators (Marinova and Balaguer 2009). In 2011, Sharp expanded into the installation maintenance market in Asia and a completed new thin-film production facility investment in Italy (IEA 2003a, 2007b, 2012e). Japan also has well-developed supply chains of inverter and polysilicon producers (in 2011, there were four in the nation), and a dense network of R&D centres in both the public and private sectors, where second- and third-generation solar PV technology is being developed.

Solar park development in Japan started to establish itself in the late 2000s. In 2009, Japan's power companies announced plans to construct a number of LS-PV installations, with a total capacity of 140 MW by 2020 at 30 different locations (IEA 2010a). The government's post-Fukushima energy policy further spurred the process. By 2012, solar parks' share of total installed PV capacity had risen to 15 percent of the national total, up from around just 1 percent in the late 2000s (IEA 2013b). By early 2014, Japan had three LS-PV installations in the 50+ MW range (Table 5.3). Most of the country's regional energy companies were intending to develop many more solar parks in the future (EPIA 2012a, Kirkegaard et al. 2010).

5.6.3.2. Solar water heaters and concentrating solar power

Although Japanese companies are active in the manufacture of SWH and CSP equipment, these two sub-sectors do not figure highly in the nation's solar energy mix. The Japanese mainland's relatively northerly position above 30 degrees latitude means that only Kyushu and the Ryukyu islands have any concentrating solar power potential. By the mid-2010s, Japan had no CSP plants in operation, nor were there plans to develop one. The country's SWH market and industry started to develop in the mid-1970s in response to the 1973–1974 oil crisis. However, sales of SWH systems peaked early, in 1980, with more than 800,000 units sold. Thereafter, sales declined, falling to just 60,000 units by 2008 (Japan Renewable Energy Policy Platform 2010). At this time, Japan represented only 0.5 percent of the global market (REN21 2009). In 2010, Japan reportedly had a cumulative 7 million m^2 of rooftop SWH units installed.[7]

5.6.4. Main challenges ahead

Japan has spent decades nurturing a 'national innovation system' for solar PV sector development, yet it has lost its one dominant position in the industry. The country's dense network of R&D centres working this sector remains strong, and Japanese companies such as Sharp remain among the technology leaders in the field. At the macro-strategic plan level, solar PV was a key industry in both the New Growth Strategy and the Cool Earth 2050 plan for achieving national low carbon development. Hence, the government appears committed to the continued expansion and general development of the PV sub-sector, especially as part of its new developmentalism. Yet, as this is an emerging strategic industry for many of Japan's main economic rivals, competition will intensify.

In the aftermath of the 2011 Fukushima disaster, solar PV offers Japan one of the best alternative energy options to nuclear power. However, the government's position on the long-term future of nuclear remains unclear. Solar PV is an intermittent energy source, and Japan's high population density and limited available land for solar park development are notable constraints on scaling up installed capacity. We may expect the prime emphasis on future expansion will be rooftop PV systems and techno-innovatory advances to improve efficiency ratings, as well as accelerating the commercialisation of second- and third-generation PV. This will require a commitment to strengthening policies support and developing a stronger prosumer culture in Japan.

5.7. South Korea

5.7.1. Introduction

The territory of South Korea lies between 33 to 38 degrees latitude north, in between Japan and northeast China, and on a similar parallel to southern Spain and north Morocco. It receives good levels of solar radiation, and thus it has potential for all three types of mainstream solar energy technology. Similarities may be drawn with Japan. High-level population density and mountainous terrain pose space constraints on developing LS-PV and CSP plants, and dense urbanisation makes South Korea most suitable for rooftop applications. Well-established Korean firms have strengths in foundational industries such as electronics, semiconductors, and engineering science, thus providing the country with technological advantages in solar PV equipment manufacture. Boosted by the Lee Myun-bak government's Green Growth Strategy (GGS) and scaled up investments of *chaebol* companies, such as Samsung, South Korea's solar energy sector has expanded significantly since the late 2000s, although this has been almost entirely confined to solar PV.

5.7.2. Policy development

South Korea's government started to sponsor test research on off-grid solar PV applications in the early 1970s. The first devices were installed in unmanned lighthouses. In 1989, the Ministry of Energy and Resources (later to become the Ministry of Commerce, Industry and Energy) initiated the National Photovoltaic Project, which was essentially a long-term R&D plan (1989–2001) for establishing South Korea's PV industry (Song 1993). As this plan came to an end, other ministerial-level policies were introduced to promote solar PV development. The first arose from the 2001 Basic Plan for New and Renewable Energy Development, Utilisation, and Promotion, in which solar PV along with wind energy were priority technology areas for US$800 million R&D support. A new FiT scheme launched the same year helped to kickstart deployment of rooftop PV systems on residential and commercial buildings. This was soon followed by the 2002 Solar Land Programme, an R&D and export-promotion scheme aiming to install 30,000 new rooftop PV systems by 2008. A year later, the 2003 Second Basic Plan for New and Renewable Energy Development, Utilisation and Promotion offered very favourable low-interest loans and grants for investment in solar PV projects, support for CSP dissemination projects, set equipment efficiency targets, and established a goal of capturing 7 percent of the world market by 2012 (Avril et al. 2012). By the mid-2000s, the government had set in motion a series of policies to foster solar PV development in the country: the General Deployment Programme (capital cost support), the Rooftop Programme (aimed mainly at residential and commercial sectors), the FiT scheme, the Public Building Obligation Programme (renewable energy requirements for new buildings), and the Local Deployment Programme, which included RE technology applications in 'green village' development (IEA 2007c).

In 2008, the revised FiT scheme reduced tariff levels for solar PV as the industry moved toward deeper commercialisation. Soon after, the 2009 One Million Green Homes Programme, stemming from the GGS, provided extra subsidies for rooftop PV installations. By the end of 2011, the programme had led to 85.2 MW of new capacity added in approximately 96,000 homes (IEA 2012f). In the same year, the government conceded that its revised FiT scheme was yielding disappointing results on PV installation growth (Table 5.1) and decided to replace it with a new renewable portfolio standards (RPS) scheme, as well as promote LS-PV development in coordination with the country's main power companies (IEA 2013c). Furthermore, under the GGS and New Renewable Energy Medium-Term Plan (2010–2015), the government aimed to create KRW20 trillion (US$18 billion) of new public and private investment in the solar PV sector by 2015, with a target of 1.2 GW of installed capacity by that year and 1,971 GWh of PV electricity generated by 2030 (its 2011 level was 917 GWh).[8]

5.7.3. Domestic market and industry development

South Korea's solar PV market has grown relatively slowly. National installed capacity amounted to just 0.6 MW off-grid by 1989 (Song 1993), and the first grid-connected PV system was installed in 1996. Off-grid PV remained ahead of grid-connected capacity until 2005; by this year, South Korea's total PV capacity had gradually crept up incrementally to 13.5 MW (IEA 2007c). New policy support measures from the 2003 Basic Plan provided fresh impetus to the sector, setting it on a new growth trajectory. Installed PV capacity rose from 35.8 MW in 2006 to 357.5 MW by 2008. By 2013, the level had reached 1,467 MW (Table 5.1). The country's larger BIPV and BAPV systems and first solar park projects began to emerge in the mid-2000s. By 2007, more than 50 commercial PV power plants (3 kW to 1 MW) of 9.2 MW total capacity were already in operation (IEA 2007c). The following year, the German company Conergy completed construction of the 24 MW SinAn solar park (Table 5.3), which was at the time East Asia's biggest LS-PV installation, producing electricity for approximately

10,000 Korean households and achieving around 25,000 tonnes of annual CO_2 emissions.

The country's solar PV manufacturing base has gradually expanded since the 1990s, albeit experiencing phases of cyclical development. For example, its only PV cell producer, High Solar (formerly part of the LG group) stopped production in 2000, leaving South Korea with no manufacturing capacity in this part of the supply chain the following year. However, in 2002, two companies (KPE and Neskor Solar) started production of first-generation PV cells, exporting over half of their total combined 300 kW output. By then, there were five Korean PV module producers with 0.8 MW collective output, but which had to import the majority of their PV cell components (IEA 2003b). By 2006, Neskor Solar has ceased production of PV cells, leaving South Korea with only KPE as the sole domestic producer of this component, although it had expanded annual output to 18 MW. Meanwhile, there were now eight PV module manufacturers with a collective 16.9 MW output, despite their combined production capacity of 130 MW. This was mainly due to the high polysilicon prices noted earlier in the chapter, which were adversely affecting PV cell supply levels globally (IEA 2007c).

New impetus in South Korea's PV cell production came from the entry of three new companies to the industry in 2008: Hyundai Heavy Industry, Millinet Solar, and Shinsung. Output consequently jumped that year to 106 MW, and output capacity reached 300 MW (IEA 2009d). Manufacturing capacity in all parts of South Korea's PV industry value chain (including polysilicon and wafer production) accelerated still faster into the early 2010s, in no small part due to strengthening state support. By 2012, there were 18 Korean PV cell and module producers, with a collective 1.7 GW output on modules and 1.0 GW on cells (IEA 2013c). Relatively new entrant Hanwha Solar One had moreover broken into the world's top-ten manufacturers (Table 5.2). However, the country had no thin-film PV manufacturer after two firms had stopped minor-level production in 2010. Like their Japanese rivals, Korean firms were still producing mainly for export, at approximately 70 percent of total output by the early 2010s (IEA 2012f, 2013c).

5.7.4. Main challenges ahead

Although more Korean firms have entered different supply chain stages of the solar PV industry and production capacity levels have increased, even the more advanced and larger *chaebol* players such as Samsung, Hyundai, and LG still lag behind their European, Japanese, and American rivals on the technological front. While they are investing more in second- and third-generation PV technology, South Korea still lacks a production base in these areas, and many of its module manufacturers are still import-dependent when sourcing more technically advanced PV cells and other components due to the lack of domestic supply. Samsung and LG in particular are making notable R&D investments in second- and third-generation PV technologies, assisted also by public research institutes such as the Korea Institute of Science and Technology and the Electronics and Telecommunications Research Institute (Wu and Mathews 2012). Such continued efforts in both public and private R&D will help close techno-innovation capacity gaps between Korean and advanced foreign firms. Another obstacle to solar energy development in South Korea concerns its population density (503 people per km^2), one of the world's highest for a major territory, and considerably higher than Japan's (337 people per km^2). Like Japan, it is also mountainous and there is intense competition for scarce land resources, making it inherently difficult for solar park and CSP development.

5.8. Taiwan

5.8.1. Introduction

Taiwan is a sub-tropical island that lies between 22 to 26 degrees latitude north, situated off southeast mainland China. Annual sunshine rates range from 1,500 to 2,200 hours for most parts of Taiwan's territory, and reaches 2,500 hours in its far south (Chen et al. 2010a, 2010b). It thus has abundant solar energy potential, although the combination of frequent cloud cover and lack of large flat arid areas make it unsuitable for CSP development. Taiwan also faces the same spatial and demographic predicaments as its Northeast Asian neighbours of Japan and South Korea with regard to high population density and mountainous topography. On the positive side, Taiwanese firms also share strengths in the same foundational industries as their Korean and Japanese counterparts, particularly in electronics (including semiconductors) and ICT, which has enabled Taiwan to become a major producer and technology player in solar PV.

5.8.2. Policy development

The Taiwanese government's first solar energy policy initiative was a 1986 subsidy support measure on SWH installation. Incentives on solar energy R&D activity were introduced in 1987, but it was not until the 2000s that significant state support for the sector was offered (Liou 2010, Su 2013). Subsidy schemes for installing solar PV, SWH, and solar thermal applications were introduced over the 2000–2003 period. The 2002 Renewable Energy Promotion Plan's measures included funding for PV projects in Taiwan's offshore islands, R&D investment, thin-film PV, export promotion, improved polysilicon production capacity, and the aim to install up to 120,000 BIPV systems as a longer-term objective (Chen et al. 2010a, Huang and Wu 2007, Hwang 2010). The following year, in 2003, the first quantitative target on installed solar PV capacity was set to achieve 1 GW by 2020. The government also introduced a number of location-specific projects, starting with solar roofing (from 2000), solar city (2004–2006), solar-powered government institution building standards

(2002–2004), a remote island emergency support system against natural disasters' (2005–2006), classic building (2006–2009), and solar campus, solar community, and agricultural installations, all from 2006 (Liou 2010b).

In February 2006, the Ministry of Economic Affairs (MOEA) convened a Silicon Conference, bringing together firms and policy-makers in partnership to establish Taiwan's first polysilicon factory a year later, with more plants created thereafter. Also in 2006, the MOEA organised two photovoltaic development strategy forums in a similar exercise of state–business partnership on the future development of Taiwan's solar energy industry. The resultant strategy comprised eight main elements, including polysilicon production, R&D, new material science, testing and certification systems, boosting domestic demand, and strengthening legislation. In August 2006, the government's Photovoltaic Industry Action Plan was published in association with the government's Green Energy Industry Sunrise Plan (GEISP; see Chapter 3). Soon thereafter, the MOEA hosted the Conference on Accelerating the Development of PV Industry in October 2007, based on the same aims, discussion areas, and consultative stakeholder principles as the previous year's forums (Liou 2010b). The landmark Renewable Energy Development Act (REDA), passed into law in 2009, gave further boost to Taiwan's solar energy sector, especially from the introduction of Taiwan's inaugural FiT scheme—the tariffs of which were revised upward in January 2011. REDA also included a new subsidy mechanism for solar thermal development.

From the late 2000s, policy-makers made greater efforts to strengthen Taiwan's techno-innovative capacity in the sector. The government's Industrial Technology Research Institute and Taipower, the public grid company, were among partner agencies engaged in the 2009 National Science and Technology Conference that led to the National Science and Technology Programme for Energy (2009–2013), in which solar energy was made a priority industry for R&D support (Liou 2010b). Later, the government's 2011 New Energy Policy to 2030 launched a new Million Solar Roofs project, with the prime aim of boosting Taiwan's domestic solar PV market based on targets of increasing total installed capacity to 610 MW by 2015 and to 3.1 GW by 2030. Annual targets have been set in motion as part of the government's renewable energy strategy, embedded within both the New Energy Policy and GEISP. In July 2012, the country set a target of an additional 100 MW of PV capacity to be installed in 2013, which was revised in November 2012 to 130 MW and then again in June 2013 to 175 MW after it became clear that domestic production levels had been underestimated.[9] This mirrored the experience of mainland Chinese policy-makers. In the SWH sub-sector, an annual installation area target of 140,000 m^2 by 2020 has been set by the government (Chang, P.L. et al. 2013).

In January 2013, the Bureau of Energy launched its Action Plan for Overseas Expansion of the PV Market, which involved the implementation of three main strategies (providing market intelligence, matching business opportunities, facilitating financing and funding) and 16 practices aimed at improving Taiwanese firms' international competitiveness at a time when the industry was suffering from over-supply and intensifying competition from mainland Chinese companies. The Loan for Green Energy and Industrial Equipment Export element of the state National Development Fund, as well as the Small and Medium Enterprise Credit Guarantee Fund of Taiwan, provided the main financial instruments for this policy initiative.[10] One of its key aims was to help Taiwanese businesses secure LS-PV project contracts overseas and expand PV services operations.[11] A target was set for achieving 10 percent annual growth in exports. In March 2013, a new Solar Community Subsidy Guidelines scheme was introduced, which provided a minimum NT$200,000 subsidy for community solar PV projects of at least 50 kW capacity.

5.8.3. Domestic market and industry development

5.8.3.1. Solar PV

As Table 5.1 shows, Taiwan has lagged well behind Japan, South Korea, and mainland China on installed PV capacity, even when its relatively smaller GDP is taken into account. By 2013, this capacity level stood at 392 MW (Table 5.1). The fact that Taiwan has maintained its position in the world's top-four PV manufacturing economies for some years makes this rather disappointing domestic market performance all the more surprising. Cell production is the core value-chain element of the Taiwanese PV industry, accounting for an estimated 70 percent of revenue and output (Su 2013). By 2013, the industry comprised more than 200 Taiwanese firms, which was more than double the 80 or so that existed just 4 years previously. As in other parts of Northeast Asia, deepening vertical integration has been evident (Chen et al. 2010b, Photovoltaic Science and Engineering Conference 2013).

The foundational strength of Taiwan's solar energy sector is based on internationally competitive electronics and ICT industries. Its wafer foundry, semiconductor, and integrated circuit design firms have been competitively ranked first or second in the world for some time. Many Taiwanese electronics companies—such as Taiwan Semiconductor Manufacturing Company, which operates the world's largest semiconductor plant—have successfully diversified into the solar PV sector, particularly from the late 2000s onwards. There is furthermore an important geospatial dimension to the technology cluster and economies of scope advantages enjoyed by Taiwan's solar PV industry in that key firms are geographically concentrated in three large techno-industrial parks: Hsinchu Science and Industrial Park (which accounts for around 40 percent of Taiwan's solar PV production), Southern Taiwan Science Park (around 20 percent of the total), and the Central Taiwan Science Park (specialising in thin-film PV). This has helped to foster networked collaboration among partner firms and the pooling of skills and services provision. All three techno-industrial parks have developed vertically integrated value-chain configurations across the sector (Su 2013). Both the Southern and Central Taiwan Science Parks were established as part of the government's Green Silicon Island programme, which was introduced from 2001 with the primary aim of industrial cluster development in green economy sectors (Liou 2010b). This relevant to Chapter 2's discussion about the industrial cluster aspects of state support for renewable energy development. Approximately 80–90 percent of PV cells manufactured in Taiwan are exported. Moves toward developing second-generation thin-film solar cell design and production capacity began after the 2008 price spike in polysilicon. The introduction of the GEISP in 2009, with its supporting measures to develop a robust and complete PV industry cluster by 2015, provided a boost across the whole production value chain.

5.8.3.2. Solar water heaters and concentrating solar power

Although Taiwan had no CSP installation operating in its territory by the early 2010s, there were Taiwanese firms helping develop CSP projects overseas (e.g. Ya-Fei Green Energy in Spain),[12] as well as a number of foreign CSP technology firms that were investing in joint ventures with domestic firms in Taiwan's techno-industrial parks. The SWH industry is the longest-standing of all three solar energy sub-sub-sectors in Taiwan, with the first units being produced and used from 1978. Subsidy support measures introduced in 1986, 2000, and 2009 had significant positive effects on SWH manufacture and utilisation (Chang, K.C. et al. 2008, 2009, 2011, 2013; Chang, P.L. et al. 2013; Chen et al. 2010a). For example, the original 1986 subsidy scheme resulted in a six-fold increase in the annual SWH installation rate by the end of it by 1991, and the 2000–2009 scheme led to a more than doubling of the rate over the period, to almost 120,000 m^2 a year. In 2013, there were an estimated 520,000 SWH systems in operation (approximately 2 million m^2), representing around 7 percent of all households on the island, and globally ranking Taiwan fifth in the sub-sector. There were, by this time, around 70 Taiwanese SWH manufacturers. The vast majority of Taiwan's SWH production is for the domestic market rather than for export (Photovoltaic Science and Engineering Conference 2013).

5.8.4. Main challenges ahead

Up to the late 2000s, Taiwan's industry had invested relatively little in solar energy R&D. The relative lack of state support on techno-innovation was partly to blame. It could be argued that Taiwan's ICT firms had sufficient funding and techno-innovatory capacity to make such investments, and therefore the government was reluctant to direct scarce public funds toward them. However, after the global takeoff of the solar PV industry, Taiwan's government and business community saw an opportunity to develop an emerging strategic industry based on the economy's existing techno-innovatory strengths. Yet, Taiwan's solar PV firms still remain significantly dependent on foreign technology partnerships, especially with American and Japanese firms

(Wu and Mathews 2012). Despite repeated calls made by the government for the private sector to invest more in solar energy R&D, there has been very limited direct public support here. The landmark REDA legislation did not include any direct provisions for R&D activity. More efforts will be required on improving the techno-innovation capacity of Taiwan's solar energy sector to compete at the qualitative end of the first-generation PV market, and also in emerging second- and third-generation PV markets.

At 645 people per km² recorded in 2012, Taiwan is one of the world's most densely populated territories; this rate is higher than both South Korea's (503 people per km²) and Japan's (337 people per km²). Moreover, like both of its Northeast Asian neighbours, Taiwan's terrain is highly mountainous. This has created a number of spatial constraints on the development of solar energy installations. Opportunities to develop solar parks are obviously limited, at least on land. Limited space for housing has compelled the construction of relatively high-rise apartment blocks, thus restricting potential rooftop space for solar PV and SWH installations. However, on this issue, research conducted by Chen et al. (2010b) and Chang, K.C. (2013) suggests that Taiwan has only reached around a tenth of its full potential for both roof-based systems.

5.9. Southeast Asia

5.9.1. Introduction

Spread over the tropical zones on both sides of the equator, Southeast Asia has abundant solar energy potential. In the region's northerly reaches (parts of Laos, Myanmar, Thailand, and Vietnam), there are suitable sites for CSP technology. Southeast Asia is also a significant producer of solar energy equipment. However, installed capacity levels are low, primarily due to lower levels of economic development, infrastructural provision, and policy support (Urmee and Harries 2012). Thailand has the highest level of installed PV capacity in Southeast Asia, with Malaysia following a distant second. The sub-region's SWH industry is rather underdeveloped and has received surprisingly limited government support.

5.9.2. Policy development

For most Southeast Asian nations, solar has been one of the most important sectors in government strategy-making on renewables, energy security, and economic development generally. The sub-region's earliest solar energy policies were directed at providing electricity to remote communities. Thailand's Ministry of Public Health and Medical Volunteers Foundation began to install solar PV equipment in rural health stations in 1976 (Green 2004). New state-funded projects for PV-powered water pumping stations in northeast Thailand and solar battery-charging stations in remote areas nationwide were introduced in the 1980s (Kirtikara 1997). Malaysia commenced similar programmes in the early 1980s for rural electrification under the Seventh Malaysia Plan (Chua and Oh 2012, Mekhilef et al. 2012). Indonesia later followed suit with its 1992 Solar Villages initiative, part of the government's Rural Electrification Assistance Programme (Dasuki et al. 2001). A few years later, the country received US\$44 million of special loans from the World Bank and its Global Environment Facility over 1997 to 2001 to expand solar PV installations in rural communities (Miller 1998, Miller and Hope 2000). The policy focus then shifted somewhat in the 2000s, when solar PV was identified as a key emerging strategic industry in Malaysia's Third Industrial Master Plan (2006–2020), Indonesia's National Energy Blueprint (2005–2025), Philippines Renewable Energy Policy Framework (2003–2013), Singapore's 2007 Comprehensive Blueprint for Clean Energy, Thailand's Strategic Plan for Renewable Energy Development (2004–2011), and Vietnam's Renewable Energy Action Plan (2001–2010). The Philippine, Malaysian, and Singapore governments in particular sought to use their own national electronics industry as a platform for establishing a solar PV manufacturing hub.

Singapore has especially focused on high-end, R&D driven solar PV production. As part of its Comprehensive Blueprint for Clean Energy programme, the Singapore state's Solar Capability Scheme and the Clean Energy Research and Test-bedding platform introduced in 2007 aimed to advance techno-innovation in the field. A year later in 2008, the creation of Solar Energy Research Institute of Singapore (SERIS) further consolidated the city-state's position as Southeast Asia's primary hub for applied solar energy research, with a special focus on tropical climate applications. More recently in 2011, the Floating Solar PV programme experimented with test-bedding 2 MW systems on Singapore's reservoirs in an innovative approach to solving the city-state's spatial constraints. Since the early 2010s, the Thai government has also endeavoured to develop indigenous solar PV innovation capacity based on fostering university-industry collaborative research (Department of Alternative Energy Development and Efficiency 2012, Solar Club 2012, Sugandhavanija et al. 2011).

The introduction of FiT schemes in many Southeast Asian nations during the early 2010s helped to notably expand installed PV capacity, particularly in Malaysia and Thailand (Table 5.1). The Thai government has additionally provided financial support to the country's PV industry via the Energy Service Companies Venture Capital Fund and soft loans from the Department of Alternative Energy Development and Efficiency. Furthermore, its 2008 solar hot-water hybrid system promotion project is one of Southeast Asia's very few policy measures introduced to support SWH development. Aside from its FiT scheme, Malaysia's Building Integrated Photovoltaic Technology Application project (2005–2010), partly funded by the United Nations Development Programme, aimed of reduce the long-term BIPV technology costs and was deemed generally successful (Chua and Oh 2012, Kadir et al. 2010, Muhammad-Sukki et al. 2011). Later, in 2010, Malaysia's Green Technology Financing Scheme provided important financial incentives for solar PV investor companies (Muhammad-Sukki et al. 2012). Only the Philippines and Thailand have introduced RPS schemes relevant to solar energy. Meanwhile, the Singapore government has deployed a Solar Leasing scheme to expand BAPV and BIPV applications in Housing Development Board (HDB) apartments, where 80–90 percent of residents live.[13]

5.9.3. Domestic markets and industry development

Solar PV has been the only solar energy sub-sector to have substantially developed in Southeast Asia. Its market and industry development can be dated back to the 1970s, and it has closely followed government policy priorities and targets. Thus, off-grid PV dominated well into the 2000s, with on-grid systems only starting to emerge in leading countries, such as Malaysia and Thailand, from the late 1990s. Development here has been slow: by 2006, Malaysia had only developed just under 0.5 MW of total grid-connected PV, while off-grid accounted for 90 percent of the national total of 5 MW. It was not until 2012 that on-grid PV overtook off-grid, largely due to the effects of the aforementioned then-new FiT scheme (IEA 2007d, 2013d). By 2013, Thailand had established itself as the dominant player in Southeast Asia's installed PV capacity with 704 MW, while Malaysia had 57 MW, Indonesia had 20 MW and Singapore had 15 MW (Table 5.1). Thailand's position of dominance owes much to the country's burgeoning solar park development, with its first LS-PV installation–the Mae Hong Soen 0.5 MW plant–becoming operational in 2004. By 2012, there were 16 solar parks in Thailand, with the largest being the 84 MW Lopburi and 44 MW Ayutthaya installations (Table 5.3). Malaysia has tried to follow with the construction in 2013 of its first LS-PV plant, a 10 MW Gemas installation in Negri Sembilan province.

Southeast Asia has CSP potential, but only Thailand has seriously explored it; a number of site feasibility tests have been recently undertaken in its northern provinces (Janjai et al. 2011). A 5 MW installation at Kanchanaburi became operational in 2011, constructed by German firm Solarlite in collaboration with the Thai Solar Energy company.[14] Up to 14 other such plants are in the pipeline; if completed, they would make Thailand the leading CSP electricity generator in East Asia. According to Solarlite, Thailand's national grid is not sufficiently robust to handle power produced from large CSP plants located in remote areas.[15] This is a problem facing much of Southeast Asia and China generally for RE sector development. While SWHs are widely used across the region, there were no Southeast Asian countries listed in the world's top-12 installers or producers by 2013 (REN21 2014).

Southeast Asia's solar PV industry is still highly dependent on foreign investing firms and imported components. The Malaysian government has made the greatest efforts of all in the sub-region through its strategic plans to foster indigenous production capacity and business development, helping to form the Malaysian PV Industry Association in 2006. However, Malaysia's efforts have made limited success on this front: by the early 2010s, FDI accounted for more than 90 percent of all private sector investment in Malaysia's solar PV industry (IEA 2013d). Yet, the Malaysian government has directed inward FDI to help strengthen and extend the country's PV production value chain, whereas in contrast the Thai government has opted more for imported supply sourcing.[16] Like Thailand, Singapore has main concentrated on attracting foreign PV producers (Solar Club 2012). Singapore has meanwhile become a leading regional hub for solar energy R&D as a consequence of the aforementioned state policy initiatives, with many foreign MNEs working in close collaboration with SERIS.

5.9.4. Main challenges ahead

Given the high rates of distributed generation involved in the PV and SWH sub-sectors, important questions arise concerning shares of the population in low-income places, such as Southeast Asia, that can afford to install solar energy systems even with government subsidy support. Furthermore, roof structures in remote communities may not be structurally robust enough for installations to be fitted. Mekhilef et al. (2012) found that at least 20 percent of residential buildings in Malaysia are not suitable for rooftop solar due to this reason. Survey research conducted by Muhammad-Sukki et al. (2011, 2012) in Malaysia revealed strong price sensitivity to electricity prices but not an aversion to new-generation technologies such as PV, although deficiencies in communications infrastructure could be blamed for the reported low level of public awareness concerning policy incentives, wuch as the government's FiT scheme. The survey research further found that after being informed of the FiT scheme and expressing a positive attitude to it, just under 80 percent of respondents decided not to invest in the programme for various financial reasons, such as the initial capital costs of installation, long payback period on personal investment, and debt issues. In earlier research, Green (2004) found that many early installed off-grid PV systems were failing due to lack of operational and maintenance training as well as an insufficient understanding of socio-cultural factors in remote communities concerning system management issues. These failings could be attributed partly to low-level state funding and technical capacity deficiencies. Similar experiences have reportedly applied to Thailand,[17] and the difficult challenges of expanding rooftop PV have led some governments to promote solar park development as an alternative strategy.

Even in Singapore, the sub-region's highest-income and most tech-savvy society, solar PV technology is relatively new and has yet to be proven as a dependable power generation method. The intermittency problem of solar energy is especially a concern for Singapore given the historic supply security challenges it faced over time.[18] Yet, based on research conducted by Wong (2011), if all available rooftop space on HDB buildings was used for solar PV modules, a total of 800 MW installed capacity would yield 960 GWh of electricity generated annually, far above the estimated 2004 figure of 500 GWh of electricity consumed by all HDB blocks that year.

Most of Southeast Asia lies in the planet's tropical zone, posing certain meteorological challenges for solar PV in general and CSP technology in particular, such as dealing with high-level humidity, heat and extreme weather events (e.g. monsoon rains). Only a very small fraction of PV systems worldwide are located in tropical locations, with the vast majority situated in the temperate zones of Europe, North America, and Northeast Asia. SERIS work on developing solar PV installations that can withstand the challenging climate conditions of the tropics could have beneficial effects for all Southeast Asian countries looking to expand their solar energy sectors.

5.10. Conclusion

Solar energy–and especially solar PV–is viewed by most East Asian governments as an important emerging strategic industry of the 21st century. The region has substantial solar energy resources, and moreover East Asia's strong global competitive position in related industries such as electronics and semiconductors has enabled its firms to successfully extend their existing techno-industrial competences into this renewable energy field. Its dynamic recent growth, high rates of techno-innovation, broadening scope of techno-logical applications, and ability to now compete on grid parity terms with fossil fuel–based power in many countries have led most analysts to conclude that solar will only become a more crucial part of our low carbon future. Solar energy's ability to deliver power generation in a highly distributed manner, in urban and rural communities alike, gives it a somewhat unique position to spread the renewables revolution throughout the various parts of East Asian society. In sum, the sector can be expected to play a vital role in the region's new developmentalism.

This chapter examined how East Asian governments have promoted solar energy development through the use of different policy mechanisms and stra-tegic plans, which are often integral to promoting renewables. Over time, solar's deployment on power generation has shifted in the region from primar-ily serving remote off-grid areas to becoming an established grid-connected power source, due mainly to policy incentivisation and direction. State-business collaboration regarding this development has varied considerably. Although the solar PV manufacturing industry is the most privatised RE sector element in China, state-owned power companies dominate on the installation side, working closely with state ministries and other SOEs on solar park and distributed grid infrastructure development in particular. However, like in other East Asian countries, solar energy R&D activities in China have involved deepening, co-funded private-public partnerships. In Taiwan, the government has created various institutionalised frameworks to work along-side private firms and business associations in developing all aspects of the PV production value chain. Since the late 2000s, the South Korean and Japanese governments have developed more functionally effective relationships with business on advancing solar energy techno-innovation, based on the goal of becoming industry and technology leaders. Their counterparts in Southeast Asia–most notably, Thailand, Malaysia, and Singapore–are also working with domestic and hosted foreign firms alike to secure key positions in the global value chain. Moreover, ambitious targets have been set, especially for solar PV, for expansion of low carbon macro-development and renewable energy strategies of East Asian states.

At the same time, it was noted that solar energy's future in the region faces some crucial challenges ahead. The intermittent power generation of solar PV means that it will be primarily viewed only as a complementary energy source, unless significant breakthroughs in storage technology are made. Spatial constraints in East Asia's most densely populated societies also pose difficulties for solar park development. Only few areas of the region are suitable for CSP installations. The sector has too been subject to intensifying trade disputes, with US and EU restrictions placed on imported Chinese PV equipment. Despite the sustained efforts of East Asia's government to foster stronger indigenous techno-innovatory capabilities and comprehensive production value-chain capacity, significant gaps on fronts still persist. These last challenge areas are likely to be the principal focus on state policy and strategic planning on solar energy development in the region over future years.

Notes

1　In 2004, this was around US$40 per kg, but by early 2008 had risen sharply to a peak of over US$470 per kg. By early 2012, the market price of polysilicon was around US$27.
2　http://www.greentechmedia.com/articles/read/Update-Solar-Firms-Setting-New-Records-in-Efficiency-and-Performance/?utm_source=feedburner&utm_medium=feed&utm_campaign=Feed%3A+greentechmedia%2Fnews+%28Greentech+Media%3A+News%29
3　http://www.pv-tech.org/news/china_calls_for_domestic_pv_and_bipv_applications
4　http://cleantechnica.com/2012/09/18/china-to-double-2015-solar-pv-target-to-40-gw
5　http://cleantechnica.com/2012/03/30/china-to-have-3-gw-of-concentrated-solar-thermal-power-csp-by-2020 (accessed November 29, 2012).
6　Sanyo and Mitsubishi are the country's other large-scale vertically integrated solar PV firms.
7　http://grist.org/article/on-rooftops-worldwide-a-solar-water-heating-revolution
8　Also, solar thermal of 2,046 GWh by 2030.
9　http://asian-power.com/environment/in-focus/taiwan-increases-installed-solar-power-target
10　Bureau of Energy press release (May 28, 2013).
11　Bureau of Energy press release (January 14, 2013).
12　http://www.renewableenergyworld.com/rea/news/article/2011/02/country-profile-taiwan
13　Research interview, Housing Development Board Singapore, January 2013.
14　http://www.rechargenews.com/solar/article1293523.ece
15　http://social.csptoday.com/emerging-markets/solarlite-powering-southeast-asian-csp
16　Research interview, Ministry of Energy, Bangkok, December 2012.
17　Research interviews, Ministry of Energy, Bangkok, December 2012.
18　Research interview with Housing Development Board, Singapore, January 2013.

6 Hydropower

6.1. Introduction

Hydropower is the oldest and most established mainstream renewable energy (RE) sector and by far the most dominant in power generation. As a mature energy technology, it is not generally viewed in the same dynamic innovatory terms as wind and solar energy, and it has experienced only modest growth in recent years globally. Furthermore, hydropower is strongly associated with modernisation-oriented development policies of the 20th century in East Asia and elsewhere. Large hydroelectric dams particularly encapsulated the idea of modern science and technology's ability to subdue powerful forces of nature to the material needs of humankind. In this sense, these huge constructions became synonymous with modernisation itself, and moreover compatible with early notions of ecological modernisation. Although hydropower still plays—and will continue to play—a vital role in mitigating climate change and delivering cleaner energy in relation to fossil fuel alternatives, the considerable environmental and socio-economic costs arising from large-hydro especially make this the riskiest and most controversial RE sector of all. Consequently, increasing emphasis has shifted to small-hydro technologies and applications, which are considered to have a far less adverse impact.

6.2. Hydropower: an overview

6.2.1. Hydropower explained

Hydropower captures the energy of water flowing from higher to lower elevations, where plant turbines mechanically convert the kinetic energy of water passing through them into electrical power. The amount generated depends on turbine size, volumes of flowing water, and vertical falling distance (known as the 'head'). It has been used as a source of power by many ancient civilisations, mainly for agricultural and food milling purposes. During the Industrial Revolution of the late 18th and early 19th centuries, mechanical hydropower was used in the early development of the textile industry, and a hundred years or so ago it was hydropower rather than fossil fuel sources that emerged as the world's first main source of electricity[1]

(IEA 2012g). The first hydroelectric power plants were constructed in the late 1870s and early 1880s in Europe and the United States,[2] and many hydropower dams built in the early 20th century are still operational today. Typically, though, most have a lifespan of between 40 and 80 years (IPCC 2011). Hydropower is by far the largest renewables sector, accounting for 16.9 percent of global electricity generation capacity in 2013–and around the same share in actual power output–spread over 160 countries; around two-thirds of the renewable energy contribution to this total (REN21 2014). In over 60 nations, hydropower provides the bulk of their electricity needs, and it remains the most proven, efficient, flexible, and reliable source of electricity in the renewables spectrum. Furthermore, it is a mature energy technology that competes effectively on price with fossil fuels and nuclear, and therefore does not usually need the same level of market support measures from governments as most other RE sectors. Most hydropower dams can convert up to 90 percent or more of energy inputs from flowing water into electricity, and thus they operate at very high efficiency levels and capacity factor ratings.

As noted in Chapter 1, hydropower does not suffer from the intermittency problem. It can respond quickly to meeting different levels of electricity demand as this fluctuates within daily cycles. In contrast to thermal plants (i.e. mostly fossil fuels and nuclear-based), where start-ups can take several hours or more, hydro plants usually can start producing electricity almost instantaneously (IRENA 2012d). These attributes enable it to deliver both baseload power and meet peak demands, and they can act as an essential complement to intermittent sources of power offered by wind and solar PV as they become scaled up. However, hydropower output can be constrained during periods of drought in plant catchment areas. In many countries, this is seasonally determined and corresponds to rainfall level patterns through the year.

The sector can be broadly categorised into three technology types:

1. *Run of river.* In this type, power is generated by river flow water, and thus is most dependent on rainfall precipitation and runoff levels. These plants may include a short-term storage facility (or 'pondage') that allows for some hourly or daily flexibility in meeting variable load demands. A series of run-of-river dams may operate in a cascade along the same river, requiring a coordinated regulation of water flow. This is especially relevant in China and Southeast Asia.
2. *Reservoir.* This technology is based on released stored water; hence it offers more flexible power generation options due to its reduced dependence on variable inflows. Large-scale reservoirs are able to retain up to years of average inflows and thus offer good energy security options. They also perform multi-functional services, primarily on flood and drought control, irrigation, freshwater supply, improved river navigation, and tourism.

3. *Pumped storage:* Here, stored water is recycled by moving between upper and lower reservoirs. In these plants, water is pumped from the lower to upper reservoir when electricity costs or prices are lowest (usually at night, when demand is low) and then released in reverse to generate hydropower at times of peak demand when costs are high. Because the pumping process normally involves using at least 15–20 percent more electricity than is produced from the down-flow power generation process, it does not fully qualify as an energy source as such; rather, it is a storage facility. It also cannot be strictly categorised as renewable energy because of its dependence on electricity itself in the pumping process, which may often be sourced from fossil fuel power. While this technology is a net electricity consumer, it accounts for around 99 percent of the world's stored on-grid electricity generation capacity (Electric Power Research Institute 2010).

These three technology types may be combined in the same installation. For example, run-of-river and reservoir plants can be used in cascading river systems, whereas pumped storage plants can be integrated with reservoir hydropower dams using its water storage catchment facility. The less mature in-stream 'hydrokinetic' technology is often cited as a fourth category (IPCC 2011), which operates along run-of-river principles and is also closely associated with ocean energy technologies (Chapter 8).

Environmental and socio-economic problems arising from large-scale dam projects have led to greater impetus behind the development of small hydro. Although there is no agreed-upon international standard of what power-rated plants qualify under this classification, small-hydro plants can range from up to 1.5 MW capacity (e.g. in Sweden) to 50 MW (in China). It also includes the even smaller sub-categories of micro-hydro (typically up to 100 kW) and pico-hydro devices (normally up to 5 kW), but these ranges can also vary significantly (IEA 2012g). All small-hydro applications are invariably used to serve remote communities without grid access. For example, there are thought to be more than 45,000 small-scale hydropower plants operating in China, with a collective 62 GW capacity producing electricity for over 300 million people (Chen et al. 2013a). The recent trend in hydropower policies toward favouring minimum land-to-power ratios has led to growing preference for both run-of-river and small-hydro schemes in most countries (IHA 2010, REN21 2013).

Overall, the hydropower sector has achieved relatively modest growth in global installed capacity and power output terms, at around 3–4 percent over the last decade. Total worldwide capacity over 2000–2013 has increased 45.0 percent, from 689.5 GW to 1,000 GW (Table 6.1). Over this period, China accounted for over half this growth in global capacity and is by some distance the world's largest hydropower producer, with 260 GW of capacity by 2013, followed by Brazil (86 GW), the United States (76 GW), Canada (76 GW), Russia (47 GW), and India (44 GW). Japan is the second largest

Table 6.1 Hydropower development, East Asia and global, 1980–2013

Total installed capacity (MW)

	1980	1990	2000	2002	2004	2006	2008	2010	2012	2013
China	20,320	36,046	79,352	86,075	105,242	128,570	171,500	199,130	229,140	260,200
Japan	18,276	20,825	20,019	21,697	22,048	22,199	21,852	22,362	22,144	22,300
South Korea	802	1,340	1,549	1,576	1,579	1,585	1,605	1,625	1,718	1,718
North Korea	4	5	5	5	5	5	5	5	5	5
Taiwan	1,387	1,566	1,820	1,909	1,910	1,910	1,937	1,977	2,081	2,081
Northeast Asia	40,789	59,782	102,745	111,262	130,784	154,269	196,899	225,099	255,088	286,304
Cambodia	18	10	10	12	13	13	13	13	224	224
Indonesia	1,313	3,145	4,393	4,383	4,566	4,897	4,872	4,878	4,879	5,074
Laos	155	230	622	624	672	681	681	2,520	2,973	3,212
Malaysia	649	1,457	2,054	2,106	2,095	2,120	2,120	2,115	3,317	3,917
Myanmar	169	258	340	390	746	771	771	800	2,660	2,780
Philippines	961	2,168	2,304	2,518	3,217	3,257	3,291	3,400	3,475	3,475
Thailand	1,361	2,455	2,923	2,936	3,476	3,476	3,481	3,488	3,489	3,489
Vietnam	280	675	3,292	4,155	4,155	4,580	5,500	5,500	10,037	14,000
Southeast Asia	**4,906**	**10,398**	**15,938**	**17,124**	**18,940**	**19,795**	**20,729**	**22,714**	**31,054**	**36,171**
East Asia	**45,695**	**70,180**	**118,683**	**128,386**	**149,724**	**174,064**	**217,628**	**247,130**	**286,142**	**322,475**
% share of world total	*9.9*	*12.2*	*17.2*	*18.0*	*19.9*	*21.9*	*25.5*	*26.9*	*28.9*	*32.2*
European Union	**95,856**	**93,654**	**99,772**	**103,699**	**100,824**	**100,699**	**101,767**	**103,533**	**102,000**	**102,000**
% share of world total	*20.7*	*16.3*	*14.5*	*14.5*	*13.4*	*12.7*	*11.9*	*11.3*	*10.3*	*10.2*
United States	**81,700**	**73,923**	**79,359**	**79,356**	**77,641**	**77,821**	**77,930**	**78,825**	**78,900**	**78,000**
% share of world total	*17.7*	*12.9*	*11.5*	*11.1*	*10.3*	*9.8*	*9.1*	*8.6*	*8.0*	*7.8*
Brazil	**27,500**	**45,558**	**61,063**	**65,311**	**68,999**	**73,434**	**77,870**	**80,703**	**84,000**	**86,000**
% share of world total	*6.0*	*8.0*	*8.9*	*9.1*	*9.2*	*9.3*	*9.1*	*8.8*	*8.5*	*8.6*
Canada	**47,919**	**59,195**	**67,230**	**69,029**	**70,680**	**72,661**	**74,230**	**74,901**	**74,900**	**76,000**
% share of world total	*10.4*	*10.3*	*9.8*	*9.7*	*9.4*	*9.2*	*8.7*	*8.2*	*7.6*	*7.6*
World	**461,898**	**572,903**	**689,521**	**714,521**	**751,767**	**793,834**	**853,812**	**917,544**	**990,000**	**1,000,000**

Sources: EIA (2014), CNREC (2013), IHA (2013); several of the 2013 figures come from communications with the IHA based on national reporting.

Note: Figures exclude pumped storage.

East Asian hydropower producer, with 22.3 GW capacity by the same year and an additional 25.3GW of pumped storage installed–one of the highest levels in the world. Table 6.1 shows that the next largest producers in the region are Vietnam (14.0 GW), Indonesia (5.1 GW), Thailand (3.5 GW), Philippines (3.5 GW), Laos (3.2 GW), Taiwan (2.1 GW), South Korea (1.7 GW), and Myanmar (2.8 GW). As a region, East Asia has accounted for a gradually rising share of the sector's global capacity, from 17.2 percent (118.7 GW) in 2004 to 32.2 percent (322.5 GW) by 2013 (Table 6.1). This was almost three times the level of the European Union and almost four times the level of the United States.

While the sector has contributed a steady 16–17 percent to total global electricity supply over recent years, this share was notably higher during the 1970s and 1980s at just over 20 percent (Energy Information Administration 2013, Sternberg 2010). Without China's recent hydropower expansion programme, the sector's global share would have notably fallen further still. Dam construction reached an historic peak in the 1970s but then fell to around half this level by the end of the 1990s (World Commission on Dams 2000). After being long viewed as 'shining icons of prosperity and modernity' (McCully 2001: xvi), public opposition to hydropower gathered strength as its negative socio-economic and environmental impacts became increasingly apparent. The World Bank–up to this point the industry's largest funder–and other international development banks gradually reduced their finance streams for hydropower projects during this period. Many developed countries had meanwhile exploited their near or full hydropower potential, or had chosen to meet their Kyoto Protocol carbon reduction targets via other clean energy options. However, the new political ecology of addressing climate change in the 2000s led to a number of developing nations investing heavily in hydropower to help satisfy fast-growing domestic electricity demand (Fletcher 2010).

6.2.2. Hydropower: techno-innovation and production

Large-hydro dams involve a substantial construction effort that typically uses local sourced materials, such as concrete (e.g. 27 million cubic metres for the Three Gorges Dam's 181-metre long central wall), as well as generation machinery that can be sourced domestically or internationally. Production of hydropower turbines, generators, and other related equipment has a long history, and they are produced in various countries. Three of the sector's top four manufacturing companies are European (Alstom from France, Andritz from Austria, and Voith from Germany), and the fourth, IMPSA, is from Argentina; together, they account for around half the world market. Japan's Hitachi and Toshiba, and Chinese firms Dongfang Electric (which also makes wind turbines), Harbin Electric, and Zhejiang Machinery & Equipment are also global players in the industry (REN21 2011, 2013). Alstom supplied 700 MW turbines installed at the Three Gorges Dam, as well as 812 MW turbines (the world's largest unit produced by the early 2010s) to China's sixth largest dam project at Xianjiaba. Voith meanwhile supplied 784 MW turbines for the Xiluodu dam, the world's third largest hydropower plant and power station. The electromechanical works of Malaysia's 2.4 GW Bakun dam were contracted to IMPSA and Alstom, and Russian firm Hydroproject designed Vietnam's 2.4 GW So'n La dam.

Despite the importance of foreign firms in East Asia's hydropower development, Chinese firms in particular are supplying an increasing share of its national market. Dongfang Electric now claims to hold 40 percent of this market and moreover exports to over 20 countries. Harbin Electric claims a similar domestic market share, with exports to more than 30 countries and established collaborative partnerships with Alstom, Siemens, Hitachi, and other developed country companies.[3] East Asian governments and companies are also active in developing hydropower projects in other parts of the region and worldwide. As we later examine, the China Power Investment Corporation is the main investor in Myanmar's 3.9 GW Myitsone dam, and Thailand has co-developed Laos's Nam Theun 2 dam. Japanese, Korean, and Chinese power companies have become increasingly engaged in various dam projects in Southeast Asia, South Asia, and Africa as part of overseas aid policies (IHA 2013). Regionalised or international cooperation on hydropower development is essential for major transnational rivers where numerous dams are constructed in upstream and downstream countries, such as the Mekong/Lancang in Southeast Asia and China. Given the ability of hydropower to often generate GW-level power, international trading of its electricity is often commercially viable, as in the Thailand-Laos case, and thus cross-border grid connectivity is another area of international collaboration in the sector.

Large-hydro especially involves high upfront investment costs that can normally only be covered by big companies or a consortium of investing firms, usually backed by national governments or international development banks. The long-term planning and construction process can also be a major deterrent for many investors. In sum, hydropower has been a sector requiring strategic investments where the state is often the key development player. In 2013, an estimated US$35.5 billion was invested in the hydropower industry worldwide—the third highest for any renewables sector (behind wind and solar), and many analysts believe the actual investment level was probably much higher (REN21 2014). The Kyoto Protocol's Clean Development Mechanism (Box 1.2, p. 32) has too been a common funding source for hydropower projects in developing countries, including many East Asian countries.

Hydropower's main civil construction cost elements include planning and licensing, plant construction, environmental and socio-economic impact management, and water quality monitoring. There are additionally the costs of procuring and installing power generation machinery. Civil construction costs are normally by far the greatest and will be largely determined by local source price factors, whereas generation machinery costs are determined more by international market prices. Typically, turbines, generators, and other major equipment elements will have a 30- to 40-year lifespan, and more technologically advanced upgraded equipment will improve plant efficiency, normally anywhere between 10 and 30 percent (IEA 2012g). A third cost element–operating and maintenance–is usually a fraction of total capital investment costs, and hydropower's high capacity factor ratings make it a high-performing, high-efficiency energy source with very competitive levelised cost of electricity rates.

Although hydropower is largely a mature energy technology sector, it still attracts significant levels of investment in new technology development and innovation. Many turbines installed in dams run at efficiency ratings well exceeding 90 percent, so there is relatively limited scope to achieve further technical advances on this front. New turbine design has thus focused more on functionality issues, such as variable speed and matrix technologies, fish-friendly turbines, hydrokinetic turbines, and abrasive-resistant turbines. Improving electro-mechanical performance for installations with low (less than 15 metres) or very low (less than 5 metres) head has been another key area of innovation that could open up new opportunities for sites hitherto not suitable for conventional hydropower technology (IPCC 2011, IHA 2013). Generally speaking, the sector's equipment manufacturers are striving to achieve higher levels of efficiency, reliability, and longevity through the use of computational fluid dynamics design, advanced manufacturing processes, new material science techniques, and state-of-the-art information and communications technology in plant design and operations (IEA 2012g).

6.2.3. Obstacles to the expansion of hydropower

Of all the renewable energy sectors, hydropower–and especially its dominant large-hydro variant–is the most controversial, due principally to the adverse environmental and socio-economic impacts caused by dam constructions. Growing public awareness of these impacts and corresponding opposition to dam development has constrained hydropower development in many parts of the world, and it looks set to become an increasingly significant obstacle to the sector's future expansion. The main adverse environmental and socio-economic impacts in question are as follows (IEA 2012g, IHA 2013, International Rivers 2014, IPCC 2011, REN21 2013):

- *Cumulative riverbed silting:* River transport sediments (sands, gravels, silt, clay particles) are deposited when they reach a reservoir or still water area, bringing about geomorphological and thereby ecological changes. This is obviously more of a problem for reservoir-based dams. Unless flushed through lower outlets, the cumulation of riverbed silts will mean dams may have definable lifespans, thus raising issues of their renewability. Flushed-through sediment flows can cause downstream ecological damage if not properly regulated.
- *Carbon and methane emissions:* Both natural and human-made freshwater systems emit some level of greenhouse gases (mainly methane and

carbon dioxide) from decomposing organic material, as well as capture carbon in their sediment. Critics of hydropower argue that dams make significant net contributions to such emissions, especially in reservoir constructions that have flooded dense vegetation in relatively high temperature areas, and when reservoir draw-downs (deployed for maintenance purposes) or naturally falling water levels regularly expose rotting organic matter. Methane emissions can be significant at these times, and are 20 times more potent a greenhouse gas than carbon dioxide. The dam construction process additionally causes emissions. However, lifecycle assessments on the net greenhouse gas effects of hydropower dams are generally inconclusive, and no firm scientific consensus yet exists on the matter.

- *General damage to ecological systems*: Dams can have notable deleterious effects on and local ecosystems and wildlife, including fish, mammals, birds, and invertebrates. Older dams were often designed with no consideration of migrating fish but more modern plants allow for both upstream and downstream migration. Riverside ecosystems may too be subject to the damage caused by hydropower plants, such as soil erosion due to hydrological changes.
- *Displacement of communities and disrupted or destroyed livelihoods*: This is especially relevant to large hydropower installations serving the energy needs of load centres some distance away, but where the impacts of dam construction on local communities (often in remote, poor areas) are not taken into consideration. Up to 80 million people worldwide are thought to have been displaced as a consequence of dam construction, and a reported 23 million of these were in China alone by 2007 (International Rivers 2012). Psychological traumas caused from forced movements and losses of local cultural heritage in dam-flooded areas are other relevant social costs. Displacement can also often cause loss of both short-term and long-term earnings for affected populations.
- *Reduced water quality*: Hydropower dam impact on water quality is very site-specific. This is particularly a challenge for reservoir dams where there is significant organic and human-made material waste cumulating from upstream sources. Furthermore, low dissolved oxygen levels tend to occur in the reservoirs bottom waters due to heavy organic sediments, causing ecological damage downstream when flushed through lower outlets. This can be mitigated by new turbine designs and multi-level water intakes in more modern reservoir plants. Run-of-river plants can actually improve dissolved oxygen levels of water and also capture floating debris for disposal.
- *Other potential adverse impacts*: Earthquake risk has been associated with reservoir dams in particular, due to the extra stress that rising and falling water levels exert on rock formations. The construction of large dams in zones of seismic activity also poses a similar risk. Flooding risk is relevant when dams cannot cope with regulating sudden increases

in rainfall or run-off water. Conversely, droughts can occur from poor management of water flows at dams in the same river system. Toxic human-made waste transported downriver cumulating in dam areas can cause health problems to local communities. Industrial waste materials not removed before they become submerged in upstream dam areas can contaminate water supplies.

Smaller hydropower plants and community-based hydropower projects could avoid many of the negative externalities noted above. However, the aggregated environmental and socio-economic impacts of many small-hydro devices might be actually equivalent to, or even more significant than, those from a single large dam generating the same total power level (IEA 2012g). Addressing the various problems associated with hydropower development has led generally to more careful planning, greater stakeholder engagement, and improved monitoring procedures over the plant's lifecycle. Codes of conduct have been introduced that enshrine these principles at the national and international level. The IHA's Hydropower Sustainability Assessment Protocol was published in 2010 after a long stakeholder consultation process and is the best known and most widely adhered to of these, based on up to 23 sustainability assessment criteria that include biodiversity, indigenous peoples, resettlement, water quality, downstream flow regimes, erosion, and sedimentation. The World Commission on Dams (2000) report was an important precursor to the IHA protocol, which helped to establish a number of standards, principles, and guidelines on good sustainable hydropower practice.

There are other obstacles to the expansion of hydropower to consider. The extent of a country or region's river or freshwater supply generally presents fundamental resource constraints on hydropower development. This has become increasingly relevant in an era of mounting demands on freshwater supplies, where water security risks have intensified worldwide. Furthermore, many countries have already exploited much of their hydropower potential already, mainly in countries with long histories of dam construction, such as Japan.

6.3. China

6.3.1. Introduction

With an installed generation capacity of 260.2 GW by 2013, hydropower in China is the largest singular renewables sector in the world by some considerable distance, accounting for well over half of all RE power capacity in East Asia. The socio-economic and environmental costs arising from the country's mega-scale dam construction programme have also been of considerable domestic and international concern. China has over 50,000 rivers with a basin area of 100 km^2, and almost 4,000 of these have hydropower potential of over 10 MW each. Most of this potential is concentrated in the central, south, and southwest provinces, where the great majority of the government's 15 hydropower bases are located in cascade configurations along large rivers, such as the Yangtze, Jinsha, Yalong, Nu, Dadu, Hongshui, Nanpan, Wujiang, Lancang (Mekong), Yuan, and Yarlung Zangbo (Fang and Deng 2011, Zhao et al. 2012d).

6.3.2. Policy development

The Nationalist government (1911–1949) operated a somewhat ad hoc hydropower policy, building a number of relatively small dams across the country during this period. Not long after the Chinese Communist Party came to power, the new government under Mao Zedong introduced a large-hydro programme as part of its national modernisation and industrialisation strategy operationalised through early Five-Year Plans (FYPs). Large dams came to symbolise the Maoist period slogan of 'man must conquer nature,' and decisions on each dam project were made at the highest political level. All political and public opposition was strictly suppressed (Shapiro 2001). Han (2013: 316) noted that this was particularly the case during the Great Leap Forward (1958–1961) and the Cultural Revolution (1966–1976) when 'any dissent against building large dams was considered tantamount to a direct threat to the regime.' Misconceived and poorly designed policies implemented during this period created a series of local environmental disasters and numerous failed dams. This was compounded by significant socio-economic problems arising from forceful resettlements and inadequate compensation for millions of displaced people.

With the advent of the economic reform period in 1978 came another phase of dam-building in China. Fewer but larger dam projects were initiated, and their main purpose focused more on hydropower generation than on irrigation and flood control. New liberalisation policies opened up the economy to foreign investing companies, enabling hydropower projects to benefit more widely from foreign technology, equipment, finance, and engineering techniques, leading to significant improvements in build quality and technical performance (McDonald et al. 2009). As is later discussed, in the 1980s and 1990s Chinese state agencies began to adopt a more systematic approach to cascade hydropower development along the nation's largest rivers. The 1996 Brightness Programme, the Chinese government's first significant multi-sector policy initiative on renewables (see Chapter 3), offered some incentives for dam project developers, although these had minimal impact.

The current phase of China's hydropower development started from the early 2000s, when a number of factors converged to spur the sector's expansion. The first was the introduction of the government's regional masterplan, the Western Development Strategy (WDS) in 2001–2002, which was aimed at addressing core-periphery divides between the country's rapidly prospering coastal southern and eastern provinces on the one hand, and the interior western provinces on the other whose incomes and development still languished in comparison. The WDS involved substantial new investments in energy infrastructures where hydropower was assigned an important role. The construction of three new ultra high-voltage transmission grid lines running from west to east, combined with low-interest finance from state banks, kickstarted many large-hydro projects around this time, mainly in southwestern and central regions (Vermeer 2012). Meanwhile, the 10th Five-Year Plan (2001–2005) was the first FYP to prioritise renewables such as hydropower for strategic development. Small-hydro project developers received policy support from the 2001 Township Electrification Programme, this was superseded by more generous assistance offered by the superseding 2006 Village Electrification Programme. After the 2002 'corporatisation' reforms of China's energy industry (see Chapter 3), the nation's five large power companies looked to consolidate their competitive positions in the new domestic market environment by initiating new hydropower projects (Han 2013). It is worth noting that Hu Jintao, who became president in 2002, began his career in the hydropower industry and previously worked for Sinohydro Corporation. Additional momentum on hydropower development was provided by the 11th FYP (2006–2010), which announced the construction of 70 GW new capacity that would be approved by the plan period's end. The strategic intent on greatly expanding the sector was further revealed in the 2007 Medium and Long-Term Development Plan for Renewable Energy.

At the same time that China was undertaking the world's most ambitious hydropower programme to date, important changes were occurring in the sector's policy-making apparatus and implementation mechanisms. This could be understood in the context of how China's leadership began to adopt ecological modernisation ideas in development strategy thinking from the early-to-mid 2000s onwards (Chapter 2). As the 'scientific development' and 'harmonious society' concepts started to influence economic policy-making generally, the government introduced new laws and policies on strengthening environmental and socio-economic impact assessments on hydropower projects, enhancing local stakeholder engagement processes, and improving resettlement and compensation packages. These resulted in relatively more inclusive planning and stricter approval procedures, and more accurate calculation of adverse impacts incurred from large-hydro dams in particular. The most important new legislation brought into force was as follows:

- *2003 Environmental Impact Assessment (EIA) Law*: Required hydropower developers to secure EIA approval from the State Environmental Protection Agency (SEPA) under stricter approval criteria than previously applied. EIA findings were to be made publically available.
- *2006 Regulations on Land Acquisition Compensation and Resettlement of Migrants for Hydropower Construction*: Revision of the preceding 1991

legislation that set much higher standards on remuneration and safe-guarding sustainable livelihoods for displaced or other affected peoples, setting out more clearly defined rights and responsibilities for all stake-holders, and most importantly stronger obligations on hydropower companies, although local governments are primarily responsible for effectively implementing the policy.

- *2008 Provisional Measures for Public Participation in Environmental Impact Assessment*: Provided for wider public and stakeholder engagement in the EIA process, most significantly for environmental non-governmental organisation (NGO) involvement.

These laws brought relative improvements to accounting for the environmental and socio-economic costs of hydropower development in China, yet there still remained crucial weaknesses in both the design and implementation of these policies (Brown and Wu 2010, McDonald et al. 2009). Furthermore, while SEPA was afforded a more influential role in hydropower policy, the National Development and Reform Commission remained the ultimate arbiter of policy decisions, and thus economic development aims still had primacy over environmental goals. China's leadership was nevertheless compelled for political reasons to take into greater account growing public unrest regarding government economic development plans, especially in poorer interior provinces where much of the country's large-hydro dams were being built. Consequently, the approval of proposed new dam projects were either rejected or delayed in a rising number of cases. This included halts in dam cascade development on the Nu River in the mid- to late-2000s, where the then Chinese Premier Wen Jiabao himself personally intervened. More recent policy measures in 2011 and 2012 introduced stricter EIA rules. Faced with blocks on large-hydro project development, many developer companies reportedly turned instead to the construction of very numerous 'unlicensed' small-hydro dams (Vermeer 2012).[4]

China's hydropower policy-making has also become increasingly contested due largely to the effects of post-1978 reforms. Devolved policy competence from national to local government, as well as the associated proliferation of different state actors in a rapidly growing economy, created a more pluralised (rather than democratised) policy-making environment–what may be referred to as 'fragmented authoritarianism' (Mertha 2008). Given the huge size of the industry relative to the country's other RE sectors, this was especially relevant to hydropower. Exercises of state capacity here have become more complex than in most other areas of renewables development in China because of the vast multitude of stakeholders involved. Consensus has to be struck between different national and local government agencies working alongside a myriad of state-owned enterprises (SOEs) involved in competitive bargaining and negotiating processes over thousands of dam projects, both large and small.

To some extent, fragmented authoritarianism has created certain fissures within the state that has enabled NGOs and other non-state actors the opportunities to partially influence policy-making outcomes by mobilising support from like-minded government officials. Key NGOs include Green Earth Volunteers, Friends of Nature, Green Islands, Global Village Beijing, and Green Watershed, which in 2004 formed the China Rivers Network coalition to coordinate campaigns against mainly large-hydro projects. These organisations have in addition strengthened their network links with their Southeast Asian counterparts to bring greater oppositional pressures from civil society to bear on dam-building along shared transnational rivers, such as the Nu/Salween and Lancang/Mekong. During the 2000s, and in coordination with their Chinese partners, the Southeast Asia Rivers Network and the Salween Watch Coalition sent the Chinese government various petitions against its cascade dam development plans along the Nu, outlining the significant damage they could cause to the livelihoods of large numbers of people downstream across the China–Myanmar border (Han 2013). It is difficult to confirm what degree of impact civil society opposition has had on hydropower policy in China, but it has become clear that the state has gradually taken a more circumspect approach, especially regarding large-hydro development. The government's target of expanding the sector up to 350 GW of installed capacity by 2020 is based on the completion of the current dam construction programme. Future targets set beyond this date are likely to be notably less ambitious and based more on small-hydro development.

6.3.3. National market and industry development

In 1910, China completed construction of its first hydroelectric dam, the Shilongba plant on the Tanglangchuan River in Yunnan province. Using mostly Siemens generating equipment, the then 480 kW capacity plant became operational in 1912 and was still generating electricity over a century later. After the Shilongba dam's completion, the Nationalist government constructed another 21 medium- to large-scale dams. When the Communists came to power in 1949, China's installed hydropower capacity was just 0.2 GW. The Mao government's new dam-building programme expanded this level to 0.5 MW by 1955, 1.9 GW by 1960, and 3.0 GW by 1965 (Chang et al. 2010, McDonald et al. 2009). However, an alarmingly high number of these dams collapsed, including more than half the 110 dams built in Henan Province by the mid-1960s. The death toll from the collapse of the Banqiao and Shimantan large dams in Henan later in August 1975 was estimated at between 86,000 and 230,000 people (Han 2013). According to He et al. (2008), in addition to the above disasters there were 3,479 other dam failures between 1954 and 2003; of these, 123 were medium-scale and 3,356 small-scale dams. Many collapsed within their first year of construction due to poor design and sub-standard materials. By far the worst period was the first half of the 1970s—a legacy of the Cultural Revolution, when many skilled engineers were removed from their positions of responsibility. Millions of properties and hectares of farm land were devastated by these disasters. The last significant dam collapse occurred in August 1993 at the Gouhou plant, leading to the death of almost 300 people.

Soon after the advent of the economic reform period under leader Deng Xiaoping, the government introduced an intensive dam cascade development programme in the country's western source rivers. Thereafter, large-hydro projects started to dominate the sector (Fang and Deng 2011). New large dams not only helped power the national economy's rapid economic growth but also spurred the development of China's burgeoning hydro-power industry. Companies such as Sinohydro Corporation and Gezhouba Group later became global players, with investments in various projects worldwide. National installed capacity rose from 20.3 GW in 1980 to 36.0 GW by 1990 and then to 79.4 GW by 2000 (Table 6.1). Hydropower's contribution to China's total installed power generation actually peaked in the mid-1980s at around 32 percent. Over time, the industry has remained completely dominated by Chinese SOEs in both construction and operation. By the early 2010s, the top 10 hydropower plant operators shared around half the market, with four of the five large power companies (Huadian, Datang, Huaneng, and Guodian; see Chapter 3) being among these. In construction, state-owned Sinohydro–the world's largest dam-building company–accounted for two-thirds of China's market, Gezhouba Group accounted for a fifth, and Anneng Group was around 4 percent. Foreign firms are limited to providing generation machinery supply and technical consultancy services (IHA 2013, Solidiance 2013).

Between 2000 and 2013, China's installed hydropower capacity more than tripled (Table 6.1), overtaking the United States as the world's largest producer nation in 2001. China also accounts for many of the world's largest dam projects, and all of East Asia's top 10 either built or in progress, as well as 18 of the region's top 20 largest, although 11 of these were still in under construction in 2014 and due for completion sometime before 2020 (Table 6.2). In addition, the country is the world's leading builder of dams overseas, with Chinese companies and financiers involved in more than 300 projects across 66 nations (International Rivers 2014). The internationalisation of China's large dam-building industry, especially as its own domestic market potential diminishes, could become another area where the country impacts significantly on the global renewable energy landscape. As noted earlier, the more pluralistic policy environment combined with the government's 'going-out' investment strategy has enabled Chinese hydropower firms to have both more flexibility and greater opportunity to expand operations abroad, particularly in Southeast Asia (as later discussed) and Africa. China's hydropower equipment manufacturers also export to Latin America, South Asia, and Eastern Europe. The Chinese government's more pragmatic approach to development assistance and as well as competitive-priced dam construction options have too appealed to host low-income nations.

Table 6.2 Largest hydropower dams in East Asia (over 1,000 MW)

Rank	Plant	Country	Capacity (MW)	Year completed
1	Three Gorges	China	22,500	2012
2	Xiluodu	China	13,860	Under construction (2015)
3	Baihetan	China	13,050	Under construction (2019)
4	Wudongde	China	8,700	Under construction (2015)
5	Longtan	China	6,426	2009
6	Xiangjiaba	China	6,400	Under construction (2015)
7	Nuozhadu	China	5,850	Under construction (2015)
8	Jinping-II	China	4,800	Under construction (2014)
9	Laxiwa	China	4,200	2010
10	Xiaowan	China	4,200	2010
11	Myitsone	Myanmar	3,600	Under construction (2017)
12	Pubuguo	China	3,600	2010
13	Jinping-I	China	3,600	Under construction (2015)
14	Ertan	China	3,300	1999
15	Gezhouba	China	3,115	1988
16	Goupitan	China	3,000	2009
17	Guanyinyan	China	3,000	Under construction (2016)
18	Dagangshan	China	2,600	Under construction (2016)
19	Changheba	China	2,600	Under construction (2016)
20	So'n La	Vietnam	2,400	2011
21	Bakun	Malaysia	2,400	2012
22	Jinanqiao	China	2,400	2011
23	Guandi	China	2,400	2012
24	Lijiaxia	China	2,000	2000
25	Liyuan	China	2,000	2012
26	Shuang Jiang Kou	China	2,000	2012
27	Hòa Bình	Vietnam	1,920	1994
28	Shuibuya	China	1,840	2008
29	Xiaolangdi	China	1,836	2000
30	Baishan	China	1,800	2006
31	Jinghong	China	1,750	2008
32	Manwan	China	1,750	1996
33	Pengshui	China	1,750	2008
34	Houziyan	China	1,700	Under construction (2015)
35	Gongboxia	China	1,500	2006
36	Shuikou	China	1,400	1999
37	Dachaoshan	China	1,350	2003
38	Tianshengqiao-II	China	1,320	1997
39	Longyangxia	China	1,280	1992
40	Geheyan	China	1,240	1993
41	Liujiaxia	China	1,225	1969
42	Yantan	China	1,210	1995
43	Wuqiangxi	China	1,200	1996
44	Wujiangdu	China	1,130	1998
45	Tingzikou	China	1,100	Under construction (2014)
46	Tominaga	Japan	1,095	1980
47	Silin	China	1,080	2008
48	Wanjiazhai	China	1,080	1998
49	Nam Theun 2	Laos	1,070	2010
50	Imaichi	Japan	1,050	1988
51	Guangzhao	China	1,040	2008
52	Cirata	Indonesia	1,008	1988
53	Fengman	China	1,004	1953

Sources: Author's own research, various media sources.

Box 6.1 Small-hydro development in China

Most attention given to China's hydropower sector has tended to focus on its large dam-building programme. Studying the development of small-hydro, though, is vital to understanding important evolutionary changes that have occurred in this sector over time from multiple perspectives: industrial, policy, technology, and socio-environmental. Given the past controversies and future limitations of large dam development in the country, small-hydro is also highly relevant to the Chinese government's post-2020 hydropower strategy.

Small-hydro has long played a crucial role in providing off-grid electricity to China's remote communities. Whereas the country's large dams have primarily served its expanding urban load centres in the south and east through long transmission links, small-hydro plants (SHPs) principally serve the power needs of their locality. Around a quarter of the national population, mainly its rural county elements, depends mainly on small-hydro for its electricity supply. The classification of SHPs in China has changed over time, increasing in capacity ratings as larger dams were built. In the 1950s, SHPs of less than 0.5 MW were referred to as rural hydropower stations. In the early 1960s, dams of up to 3 MW capacity were categorised as SHPs, then raised again to 12 MW by the end of the decade and later to 25 MW when the government introduced its National Rural Electrification Programme in the 1980s. As the nation's average dam size continued to grow, it was announced in the 1990s that SHPs up to 50 MW would benefit from policies aimed at small-hydro development (Zhao et al. 2012d). Today, China classifies dams from above 0.5 MW to 50 MW as small-hydro installations, which is a very broad rating band by international standards. Meanwhile, its micro-hydro category is rated at above 100 kW to 500 kW and pico-hydro installations at 100 kW or below; again, these ratings also are much higher than the international norms (Li 2012).

When China's small-hydro industry began to emerge in the 1950s, its technology and equipment were very rudimentary. SHPs often used self-made wooden or cast-iron turbines that were highly inefficient. A limited range of mass-made factory produced equipment was manufactured toward the end of the Maoist period, but these were still of poor quality. The average SHP operating hours remained no more than 2,000 hours per annum by 1980. Economic reforms introduced in the 1980s helped to strengthen the techno-industrial capabilities of China's hydropower sector generally. Small-hydro benefitted from the emergence of new and more capable firms, general advances in techno-innovation, the more open diffusion of SHP technologies, and improved planning and site survey functions (Gu and Liu 1996). In

1983, the Chinese government launched a small-hydro development scheme in 100 pilot counties (2.5 million household beneficiaries) as part of its wider rural electrification programme. This was extended under the 8th FYP (1991–1995) to 200 counties, and then to 300 counties during the 9th FYP (1996–2000). However, this still only affected a very small part of the population, and policy initiatives such as the 1996 Brightness Programme, 2001 Township Electrification Programme, and 2006 Village Electrification Programme only had a relatively limited impact, especially when compared to the then-burgeoning large-hydro industry.

In reality, national government did not need to incentivise entrepreneurship in China's small-hydro industry because it was already flourishing without such policy support in the early economic reform period (European Small Hydropower Association 2005). By the late 1980s, there were already had around 66,000 SHPs dispersed around the country with 10.8GW of aggregated installed capacity, representing over a third of the hydropower sector total and around 10 percent of national installed generation. State-owned enterprises of various sizes–including local SOEs backed by local governments–were key players in this development (Zheng et al. 1989). By the mid-1990s, SHP numbers had fallen to around 47,000 as older, less efficient plants were decommissioned. Those remaining still contributed similar shares to these totals, with 15.8 GW combined capacity and an annual 50 TWh power output (Gu and Liu 1996). By 2011, there were a reported 45,165 SHPs in operation, with a total installed capacity of 62.0 GW generating 174.4TWh of power, managed by roughly 20,000 enterprises spread over half China's territory, and providing electricity to over 300 million people (Chen et al. 2013a).

Small-hydro's share of China's total hydropower capacity had fallen to less than 30 percent by the early 2010s, yet this was still a substantial proportion compared to most other countries (Li 2012). Furthermore, Chinese SHP developers and equipment manufacturers have become increasingly internationalised. There is a large concentration of these firms in the southern Sichuan and Guangxi provinces, and around 70 Chinese small-hydro developers now operate overseas (Zhao et al. 2012d).

Questions surrounding whether small-hydro offers China's hydropower sector a more sustainable and ethical future have become increasingly important given that large dam development in the country is likely to peak in the next few years. In their study on the subject, Kibler and Tullos (2013) found that aggregations of installed SHP capacity may actually have worse biophysical (i.e. socio-economic and environmental) impacts per MW than equivalent large-hydro capacity. Using the Nu River basin for their empirical work, their research concluded that the combined adverse impacts of small-hydro installations exceeded that for large dams for 9 of the 14 metrics evaluated, including habitat diversity, potential to modify hydrologic regimes, and overall effects on water quality. In contrast, large dams had comparatively greater impacts on total land inundation, potential to trigger seismic events, and sediment transportation. The authors recommended that more robust standards for SHP impact assessment and governance were thus needed, especially if the future emphasis of the Chinese government's hydropower policy was to switch away from large-hydro to small-hydro.

Another problem is that China's 'small' hydropower classification includes quite large plants (up to 50 MW) by international comparison, and that data on micro-hydro and pico-hydro applications are extremely limited. Although only able to realistically serve the needs of small communities and households, promoting the further development of micro-hydro and pico-hydro has the advantages of advancing greater societal control and ownership in China's hydropower sector generally. Given that very small-small-hydro devices also have a negligible environmental impact, this development would furthermore be consistent with what Zhao et al. (2012d: 6) suggested for the sector's future in the country, moving from 'pure engineering hydropower to ecological hydropower and from pure technical engineering to social engineering.'

6.3.4 Main challenges ahead

The considerable socio-economic and environmental risks associated with China's large-hydro programme present the most significant challenges in this sector (Tang et al. 2013). China is still constructing numerous large-hydro dams and many hundreds more medium- and small-scale dams. Easing tensions within Chinese society over 'blind development' policies, which take little account of their adverse environmental and socio-economic impacts, has become a growing priority for the country's leadership. It was previously noted that by the mid-2000s the Chinese government started to more seriously acknowledge the significant risks and obstacles in the sector, and they introduced new policies and regulations in an effort to address them. However, these have only been able to mitigate the adverse impacts of large-scale dam-building to a limited degree. The enormous importance of hydropower in China's renewables spectrum, which is likely to account for at least two-thirds of total installed RE capacity up to 2020, makes it critical for the government to further improve its policies and strategies in this area. A key challenge here relates as much to implementation of new legislation and plans as it does to their design, and this in turn will involve national government exercising greater influence over local governments and SOEs actively engaged in hydropower development.

Given the huge number of sizeable hydroelectric dams that have been constructed across China over recent years, it is difficult to summarise empirically what socio-economic and environmental problems and thereby challenges have arisen in its hydropower sector. Moreover, for the sake of avoiding repetition, it is instructive to note that the general obstacles to hydropower's future development outlined earlier in the chapter apply more to China than any other country. A case study on the Three Gorges Dam (TGD) project in Box 6.2 will, though, serve as a good representative illustration in this regard for at least large-hydro development. The TGD's story also connects closely with many historic challenges of China's hydropower

Box 6.2 The Three Gorges Dam: clean energy hero or villain?

China's Three Gorges Dam (TGD) on the Yangtze River is both hero and villain of the country's renewable energy development. Becoming fully operational in May 2012 with an installed capacity of 22.5 GW, the TGD has been a flagship project of Chinese economic development generally. Its annual 'clean energy' savings of 31 million tonnes of coal and 100 million tonnes of greenhouse gas emissions make a significant contribution to decarbonising China's energy sector. Yet, the huge socio-economic and environmental costs arising from both the TGD's construction and operation have made it highly controversial and a target of much criticism, inside and outside the country.

The initial idea for the TGD was first proposed by the Nationalist government in 1919. Much later, the Mao government started to conduct feasibility studies on the dam in the 1950s. However, in the 1960s, China still lacked the technical capability to build it, and in the 1970s lacked the money to finance it. Only by the mid-1980s was the government able to put together a viable plan for the project (Fearnside 1988). In 1992, this plan was approved by the National People's Congress, but with a record number of delegates (32 percent) either voting against or abstaining support for building the dam, which began in 1994 and was structurally complete by 2009. The last of hydroelectric turbines (32 of 700 MW rated capacity, plus two at 50 MW) were fitted in the 2,335-metre-long, 175-metre-high installation in 2012. The dam cost an estimated US$25 billion and created a 600-km long reservoir upstream. At 22.5 GW, it is by far the world's largest power station by capacity rating, equivalent to almost a quarter of Britain's total national grid and just below that for Malaysia's.

The TGD's hydroelectric function was the primary motive for construction, but other functions are also important. Over the last century or so, major catastrophic floods on the Yangtze River occurred in 1911, 1931, 1935, 1954, and 1981, resulting in over 300,000 deaths and millions of displaced people. A later flood event in 1998 killed over 4,000 people, inundated 25 million hectares of farm land, and caused material damage estimated at US$36 billion, providing stronger justification for the TGD. The dam is also situated in a critically important area: the Yangtze River basin is home to one-third of the national population, around 70 percent of the country's rice fields, and roughly 40 percent of Chinese industrial output (Liu et al. 2013b, Tullos 2009). Building the dam would spur economic development in this key riparian zone by improved navigation: the maximum ship size on the river increased from 2,000 to 10,000 tonnes and thereby considerably strengthened commercial transportation links between major interior cities such as Chongqing to the east coast (Jackson and Sleigh 2000).

While the TGD has delivered many benefits, it has also incurred considerable environmental and socio-economic costs. Its construction necessitated the inundation of two cities, 11 counties, 140 towns, 326 townships, and 1,351 villages covering 23,800 hectares and the resettlement of an estimated 1.5 million people (Chien and Lior 2011, Liu et al. 2013b). Although around 35–45 percent of the project's budget was allocated to resettlement and compensation for affected local communities, according to International Rivers (2012) there were reports of this process being badly mismanaged and subject to notable levels of corruption by implementing local government officials, with around 12 percent of funds being embezzled. Filling the TGD reservoir is thought to have destroyed a number of archaeological sites of important cultural interest. Ecological costs and geological risks associated with the TGD's development have included the following (Chien and Lior 2011, International Rivers 2012, Solidiance 2013):

- Undermining the Yangtze River's rich biodiversity, as the dam is located in one of its conservation areas where 25 endangered fish species exist, such as the Chinese sturgeon, river sturgeon, river dolphin, and paddlefish.
- Stagnant water forming in the TGD's long reservoir, where industrial effluents discharged by factories upstream into the river have accumulated, creating regular algae blooms. Reservoir siltation has too accumulated to notable levels. In addition, silt-free water released from the dam is causing significant erosion of river banks

downstream, thus counteracting some of the TGD's flood control benefits.

- Soil erosion is affecting more than half the reservoir area, causing frequent landslides and an estimated 178 km of river banks thought to be at risk of collapse. If this transpires, another 530,000 people will need to be relocated by 2020 as a result of this risk.
- Seismic activity in the TGD reservoir zone has increased since its filling process was substantially completed by the late 2000s.

China's national government agencies have exhibited growing concern over the high socio-economic and environmental risks associated with the dam's construction and operation, which are closely connecting with key challenges it faces generally with its large-hydro development strategy. In September 2007, Beijing acknowledged the need to strengthen EIA procedures applied to the TGD, and later in May 2011 conceded that its installation has posed notable environmental, geological, and socio-economic welfare risks (International Rivers 2012). Despite the adoption of stronger environmental protection laws and regulations introduced by the government since the early 2000s noted in the main text, weaknesses in their design and implementation (e.g. on punitive fines for non-compliance, and regarding stakeholder engagement) has given more scope for the adverse impacts of China's large dams, such as the TGD, to go unchecked. The resultant problems arising from the Three Gorges Dam project has thus caused many doubts concerning the future of other large-hydro developments in progress, not just in China but also worldwide.

development. Like other large-hydro dam projects, it has too shown the importance of fostering more effective stakeholder relations among national government, local government, developer companies, and local communities, as well as adopting a more holistic and effective approach to environmental impact assessments.

China also faces challenges related to the now-limited site options for new large-hydro projects. Box 6.1 examined the prospects for future small-hydro development and noted some problems on this front too; it was concluded that supporting a more widespread deployment of pico-hydro and micro-hydro installations had a number of advantages. Finally, there is the challenge of ensuring that the internationalisation of China's hydropower industry acts as a benign force for sustainable development in other countries. This relates to similar debates concerning its wind and solar energy counterparts, and how China is playing an increasingly effective role in advancing renewable energy development globally.

6.4. Japan

Japan was the first East Asian nation to develop a hydropower capability, dating back to the late 19th century, when the technology itself was beginning to emerge. The country started to generate hydroelectric power at the Miyagi Boseki Miisawa Power Plant in Miyagi Prefecture in 1888 for private, local community use. This was just ten years after the world's first plant began operations in Britain. A similar plant was built just four years later at Keage, Kyoto in 1892. In 1907, Japan started building its first public hydropower facility in Komahashi, Yamanashi Prefecture, supplying 4 MW capacity-rated power to the Waseda Transformer Station in Tokyo, located over 75 km away. Later, in 1914, the Inawashiro hydropower plant also began to serve the capital city from an even greater transmission distance of 225 km. Although these early plants tended to use European-made turbines and other equipment, a fledgling Japanese hydropower manufacturing industry soon began to establish itself. Hitachi especially assumed a strong position at this stage; for example, in 1917, the company suppled 750 kW rated turbines for the new Nikko Daiichi power station (Tokyo Electric Power Company 2003).

The two main periods of expansion in Japan's hydropower sector occurred from the 1920s to 1930s, and then from the 1950s to early 1980s. By the end of this first period, 13 hydroelectric plants had been constructed, and in the 1950s alone another 20 MW scale dams were completed, which played an essential role in the post-war economic reconstruction process (Dinmore 2013). By the end of the 1980s, there were over 70 such dams in Japan operated by a range of owners, including the ten regional *Denjiren* power companies (Chapter 3), local prefecture governments, central government ministries, and the national energy company J-Power (Japan Commission on Large Dams 2011). By the mid-1950s, hydropower still contributed the majority share (around 60 percent) of Japan's power generation capacity. However, in the mid-1960s, it was overtaken for the first time by fossil fuel power; by the mid-1980s, it only accounted for 20 percent, falling further to 17 percent by the mid-2000s. Total installed hydropower capacity had reached over 10 GW by the 1960s—at the time one of the highest levels in the world—and over 20 GW by 1986 (Ushiyama 1999b). Thereafter, the sector entered a plateau, where it has remained ever since. As Table 6.1 shows, Japan's capacity level has fluctuated between 20.0 GW and 22.4 GW in recent years, and it was 22.1 GW in 2012. J-Power (2013) has estimated that the country's conventional hydropower potential is 34.7 GW, suggesting that more than 60 percent of this has already been exploited.

Despite the sector's relative decline, the government continued to conduct surveys in the latter years of the 20th century to investigate possible new sites for hydropower development and how to strengthen capabilities of existing plants (NEF 2007). Silting and sedimentation have been major issues in many of Japan's large-hydro plants, especially older ones. A pumped storage industry began to establish itself from the 1970s onwards and expanded into the 1990s, surpassing conventional hydropower in this decade in capacity terms, reaching 26.3 GW by 2012. Also in the 1970s, the nation's small-hydro industry began to take off around this time, mainly serving private and local community energy needs. By the early 2010s, there were more than 1,800 small-hydro devices in Japan, but the positive contributions they made to national capacity and power output were offset by mostly large-hydro plant or equipment decommissioning at older installations. Over recent years, J-Power has been actively developing overseas hydropower projects, including in many parts of East Asia such as China, Thailand, and Vietnam.

Government policy support for hydropower has been relatively weak over recent decades, with small-hydro being the only real beneficiary. The 1980 Small and Medium-sized Hydro Power Plants Project scheme offered subsidies to investors in the sub-sector, with support ending in 1999. Thereafter, small-hydro has been promoted through generic renewable energy policy measures, such as the 2003 'Green Power' Renewable Portfolio Standards scheme mandating power companies to increase their utilisation of sub-1 MW rated hydropower devices up to 2010; an initial target was set for raising installed hydropower capacity generally to 26.5 GW by this year, but this was later revised downwards to a much less ambitious 20.7 GW (NEF 2007). In addition, the government continued to offer subsidy support for development costs of hydropower projects of up to 30 MW capacity under its 1997 Support for Deployment of New and Renewable Energy programme (NEF 2007, Ushiyama 1999b). Most recently, the 2012 feed-in tariff scheme that superseded the RPS provided 20-year tariff schedules for small-hydro installations up to 3 MW capacity. According to J-Power (2013), relatively small- to medium-scale plants (10–50 MW) accounted for over 40 percent of the country's total hydropower capacity, and plants with 100+ MW capacity have around a 20 percent share of this total.

Most of Japan's yet-untapped hydropower potential is in small-hydro, as nearly all available sites for developing large-hydro have been used (J-Power 2013). In the past, the country's large dam projects have been criticised for the 'pork barrel' politics that implicated national government ministries, local government officials, and construction firms in corrupt practices, and where these projects took little heed of environmental or socio-economic impacts (Kerr 2002, Reiko 2001). The only large-hydro project recently under proposal, the Yanba Dam, has been highly controversial and delayed due to strong public opposition (Dinmore 2013). Moreover, small-hydro offers Japan more environmentally friendly options and more promising opportunities for local community stakeholdership in hydropower projects. The key issues for Japan's small-hydro future are locating commercially viable sites with stable water flows, strengthening government policy support for development projects, and streamlining planning permission processes. The country lags behind many other advanced industrial nations in this sub-sector, which is somewhat surprising given Japan's long history of hydropower development and the competitive strengths of Japanese manufacturers in the hydropower equipment industry.

6.5. South Korea

The first hydropower dams in South Korea were constructed toward the end of the Japanese colonial period in the late 1930s and early 1940s. Initial plants completed at Boseong-gang (4.5 MW in 1937), Cheongpyeong (140 MW in 1943), Hwacheon (108 MW in 1944), and Seomjin-gang (35 MW in 1945) principally served the energy needs of Japan's war effort–for example, powering Korean factories that produced military equipment and munitions. All four dams were still in operation by the mid-2010s and were owned by Korea Hydro and Nuclear Power (KHNP), a previous division of Korea Electric Power Corporation (KEPCO). The Korea Water Resources Corporation (K-Water) is the other company that, together with KHNP, dominates the country's hydropower sector. Both are also major players in the ocean energy industry.

In 1957, KHNP completed the installation of the country's first hydropower plant built after the Korean civil war at Goesan, which was a small 3 MW plant. Hydropower development thereafter become intrinsically linked to the new Five-Year Plan macro-development strategies introduced from the early 1960s onwards. KHNP constructed two dams in this decade at Chuncheon (1965, currently 62 MW) and Uiam (47 MW, 1967). Four new dams were completed in the 1970s: Paldang (120 MW in 1973, KHNP), Soyang (200 MW in 1973, K-Water), Andong (80 MW in 1976, K-Water), Anheung (0.5 MW in 1978, KHNP). In the following decade, K-Water constructed a further three plants, at Daecheong (90 MW in 1980), Chungju (400 MW in 1985), and Hapcheon (101 MW in 1989), and South Korea's last large-hydro dams were completed in the early 1990s at Gangneung (82 MW in 1991, KHNP) and Imha (50 MW in 1991, K-Water). All 16 dams were still in operation by the mid-2010s, some slightly increasing capacity levels over time through equipment upgrading while others experienced partial decommissioning. The country's total hydropower capacity level has reflected sudden periodic increases over time as new plants became operational. However, since the last of the large dams were completed, this level has increased only marginally, from 1,445 MW in 1991 to 1,718 MW by 2012 (Table 6.1).

South Korea's hydropower dams have served multiple functions. Flood control and freshwater supply management have been afforded the same priority as power generation. Freshwater resources became increasingly precious in the country due to the demands of rapid industrialisation and urbanisation (Kang and Park 2002). Moreover, the dam construction projects of the 1960s to 1980s had weak environmental and socio-economic impact assessment processes. As South Korean society became more democratic and prosperous, public opposition to future hydropower projects deepened (Han et al. 2008). In 1996, the government announced its plan to build a new dam at Yongwol on the Tong River, but it was later ditched in June 2000 due to the strength of such opposition. However, six relatively large pumped storage plants were built in the mid-1990s by KHNP and K-Water, with their collective installed capacity of 4.7GW being almost three times that of the conventional hydropower dam total.

Small-hydro projects have also recently flourished in South Korea. Test plants of between 0.9 and 6.0 MW have been developed at Yipo, Seungcheon, Hapcheon, Nakdan, and Gangjeong since 2009.[5] More generally, government policy on managing hydro-resources has shifted away from hydropower toward river engineering activities. The flagship Four Major Rivers development project of the Green Growth Strategy (GGS; Chapter 3) is indicative of this shift, where river restoration, channelization, and canalisation rather than damming now serve the policy objectives of flood control and freshwater management, as well as create new eco-recreational spaces (river parks) in city centres for the benefit of South Korea's now highly urbanised society (Woo 2010). It is therefore unlikely that any significant expansion in large-hydro will occur in the country; rather, efforts will focus on improving the operational power efficiency of existing dams as well as utilising their power generation capacity more fully.

Indeed, power output levels from South Korea's dams have fluctuated wildly over the last two decades due to electricity generation serving a somewhat auxiliary function at many installations where flood control and water supply management have been deemed a higher priority. Total power output from large-hydro peaked in 2003 at 4,757 GWh, more than double the 2001 output level but from more or less exactly the same installed capacity, and then dipped to 2,596 GWh in 2009–the same year that the GGS's National Energy Plan set a large-hydro output target of 3,860 GWh by 2030. This therefore relates more to maintaining a generally average level of output from already constructed large dams. Meanwhile, the country's small-hydro sector has been operating at close to full capacity levels, with its collective output gradually rising over the years to reach 529 GWh by 2012 (KEPCO 2013). The GGS set a 2030 target of 1,926 GWh for small-hydro, which would require well over a three-fold increase in the 2012 level if realised.

6.6. Taiwan

Like South Korea, the first hydropower plant in Taiwan was built during its Japanese colonial period, in 1905 at Guishan on the Xindianqi River. This was a relatively small dam, and much larger twin installations were built at Wujie on Sun Moon Lake from 1919 to 1934. When they became operational, these dams tripled the island territory's power generation capacity and thereby played a crucial role in Taiwan's early industrialisation. The Japanese colonial administration had plans to build a number of additional hydropower dams in the 1930s but were cut short by events of World War II (Tung 1998). By the mid-1950s, hydropower was responsible for over 90 percent of the island's power needs and over 80 percent of installed power generation. The sector continued to dominate all others up until the mid-1960s. By then, four dams had been built, at Tienlun (195 MW in 1956), Wushoh (77 MW in 1959), Kukuan (180 MW in 1961), and Shihmen (130 MW in 1964); the last dam was built with the assistance of the United States aid programme. This second historic phase of large-hydro coincided with the establishment of the Taiwan developmental state.

A further expanded dam construction programme was linked to the government's Six-Year Plan framework. In 1970, the island's largest dam then built was completed at Qingshan (360 MW), joining others in a grow-ing cascade of installations on the Dajia River. Another relatively large dam was completed soon after at Techi in 1974 (234 MW; otherwise known as Tachien) by SOE Taipower, which has historically dominated the sector (Chen et al. 2010a). A 50 MW plant was installed at Zengwen in 1973 but, like many other smaller capacity units, it principally served irrigation and other freshwater supply functions. Only two hydroelectric dams were built in the 1980s, at Junghua (40 MW) and Feitsui (70 MW), both in 1984; Feitsui was a redevelopment of Taiwan's very first plant at Guishan. Also during this decade, the Taiwan government commissioned the building of the island's first pumped storage hydropower stations, located at Sun Moon Lake at Minghu and Mingtan with a combined power capacity of 2,610 MW. The last large-hydro project to be completed was the 133 MW Ma'an dam in 1998.

Hydropower began to lose its dominance of Taiwan's power generation industry in the 1960s as the government turned increasingly to fossil fuels and nuclear power. By the 1990s, hydropower's contribution was around a sixth of Taiwan's total power output (Tung 1998). Many installations do not have any power generation facilities at all, performing just flood control functions and acting as reservoirs to supply water to nearby urban popula-tions. Thus, like South Korea, their power generation units in many of Taiwan's large dams serve an auxiliary function, which explains why many rarely operate close to full capacity output (Chen et al. 2010a). Furthermore, silting has been a problem in many of its hydropower plants, both old and new alike, due mainly to insufficient investments in maintenance. Most plants are also located in remote mountain areas that have involved high development costs and investment risks—a predicament faced also by Japan and South Korea given their similar geographies. Seismic activity (e.g. the September 1999 earthquake) and extreme weather (e.g. the 2001 Nari typhoon) have additionally caused material damage to installations and often halted their operations for long periods (Hwang 2010).

From the mid-1980s to mid-1990s, Taiwan's installed hydropower capac-ity level remained at a plateau of between 1,550 and 1,590 MW. In 1995, the Water Resources Agency of the Ministry of Economic Affairs together with

Taipower completed its survey report on hydropower potential to assess the sector's long-term future, concluding that Taiwan theoretically possessed 11,730 MW of potential total capacity but only 5,000 MW was thought to be feasibly exploitable. Later studies yielded far more optimistic estimates of 22,725 MW theoretical potential (Chen et al. 2010b). The completion of the Ma'an dam in the late 1990s, equipment upgrades to existing plants, and the emergence of a small-hydro sector since then has gradually boosted Taiwan's installed hydropower capacity over recent years, from 1,820 MW in 2000 to 2,081 MW by 2012 (Table 6.1). The government has set a target of increasing this level to just 2,502 MW by 2030 under its New Energy Policy launched in 2011. As with Japan and South Korea, hydropower's future in Taiwan would appear to be confined to small-hydro and upgrading its existing large-hydro dams.

6.7. Southeast Asia

6.7.1. Introduction

Various complex and controversial issues have arisen from hydropower development in Southeast Asia. The first MW-scale dams were not built in the sub-region until the 1950s, coinciding with the advent of industrialisation in large economy countries such as Indonesia, Malaysia, Philippines, Thailand, and Vietnam. As with their Northeast Asian counterparts some decades earlier, the construction of these plants were seen as symbols of both economic modernity and the state's role in accelerating 'development.' While the expansion of hydropower continues to make significant contributions to powering Southeast Asia's industries (Box 6.3), it has also exhibited the worst excesses of cronyism and corruption, as well as weak state capacity to deliver good governance outcomes. The environmental and socio-economic costs of hydropower development in the sub-region have been considerable, and growing public opposition from a strengthening civil society has proved to be increasingly effective in delaying or halting many dam projects. Hydropower in Southeast Asia is also arguably the most internationalised (or transnationalised) renewables sector in East Asia generally. Foreign governments (most notably China), multinational business, international aid agencies, and multilateral development banks, such as the World Bank and Asian Development Bank, have all been active players. This is especially evident in the Mekong River basin, a vast transnational area where most of the sub-region's larger hydropower dams are concentrated and where most its hydropower potential lies. These dams are still viewed positively by their national government and foreign sponsors as providing substantial levels of cleaner energy, and they are closely linked to macro-development strategies. However, various controversies still persist in the sector overall, and its long-term future in Southeast Asia points increasingly to small-hydro rather than large-hydro.

6.7.2. State capacity, governance, and macro-development strategies

As in China, hydropower development in Southeast Asia has attracted significant criticism and opposition over the years, not just because of the heavy socio-economic and environmental costs associated with dam projects but also owing to related corrupt government and business practices. All of the sub-region's major hydropower producer countries have exhibited some degrees of predatory statism, where state officials have exhibited more interest in seeking personal gains through developing cronyist links with business instead of pursuing transformation economic objectives that improve public welfare (Chapter 2). Predatory state actions are thus viewed as poor or weak state capacity from a developmentalist perspective because economic decisions were made on defective political motives, leading to poor welfare and governance outcomes.

There is a relatively long history of cronyism and political favouritism in Southeast Asia's hydropower development. The involvement of Philippines President Marcos in the San Roque project up to his downfall in the mid-1980s was an early high-profile example (Kim 2010). National governments from the sub-region have also made questionable contracts with foreign parties, as we later discuss with respect to Laos' dam projects and hydropower development in the Mekong River basin. Another key issue has been limited stakeholder engagement. Other instances of weak state capacity have arisen when government agencies have underperformed on technical issues. For example, Cambodia's 190 MW rated Kamchay dam, the country's only large-scale hydropower plant that was completed in 2012, has since been operating at around only 10 percent of maximum capacity due to poor water resource management. More generally, many dams have been built with 'inadequate impact assessments, incorrect hydrological assessments, unsafe engineering, insufficient compensation and support for relocated people, and little or no consideration of their effect on the fisheries, biodiversity and livelihood in the region' (Matthews 2012: 394).

A prime reason why hydropower has attracted much controversy and criticism in Southeast Asia owes much to the sector's importance in the national energy mix, its close corresponding links with macro-development strategies, and substantial contributions to powering industrialisation. It has long been a very significant and strategically critical energy source, thus having similarities with the Northeast Asian experience historically. In 1980, hydropower accounted for 30 percent of Thailand's power capacity, 25 percent in Indonesia, 24 percent in Vietnam, 21 percent in Malaysia, and 20 percent in the Philippines. By 1990, these shares still remained high: Indonesia, 24 percent; Malaysia, 29 percent, 31 percent; Thailand, 24 percent; and Vietnam, 31 percent. This was a period when the power generation industry in Southeast Asia expanded overall. As Table 6.1 shows, hydropower continued to grow significantly into the early 2000s, especially for Laos, Myanmar, and Vietnam. At the time, all three countries had commenced GW-scale hydropower dam projects, as had Malaysia (Table 6.2). Hydropower's share of Southeast Asia's total capacity had risen slightly from 14.6 percent in 2004 to 16.6 percent by 2012.

Hydropower has been a core element of the national energy and macro-development strategies of most Southeast Asian nations, especially up to the mid-2000s, with a growing priority afforded to small-hydro (Chapter 3). The well-established hydropower producers, Indonesia and the Philippines, introduced legislation specifically aimed at promoting micro-hydro installations as early as 1990 and 1991, respectively. Since then, aside from Vietnam's 2004 National Master Plan for Small Hydropower, the main policy measures deployed to promote the sector have been embedded in strategic energy and macro-development plans, as follows:

- *Indonesia*: 2005 National Energy Blueprint (small-hydro targets of 500 MW on grid, 330 MW off grid by 2025)
- *Malaysia*: 1981 Four Fuel Diversification Strategy and 2001 Five Fuel Policy (all hydropower); 2001 Small Renewable Energy Programme, part of the 8th Malaysia Plan (small-hydro up to 10 MW); 2006 Renewable Energy Development, 9th Malaysia Plan 2006–2010 (all hydropower); 2011 Renewable Energy Feed-in Tariff (small-hydro)
- *Philippines*: 2010 Feed-in Tariffs scheme (run-of-river hydro); 2011 National Renewable Energy Programme (hydropower target of 8,729.1 MW by 2030, a 2.5-fold increase on the 2012 level)
- *Thailand*: 2004 Strategic Plan for Renewable Energy Development (special research and development support for micro-hydropower); 2011 Alternative Energy Development Plan (324 MW additional small-hydro by 2021)
- *Vietnam*: 2001 Renewable Energy Action Plan, and part of broader Vietnam Master Plan of Power Development (special emphasis on off-grid village hydro schemes as well as all hydropower generally); 2004 National Master Plan for Small Hydropower; 2008 National Energy Development Strategy (hydropower to play key role in achieving general RE target of 5 percent of total primary energy by 2020 and 11 percent by 2050); 2011 National Power Development Plan, revised the 2008 Strategy (hydropower target of 17.4 GW by 2020).

With hydropower expected to continue making a substantial contribution to the energy mixes of many Southeast Asian states, we may anticipate the sector being integral to their future macro-development strategies as well.

6.7.3. The Mekong River: international cooperation or competition?

Running through the territories of the five Southeast Asian nations of Cambodia, Laos, Myanmar, Thailand, and Vietnam as well as China, the

Box 6.3 Malaysia's Bakun dam: serving old or new
developmentalism?

The 2.4 GW Bakun dam in Malaysia serves as a useful case study of
the relationship between hydropower and economic development in
Southeast Asia. Initial proposals for the dam dated back to the 1960s,
but first construction efforts only commenced in the mid-1990s.
However, the outbreak of the 1997–1998 East Asian financial crisis
halted the project due to funding constraints, and activities were not
reinitiated until 2000 when the government recapitalised project
financing (Chong and Lam 2013). This was to be a centrepiece
element of the provincial government's Sarawak Corridor of
Renewable Energy (SCORE), one of five regional development corridors being developed across Malaysia and coordinated by central
government (IHA 2013). The Sarawak province power generation
system was hitherto largely founded on gas and coal. SCORE was
thus conceived to restructure that system toward cleaner energy, with
Bakun being the largest of three dams planned in the strategy; the
other two (the 944 MW Murum plant and the 60 MW Batang Ai
plant) were currently under in construction by the mid-2010s. Another
nine sites in the province were in addition being considered for new
dams, making 12 plants in total.

Before the Bakun dam's completion in 2012, Sarawak's generation
capacity was just 1.3 GW. The overarching aims of SCORE were to
produce up to 80 percent of electricity consumption from hydropower
(around 20 GW) by 2020, and moreover for this boost in power capacity to somewhat ironically attract US$105 billion of investment in
mostly energy-intensive heavy industries to the province to spur its
economic development. Ten specific industry sectors were targeted,
namely oil and petrochemicals, aluminium, steel, marine engineering
(predominately shipbuilding and oilrig construction, glass, palm oil,
timber, livestock, aquaculture, and tourism; Sovacool and Bulan 2012).
Thus, while SCORE's large-hydro plants supplied cleaner, less
carbon-intensive electricity in Sarawak, this was primarily aimed at
facilitating new emission-intensive industrial activity that included
fossil fuel–based sectors. Furthermore, the strategy includes ambitions
plans to expand transportation infrastructure along the 320-km development corridor zone, resulting in a loss of virgin rainforest. In a sense,
hydropower was simply perpetuating energy-intensive 'old developmentalism' in this part of East Asia.

Mekong River (Lancang River in China) is the world's eighth largest, with a basin zone that covers 800,000 km² and on which the livelihoods of at least 50 million people directly depend (Urban et al. 2013). The river serves as an important transportation route, supports the largest freshwater fish stocks in the sub-region, and is a critical source of drinking and irrigation water supply for thousands of communities. It is also where Southeast Asia's largest hydropower potential is situated. This was recognised as far back as the 1950s, but security tensions among the Mekong riparian states during the Cold War period and lack of finance impeded plans for hydropower development, until these tensions eased in the late 1980s and early 1990s (Matthews 2012). For large transnational rivers, such as the Mekong, international cooperation is imperative given the strong inter-dependent links between the water, food, and energy interests of its constituent peoples from different nations, both upstream and downstream (Grumbine et al. 2012). China and certain Southeast Asian countries also share two other smaller transnational rivers, the Nu/Salween (Myanmar) and the Yuan/Red or Sông Cái (Vietnam).

As part of new endeavours to promote regional economic cooperation and improve environmental security in Southeast Asia, the Greater Mekong Sub-Region (GMS) programme was established in 1992 with Asian Development Bank (ADB) assistance. GMS membership comprises all Mekong riparian countries, and its programme covers 11 priority sectors: agriculture, economic corridors development, energy, environment, human resource development, investment, telecommunications, tourism, trade, transport, and multi-sector cooperation. Hydropower development is closely linked to a number of these, and given that the Mekong River is essentially a transnational resource shared by a number of nations, dam construction projects in one country can have a significant downstream impact on others. To address this and further strengthen institutionalised cooperation, the Mekong River Commission (MRC) was created in 1995, with a remit to manage integrated river basin management policies among its member nations. Although hydropower development has been a principal area of the MRC's work, it has proved difficult to coordinate different national hydropower policies. At the inaugural MRC summit held in April 2010, new legitimacy was conferred to the organisation on this front, such as concerning Laos's proposed Xayaburi Dam. The MRC additionally works in tandem with the ASEAN Power Grid programme to develop cross-border power inter-connections within the Mekong River basin (Grumbine et al. 2012, Sovacool and Bulan 2012).

Bilateral hydropower deals among GMS countries have also arisen, such as Laos's separate agreements with Thailand and Vietnam. Under these agreements, Laos is also obliged to deliver increasingly amounts of power to both countries up to 2020 based on the assumption of continued substantial growth in its own hydropower capacity, supply from the already operational GW-scale Nam Theun 2 dam (Box 6.4), and the planned 1.3 GW Xayaburi dam, which started constructed in 2012 and is due for completion in 2019. This latter project is being financed by Thai banks (US$3.8 billion) and developed by the Thai construction firm Karnchang, and will sell 95 percent of its electricity to the state-owned Electricity Generating Authority of Thailand (EGAT). To meet its power export commitments, Laos is planning to construct up to 60 more dams. One could argue that hydropower has been a key source of income for the country, which is among the world's poorest. However, Matthews (2012) contended that these dam projects represent a clear example of one riparian state 'water grabbing' from another, and argued that powerful Thai private and public vested interests were able to exploit Laos's weak state capacity (e.g. lax regulations and enforcement of laws, deficient capacity to regulate development and inward foreign investment) to their advantage. He further noted that strong domestic civil society opposition in Thailand to hydropower dams was another contributing factor to EGAT seeking alternative extraterritorial options in Laos. Thai investors have also been active in hydropower development in Myanmar, where similar power import-export agreements apply (Simpson 2013).

China's involvement in Mekong hydropower dam projects in Southeast Asia has been equally controversial. According to Urban et al. (2013), by the early 2010s multiple Chinese organisations (state-owned enterprises, banks, government ministries and regulators, private contractor firms) were engaged in almost 50 medium to large-hydro projects in Cambodia, Indonesia, Laos, Malaysia, Myanmar, Philippines, and Vietnam, as well as a number of small-hydro plant developments in Thailand. Similarly, McDonald et al. (2009) reported that Chinese investors were engaged in 46 dam projects across the region by the late 2000s, with the largest numbers in the Mekong sub-region (Myanmar 17, Laos 13, Cambodia 7, and Vietnam 4). Despite China's own huge domestic hydropower programme, many of the dams it has helped fund and develop in these neighbouring countries will serve the Chinese national power grid. The China Power Investment Corporation is the principal investor in Myanmar's 3.6 GW Myitsone dam, which is Southeast Asia's largest and the 11th largest in East Asia (Table 6.2). In addition, Sinohydro Corporation constructed and operates the Kamchay dam, Cambodia's largest hydropower project to date, which constituted almost half of the US$600 million aid package the country received from Beijing in 2006 (Middleton 2008).

On the one hand, Chinese state power and finance could be viewed as using Southeast Asia's Mekong River basin as a power generation hinterland to serve its own national energy needs. On the other, China's investment in the sub-region's hydropower sector has fostered international cooperation on energy and economic development in one of East Asia's poorest areas, helped address energy security predicaments, and presented a cleaner energy alternative to fossil fuel power generation. Nevertheless, these large dam projects, like many others, have come with heavy environmental and socio-economic costs, and international cooperation among the Mekong riparian states has proved largely ineffective at addressing them.

6.7.4. Environmental and socio-economic impact controversies

The long history of civil society opposition to large dam-building in Southeast Asia closely corresponds to the earlier discussed predatory statism, where the adverse socio-economic and environmental impacts in construction zones were largely neglected. There is still widespread reported evidence of notable damage that large-hydro projects have caused or could potentially cause. Here are some examples from various sources (Jönsson 2009, Matthews 2012, Middleton 2008, Simpson 2013, Urban et al. 2013):

- *Chibwe Dam, Myanmar:* Forced seizures of land from local villagers
- *Myitsone Dam, Myanmar:* At least ten villages displaced with no recourse to appeal
- *Upper Paunlaung Dam and the Shweli 3 Dam, Myanmar:* Reported allegations that local villagers were compelled to construct new buildings and roads for dam sites without remuneration, as well as reports at the Shweli 3 Dam of the army appropriating local community assets (e.g. land, livestock, natural resources)
- *Xeset 2 Dam, Laos:* Resettlement of an estimated 20,000 people
- *Don Sahong dam, Laos (proposed):* Could potentially block the main channel of fish migration along the Laos, Cambodia, and Thailand stretches of the Mekong. More than 70 percent of the river's fish catch is dependent on long-distance migrant species. The Mekong generally has the world's most biodiverse inland fishery with over 1,200 species.
- *Tuyen Quang Dam, Vietnam:* Resettlement of about 23,000 people
- *Stung Cheay Areng dam, Myanmar (proposed):* Its reservoir would flood at least nine indigenous people villages in the unique ecosystem zone, the Central Cardamom Protected Forest, which has 31 endangered fauna species. Downstream impacts could also include deleterious impacts on riverside and coastal rice paddy fields and destruction of swamp forest fishery.
- *Tasang Dam, Myanmar:* Forced relocation of 60,000 local people.

Stronger state capacity practices aimed at delivering good public welfare outcomes and robust environmental impact assessments would considerably mitigate the adverse impacts of large-hydro development in Southeast Asia. Enhanced multilateral cooperation in Mekong River basin projects where all upstream and downstream socio-economic and environmental impacts can be taken collectively into account and addressed is also essential. Hitherto, impact assessments in the area have been invariably framed nationally rather than transnationally, leading to extraterritorial problems to solve. As

Box 6.4 Nam Theun 2 dam: foreign hydropower investment
        in action

The 1,070 MW Nam Theun 2 (NT2) dam is one of the largest in Southeast Asia and one of the most controversial. Feasibility studies for the project were conducted in the 1980s and the government green-lighted its development in 1993. Like other large-hydro projects in the sub-region, the 1997–1998 East Asian financial crisis caused a delay in the NT2's progress, but work resumed again in 1999. By 2005, nego-tiations were concluded between the Laos government, state-owned power company Electricite du Laos, and its Thai counterpart the Electricity Generating Authority of Thailand (EGAT) on an interna-tional power purchasing agreement, in which it was agreed that the dam would supply 995 MW (93 percent) of its power capacity to Thailand's national grid, with the remaining 75 MW (7 percent) supplying the Laotian grid. The World Bank was also involved in the initial feasibility studies and helping fund the US$1.3 billion project (equivalent to around 15 percent of Laos' gross domestic product, and based on 72/28 debt-equity ratio)–the first time it had invested in such a large-scale infrastructure project in over a decade. The ADB and the European Investment Bank (EIB) were too among the 26 financial institutions (including seven Thai banks and a number of export credit agencies) in total that had provided funding support. It was also the largest foreign investment in Laos as well as the world's largest cross-border power project based principally (about 85 percent) on private finance (IPCC 2011, Smits and Bush 2010). The ADB's participation in the dam project was part of the Bank's wider programme of supporting hydropower development in Asia that started in the late 1980s (ADB 2010).

The NT2 dam became fully operational in April 2010. Under the Laos–Thailand concession agreement, the Laos government will receive around US$2 billion in revenue from EGAT over a 25-year period on a contracted tariff or take-or-pay basis, with this income directed toward realising the country's development objectives via a new Poverty Reduction Fund and environment-related programmes. With the World Bank, ADB, and EIB sponsorship were commitments from all NT2 developers to comply with strict guidelines on environ-mental and socio-economic (E&S) issues (ADB 2010, World Bank 2010). This included doubling incomes of resettled villagers within 5 years through mainly farming technical assistance, improved social infrastructure provision (e.g. electricity, communications, health, education) and rehousing programmes, and undertaking robust envi-ronmental impact assessments. Various independent monitoring agencies observed how well these guidelines were adhered to, and an E&S expert panel reported annually to the Laos government and the World Bank's own international advisory group that was performing a similar function (IEA 2012g). On the whole, these official monitoring mechanisms have returned generally neutral assessments on E&S impacts. International Rivers, an anti-dam civil society organisation, alleged that NT2 developers failed to comply with set E&S guidelines on numerous fronts. This includes chronically inadequate compensation to more than 110,000 affected people downstream whose livelihoods had been compromised by the dam's deleterious effects on fishing ecosystems, riverside farming, and water quality. The dam's 450-km^2 reservoir would also submerge nearly 40 percent of the Nakai Plateau, causing significant damage to its ecosystems. The organisation also claimed that there has been poor resettlement planning, leading to new housing and infrastructure delays affecting more than 6,000 people (International Rivers 2011).

The NT2 dam is an important foundational element of the Laos government's flagship strategy to make the country the 'battery of the region,' which also includes many other major hydropower projects, most notably the even larger 1.3 GW Xayaburi dam that is due for completion in 2019. Laos is a small, low-income nation with a narrow range of economic sectors. Hydropower development is viewed by the Laos state as the principal means to transform the economy and improve levels of public welfare. State capacity for strategising economic development is thus combined here with ecological modernisation thinking, and where East Asia's new developmentalism is clearly evident. However, the NT2 experience demonstrates that even where large-hydro projects have to adhere to relatively strict guidelines, significant environmental and socio-economic costs can still arise.

Grumbine et al. (2012: 97) commented, 'In an era of rising uncertainty and declining resilience, each Mekong country must understand that sovereign security increasingly depends on co-operative environmental decision making.' Furthermore, as we have seen through this chapter, large-hydro itself is inherently problematic given its physical scale, and a more sustainable future for 'low-impact' hydropower in Southeast Asia—like in other parts of East Asia—would point to small-hydro solutions.

6.7.5. Small-hydro in Southeast Asia

Small-hydro technologies and applications have a vital role to play not just in achieving more benign socio-economic and environmental impacts in comparison to large-hydro but also in providing clean energy to Southeast Asia's many remote, off-grid communities–where, as discussed in Chapters 1 and 3, energy poverty itself has been an impediment to improving welfare (IHA 2013, Murni et al. 2013, Sawangphol and Pharino 2011). Micro-hydro (community-scale) and pico-hydro (household-scale) power devices are most relevant to the sub-region's rural electrification policies. These could make, for instance, an important contribution toward the Laos government's goal of 90 percent national electrification by 2020. However, many aspects of small-hydro remain chronically neglected in Southeast Asia. Smits and Bush (2010) noted that there is a lack of information on the current use of micro- and pico-hydro in poorer nations in particular, where the state does not have the technocratic capacity or resources to conduct national surveys.

Governments have instead have tended to devote what capabilities they have to developing large-hydro projects. Civil society organisations in these countries, such as the Lao Institute for Renewable Energy, can act as key advocates of and sources of intelligence on small-hydro in the absence of state support, promoting the development of local networks of users and suppliers, and providing training workshops and other forms of technical assistance. There are similarities here with the 'distributed' expansion of small solar PV applications at the local or grassroots level, where society has taken both the initiative and greater ownership over assimilating the use of renewable energy technologies rather than being compelled to use them as a result of top-down state directives. Yet, there is obviously a role for government in creating the institutional environment and policy incentives to promote small-hydro development, and we saw earlier that a general trend toward this has become gradually evident in Southeast Asia's larger economy nations.

6.8. Conclusion

Hydropower not only remains the largest renewables sector in East Asia and worldwide, but it also possesses many characteristics that make it somewhat unique. It is by far the oldest and most established modern renewables technology, with the earliest installations having produced electricity in the late 19th century. Hydroelectric dams played a historically important role in powering 20th-century industrialisation in virtually every part of East Asia. While hydropower has been able to deliver substantial levels of energy over time, the environmental and socio-economic costs borne from large-hydro dams have made it the most controversial RE sector of all. A further special feature of hydropower concerns its multi-functionality. It is common for dams to perform other non-energy-related services, such as flood control and irrigation, which are often deemed to be equally or more important than power generation. Finally, there is no other renewables industry where international cooperation on power generation in East Asia is more developed, as well as where East Asian governments and companies are more active in overseas RE development.

This chapter has shown how East Asia's hydropower development has been historically state-driven. Moreover, as in other countries, large-hydro dams have long been perceived as symbols of economic modernity, and more recently as expressions of East Asia's ecological modernisation. In the past, hydropower's principal purpose was framed in terms of producing substantial quantities of electrical power per se rather than a cleaner, lower-carbon method of producing it. As climate change and other environmental imperatives came to the fore, both the quantitative *and* qualitative virtues of hydropower became increasingly recognised. Yet, the case study on Malaysia's Bakun dam (Box 6.3) and its integral role played in the SCORE regional development programme is indicative of how hydropower is still powering the continued expansion of East Asia's carbon-intensive 'old developmentalism,' with similar instances also manifesting in other parts of Southeast Asia and China.

Recent hydropower policies and strategies across East Asia have varied significantly, and this can be largely understood from a historic development perspective. For the most advanced industrial economies of Japan, South Korea, and Taiwan, the sector is largely viewed as an old energy technology, first established many decades ago, where most (but not all) of its feasible potential has already exploited. They have all paid more attention recently to improving the technical efficiencies of existing installations rather than constructing new ones. Hydropower is not therefore seen as a RE sector of much future strategic industry importance or as possessing much scope for future expansion with regard to low carbon development strategies. Strong civil society opposition to dam construction is another constraint on future hydropower development in these economies.

The sector has been viewed quite differently, however, for the last decade or two in China and Southeast Asia. Here, governments have embarked on ambitious hydropower development programmes, embracing a reliable and proven energy technology easily assimilated to exploit still huge untapped indigenous resources. This generally coincided with the first-phase RE development strategies launched in the early 2000s (Chapter 3). China's hydropower strategy has had enormous impact on the region and globally, financing most of the world's largest dam projects both at home and abroad. Yet, it too is likely to peak in developing large-hydro projects by around 2020. In addition, this chapter noted the mounting and increasingly effective public opposition to large-hydro projects in China and Southeast Asia, leading to more circumspect political thinking toward them. Hydropower has also earned an unenviable reputation across East Asia for political corruption. It is a renewables sector where frequent cases of weak or misdirected state capacity have not delivered good governance outcomes, in accordance with the aims and principles of new developmentalism.

Finally, hydropower's long-term future prospects primarily lie in small-hydro technology and applications. This includes small-scale dams and even smaller micro-hydro and pico-hydro devices. As with other 'distributed' renewables, a key advantage of small-hydro is its opportunity for advancing greater societal control and ownership over energy production and consumption. The fact that large sections of East Asia's population still lack electricity grid access make this even more relevant.

Notes

1   For example, hydropower generated 40 percent of electricity in the United States in 1920.
2   The world's first hydroelectric plant was developed in 1878 at Cragside, Rothbury in northern England to power lighting systems in the property of installation. The first hydroelectric station to provide electricity to the general public (of 12.5 kW capacity) was commissioned in 1882 in Appleton, Wisconsin in the United States (IEA 2012h, IPCC 2011).
3   Company websites: Dongfang Electric (http://www.dongfang.com.cn/index. php/business/detail/?cid=31) and Harbin Electric (http://www.hec-china.com/ eng/slhd1.php), accessed January 8, 2014.
4   Vermeer (2012) further detailed that, for instance, almost 100 small-hydro dams totalling 1.7 GW in the Nu River basin have been constructed in recent years.
5   *Water Power Magazine,* June 10, 2013.

7 Geothermal energy

7.1. Introduction

Despite possessing huge exploitable resource potential, geothermal's development to date remains somewhat stunted. Like hydropower, it may be considered a mature energy technology that has experienced a relatively slow growth rate. Geothermal is not considered a dynamic renewable energy (RE) sector. Rates of techno-innovation advance and investment growth have been slow compared to most other renewables. Geothermal energy development in East Asia is very uneven, and it has occurred to any substantial degree in just a few parts of the region. However, the Philippines and Indonesia are globally ranked second and third respectively in terms of installed capacity. Geothermal is also able to provide constant energy streams and generate electricity and heat, and applications range from small household-level devices to sizeable grid-connected power stations. However, the sector's long-term future in East Asia will depend very much on state support.

7.2. Geothermal energy: an overview

7.2.1. Geothermal explained

Geothermal energy is derived from terrestrial heat stored in or discharged from fluids and rocks tapped over varying depths below the earth's surface. It is harnessed either for direct use (e.g. heated water for buildings) or power generation, and a geothermal energy system requires a combination of heat, permeability, and water (Geothermal Energy Association 2011, Goldstein et al. 2011). Direct-use geothermal accounts for around two-thirds of all energy output from the sector, and it has a very long history that can be dated back to the Middle Palaeolithic period (Cataldi 1999). Industrial applications were first developed in Italy during the early 20th century, with a 250 kW electricity-generating installation becoming operational at Larderello, Tuscany in 1913 (Burgassi 1999, Lund 2004). Japan was the first East Asian nation to establish a commercially operational geothermal plant in 1966, followed by the Philippines in 1977 and Indonesia in 1978.

The three main types of geothermal power for electricity generation are as follows:

- *Flash steam*: High-pressure steam produced by extracted underground hot water is captured to drive turbines using steam/water separators. Around two-thirds of geothermal power plants are based on this technology.
- *Dry steam* (often called *direct steam*): Steam is extracted from ground fractures, which are then directed to drive turbines inside a generator. This process is used in about a quarter of geothermal power installations.
- *Binary cycle*: Extracted hot water is passed through heat exchanges, which boil organic fluids (in so-called binary closed-loop systems) that turn the plant's turbines.

The above types are sometimes combined in the same installation to form hybrid plants, with the aim of improving versatility, thermal efficiency, and load-following capability. Cogeneration or combined heat and power plants meanwhile produce electricity and direct-use heated water. Geothermal sources are located using survey techniques similar to those used to find fossil fuel deposits. Conventional rotary drilling methods are deployed to tap into geothermal reservoirs up to 5 km deep (IPCC 2011). Exploration and drilling costs form a large element of geothermal's levelised cost of energy (LCOE) for installations over their lifetime; thus, initial investment costs are relatively high compared to most other renewables. The commercial risks faced by geothermal developers are similar to those in the oil/gas and mining sectors, especially considering that any real accurate estimation of resource potential is not known until drilling operations commence. However, once the feasibility of a geothermal resource has been established, the project's probability of commercial success is more than 80 percent. Typical LCOE rates (Chapter 3) for both geothermal power and direct use heat are generally competitive in most national electricity markets (IPCC 2011). Like hydropower, geothermal provides constant and reliable energy streams; hence it is able to complement variable, intermittent power sources such as wind and solar. Efficiency rates are also relatively high, at approximately 70–75 percent capacity factors on average, although rates exceeding 90 percent are increasingly common.

Geothermal's growth performance on power generation is similar to hydropower, at around 3–4 percent annually over recent years. Global installed capacity was 1.3 GW in 1975, rising to 5.8 GW in 1990, 8.0 GW by 2000, and to just 12.0 GW by 2013, spread across 24 countries (Bertani 2005, 2010; Goldstein et al. 2011; Lund et al. 2010; REN21 2014). As Table 7.1 shows, more than 90 percent of total worldwide capacity was concentrated in just seven countries by 2013: the United States (3.4 GW), the Philippines (1.8 GW), Indonesia (1.3 GW), Italy (0.9 GW), Mexico (1.0 GW), New Zealand (0.8 GW), Iceland (0.7 GW), and Japan (0.5 GW). China had only 28 MW of installed geothermal capacity that year and Thailand had a mere

Table 7.1 Geothermal energy development, East Asia and global, 2000–2013

	Total installed capacity (MW)						
	2000	**2006**	**2008**	**2010**	**2011**	**2012**	**2013**
Philippines	1,909	1,978	1,958	1,966	1,967	1,967	1,848
Indonesia	590	800	1,003	1,197	1,209	1,341	1,341
Japan	535	532	532	500	500	502	537
China	27	27	27	27	28	28	28
East Asia	**3,034**	**3,310**	**3,493**	**3,663**	**3,676**	**3,810**	**3,754**
% share of world total	*38.5*	*37.8*	*37.5*	*36.5*	*32.4*	*32.8*	*31.3*
United States	2,228	2,771	3,081	3,207	3,226	3,386	3,389
Mexico	755	960	960	965	887	812	1,017
Italy	785	671	671	863	863	863	875
New Zealand	437	433	593	731	769	769	843
Iceland	170	422	575	575	665	665	665
Costa Rica	143	160	160	160	208	208	208
El Salvador	161	151	204	204	204	204	204
World	**7,974**	**8,831**	**9,383**	**10,105**	**11,430**	**11,700**	**12,000**

Sources: EIA (2014), IGA (2014), and REN21 (2014).

0.3 MW capacity. East Asia's share of the global total here has gradually fallen over time, from 38.5 percent in 2000 to 31.3 percent by 2013, mainly due to the lack of both new investment and policy support in the region's leading producer countries. Meanwhile, at least 78 countries were actively deploying direct-use geothermal for heating, with an estimated worldwide total energy output of 91.0 TWh. China took the global lead (21.0 TWh in 2013), followed by the United States (18.8 TWh in 2012), Turkey (16.4 TWh in 2012), Sweden (13.8 TWh in 2010), Iceland (7.2 TWh in 2012), and Japan (7.2 TWh in 2013), which was ranked sixth (International Geothermal Association 2014, REN21 2014). Huge untapped potential exists in the sector generally, and Goldstein et al. (2011) estimated that, by 2020, global capacity for geothermal electricity generation could rise to 25.9 GW. Recent geothermal projects have apparently been hindered by, amongst other things, by the lack of available drilling rigs owing to competition for their use in the oil and gas industry (World Economic Forum 2009, REN21 2011).

7.2.2. Techno-innovation, production, and investment

Utility-scale geothermal power plants typically range between 50 and 200 MW capacity and take 5–7 years to develop, from surveying to commercial operation. The world's largest geothermal field is The Geysers in the United States (1,585 MW); the 716 MW Tongonan field in the Philippines is ranked third. By the early 2010s, there were almost 800 plants in operation. Japanese firms lead the way in geothermal steam turbine manufacturing, with Mitsubishi, Toshiba, and Fuji accounting for around two-thirds of global production (Bertani 2010, REN21 2014). However, this remains a relatively small energy industry with low business sector growth rates compared to most other renewables. In 2013, it only attracted US$2.5 billion investment worldwide (Table 1.3). By September 2013, there had been only 35 Clean Development Mechanism (CDM) projects initiated on geothermal, or 0.4 percent of the total. Of these, 22 were hosted in East Asia, with 14 of these in Indonesia (Table 1.5).

New advances in techno-innovation have primarily centred on enhanced or engineered geothermal systems (EGS), which extract heat from drilled reservoirs through rock stimulation and fluid injections techniques using artificial fluid pathways. This technology is intended for use where there is no or insufficient steam or hot water and rock permeability is low. EGS methods more specifically involve creating large heat exchange areas in hot rock, which heats up artificially created fluid (usually water) reservoirs to produce generator steam for power output. Despite being under development since the 1970s, EGS technologies are still generally at the demonstration stage and are most advanced in Europe and the United States. China is also conducting EGS tests in the northeast (volcanic rocks), southwest (volcanic rocks), and southeast (granite) regions of the country (IEA 2011c, IPCC 2011).

7.2.3. Obstacles to the expansion of geothermal energy

Environmental and socio-economic costs associated with geothermal are normally site specific and technology specific. The principal greenhouse gas discharge from geothermal plants is carbon dioxide from naturally occurring processes in generation rather than combustion, and these are moreover negligible. Because geothermal systems involve natural phenomena, there are usually discharge gases mixed with generated steam, as well as water-dissolved minerals from tapped hot spring sources. Any dangerous gas emissions are normally treated or mitigated; however, if they are discharged without treatment, they can have adverse effects on local ecosystems. There is evidence to suggest that geothermal field development can cause micro-earthquakes, ground subsidence, and hydrothermal steam eruptions. However, in the sector's century-old history, there have been no recorded incidents of damage to local communities, buildings, or other structures from this development (IPCC 2011).

While EGS technologies have helped to potentially broaden the scope for geothermal development, main zones are still limited to areas of natural seismic activity that generate commercially viable levels of hot-water resources. High upfront capital investment costs arising from exploration and plant construction remain the most significant deterrent to investors. There are also a relatively small pool of skilled companies and well-trained personnel working in the industry that are concentrated in just a few countries, posing another possible obstacle for future expansion (Brown et al. 2011).

The above reasons explain why many countries have not hitherto developed a geothermal policy or have afforded little priority to promoting the sector.

7.3. China

Geothermal is a 'Cinderella' RE sector in China. Although the country has considerable geothermal resources, it has never invested significantly in exploiting them, especially as a power-generation source. The Chinese government began geothermal exploration surveys in the 1960s, which revealed the most promising high-temperature sites to be located in the southwest (Tibet, Sichuan, and Yunnan), whereas the southeast (Guangdong, Hainan, Jiangxi, Hunan, and Fujian) possessed mid- to low-temperature geothermal resources (Finn 1979, Luo et al. 2012). During the 1970s, a number of small test plants using both flash steam and binary cycle technology up to 3 MW capacity were constructed. However, most of these were later decommissioned in the 1980s due to plant equipment scaling, other technical problems, and generation cost factors, resulting in curtailed project funding from the government (Lund 2004, Wang 2003). Market reforms in the economy helped accelerate exploration activities in the late 1980s and 1990s. However, the country had only one significant geothermal plant generating electricity by the mid-2010s–the 24 MW installation at Yangbajain in Tibet, which supplies Lhasa with around half its electricity supply. This project started in 1977, with installed capacity increasing gradually up to 1991 but no further substantive investment and development thereafter (Bertani 2005, 2010). According to Zheng (2010), the potential capacity generation of the Yangbajain field is between 50 and 90 MW. In addition to two other small grid-connected installations also located in Tibet, this gave China only a total of 28 MW installed capacity by 2013 (Table 7.1).

The number of geothermal heat pumps (GHPs) for direct use in China has expanded over the years, especially since the early 2000s, producing over 5 GW_{th} capacity by the end of the decade, up from around 0.4 GW_{th} in 2004 (Zheng et al. 2010). Research on EGS technology began in 2007 but has not made much progress since (Luo et al. 2012). The Chinese government decided to merge its only target for geothermal development with that of tidal energy, with the combined installed capacity of the two sectors to reach 110–120 MW by 2015. This linking is rather unusual given the limited technological connections between the sectors. Moreover, the government had not set a 2020 target for geothermal energy power generation, thus making it the only relevant RE sector not to be assigned one. In sum, geothermal appears to be China's most neglected renewables industry.

7.4. Japan

Japan is situated on the western frontier of the Pacific volcano belt, otherwise known as the Pacific Ring of Fire. There are around 200 volcanoes in its territory and is thought to have an estimated 70 GW of geothermal power generation potential (excluding EGS), ranking it third highest next to Indonesia and the United States (Ushiyama 1999b, Sakaguchi 2013). However, although Japan has a long history of geothermal development, it has done relatively little to exploit this potential. Installed power generation capacity has increased from 23 MW in 1970 to 164 MW by 1980, 215 MW by 1985, and 296 MW by 1994. It then accelerated to 535 MW by 2000, where it reached a plateau for many years before actually falling to 500 MW in 2010 due to decommissioning of capacity in older plants, then rising again to 537 MW by 2013 (Table 7.1). This power is generated from 20 plants across 17 sites, mainly located in Tohoku district to the north and Kyushu Island in the south; all of these commenced operations before 1999, with the exception of a small binary cycle plant established in 2004 (Sugino and Akeno 2010, Yanagisawa 2013). Despite this more or less static development over the last decade or so, the country is ranked the eighth largest for geothermal power generation globally and third in East Asia.

Japan is a historic pioneer of geothermal energy. Two experimental test plants were developed at Beppu in 1924 and Otake in 1926 on Kyushu Island to the south (Nakamura 1981). Geological surveys for prime geothermal sites continued after the Second World War in the late 1940s, and the Japan Geothermal Association (JGEA) was formed in 1960 as an alliance of Japanese firms working in the emerging industry; the group had close links with the Ministry of International Trade and Industry (MITI), the pilot agency of the developmental state (Chapter 2). In 1966, the nation's first commercial geothermal plant (23 MW, dry steam type) became operational at Matsukawa in Tohoku district, constructed by the firm Azuma Kako with MITI financial and technical assistance (Nakamura 1981). By the late 1970s, Japan had developed six geothermal plants, all located in national parks. The government's 1974 Sunshine Project laid the foundation for more robust policy support during the following two decades, such as the 1979 Project for Geothermal Power Generation, which offered state support for investment in geological surveys, research and development (R&D) activity, and plant construction (Chapter 3).

Technological developments in the industry by the 1980s allowed smaller plants (typically 20–30 MW) to operate more economically and with less environmental impact. However, the vast majority of Japan's geothermal resources potential were located in natural parks (around 80 percent) and also where recreational hot spring businesses (about 28,000 of them by 2010) were already extensively developed, which has been a notable constraint on the country's geothermal energy development (Yanagisawa 2013). As the government moved away from its traditional industrial policy approach toward a more neo-liberal policy framework in the early 2000s, state support for the geothermal R&D activity ended in 2002 (Yanagisawa 2013). A year later, however, the country's first renewable portfolio standard (RPS) scheme was introduced, albeit having relatively unambitious targets and limited support for geothermal development specifically because only binary cycle plants were effectively eligible–none of which had then been built in Japan. The 2003 RPS scheme led to the development of the nation's first binary cycle geothermal plant at Hatchobaru (of only 2 MW capacity); its test capacity generation started in 2004 and commercial operations in 2006 (Sugino and Akeno 2010).

It was not until the early 2010s that the government again turned its attention to the sector. In March 2012, the Japanese Energy and Environment Council announced the removal of regulations from the government's Natural Park Act, which had restricted geothermal energy development to just six areas of the country. MITI's successor–the Ministry of Economy, Trade and Industry (METI), in conjunction with the JGEA–began working around the same time on a plan to develop 36 new geothermal plant projects, but with a target to increase installed capacity only by 50 MW by 2020 (Sakaguchi 2013). The government's 2012 Innovative Strategy for Energy and Environment, however, had a geothermal sector development target of attaining 3.88 GW installed capacity by 2030, requiring at least a seven-fold increase in the current power generation level. Also in 2012, geothermal plant developers were eligible to apply for the new feed-in-tariff (FiT) scheme to receive financial assistance for their new projects.

As earlier noted, Japan has a large recreational hot spring industry, yet its direct-use geothermal sub-sector has not grown that significantly by international comparison. By the early 2010s, its use of the geothermal heat pumps was around 0.1 percent of the GHP capacity in the United States, and around only 0.3 percent of all direct use geothermal in Japan compared to over 90 percent in South Korea at this time (Yanagisawa 2013). On EGS technology, the government established two test sites in the early 2000s at Hijiori and Ogachi; however, due to later cuts in public funding support after 2002, these languished until more recently when government policies on geothermal development were reactivated.

Primarily due to their long historic involvement in the industry, Japanese firms are dominant global players in geothermal equipment manufacturing. Fuji Electric, Mitsubishi, and Toshiba alone account for around 70 percent of the geothermal steam turbine production worldwide (Bertani 2010). The country's largest energy company–Japan Oil, Gas and Metals National Corporation–has too become increasingly active in the sector and engaged in 16 of the aforementioned 36 new plant projects announced by METI in 2012. In sum, Japan's geothermal sector has only just emerged from a decade-long dormancy, and the country has still only just begun to exploit its considerable resources in this renewables field.

7.5. South Korea

Geological surveys for geothermal energy in South Korea began in the mid-1980s, mainly to explore potential for direct-use water heating. The country's lack of near-surface, high-temperature resources associated with active volcanic or tectonic activity has limited the prospects for conventional geothermal power generation, and by the late 2000s no such plant had been installed or was in development (Song et al. 2010). However, a rapid expansion of GHP applications and direct-use geothermal generally has occurred since the mid-2000s. In 2004, South Korea's installed direct-use capacity was 23.2 MW_{th}, of which GHPs contributed 9.7 MW_{th}. By 2012, these figures had increased to 551.7 MW_{th} and 508.2 MW_{th}, respectively, with recreational hot springs (32.6 MW_{th}) accounting for most of the remainder. For the 5 years up to this date, installed GHP capacity increased more than 50 percent annually, which Lee and Song (2013) attributed to government policy incentive measures. These coincided with the advent of the Green Growth Strategy launched in 2008–2009 and specific policy measures for promoting renewables development, such as the 2009 One Million Green Homes Programme and 2010 Mandatory Use of Renewable Energy Technology for Public Buildings scheme (Chapter 3).

Under the government's New Renewable Energy Medium-Term Plan (2010–2015), a technical roadmap was introduced for geothermal energy, with phased stages of development and targets leading up to 2030 (Korea Institute of Energy Technology Evaluation and Planning 2011). In addition to plans to strengthen South Korea's technological and industrial capabilities in GHPs and district heating applications, the roadmap contained a strategy for establishing EGS-based power generation. Installed geothermal capacity of 200MW by 2030 was additionally envisaged based on ten 20 MW individual plants. Furthermore, under the National Energy Plan (2008–2030), the government set a target of 2,803 GWh total geothermal energy output level by 2030. However, the 2012 RPS scheme excluded geothermal; thus, power companies were not obliged to increasingly generate electricity from it up to 2020. Alternatively, the government decided to launch a test EGS plant project in December 2010 at Pohang based on a 5-year public–private partnership agreement and initial aim of achieving 1 MW generation capacity by 2015 (IEA 2012h). Under the agreement, government provided a US$18.5 million share of the project's US$43.5 million budget and industry provided the remainder, with the Korean companies Nexgeo and POSCO being the principal business actors. The incorporation of geothermal into a possible later revised RPS scheme could be vital, though, for more extensive business engagement in the sector's fledging power generation industry. Meanwhile, by the early 2010s, the country's GHP industry had grown to more than 100 small firms and was represented by two business associations, the Korea Geothermal Energy Association and the Korea Groundwater and Geothermal Energy Association, each with close links to government ministries (Lee and Song 2013).

7.6. Taiwan

Taiwan's situation on the western edge of Pacific volcano belt, and in the complex collision zone between the Eurasian and Philippine tectonic plates, confers it with significant geothermal potential. However, although more than 100 recreational hot spring sites have been developed on the island, it has devoted very little attention to harnessing geothermal for power genera-tion purposes. In 1966, the Ministry of Economic Affairs' (MOEA) mining industry division began to conduct surveys for geothermal energy, and feasi-bility studies on locating the best sites for development continued into the 1970s. At the end of this process, 26 sites were identified as having a total of 1 GW power capacity potential (Cheng 1985). More recent survey work estimates this to now be between 25.2 GW and 33.6 GW for Taiwan as a whole.

In 1976, Taiwan's national oil company, Chinese Petroleum Corporation, began developing a geothermal power plant at the Chingshui site in I-lan County in partnership with the government's Industrial Technology Research Institute (ITRI), leading to the commissioning of the experimental 3 MW flash steam installation in April 1981. By November 1993, the plant was closed down due to water acidity problems that were scaling pipeline equip-ment and causing severe efficiency losses: by the end of the year, the plant's effective capacity declined to just 0.18 MW (Mao and Chan 2006). A second 0.3 MW test installation using binary cycle technology was constructed at Tuchang in 1985, but in 1994 this experimental plant also closed down due to curtailed government funding. The sector remained largely dormant from this time until the mid-2000s, when the government introduced new subsidy support measures for geothermal development in 2005, leading to renewed survey work undertaken by Taiwan's power companies thereafter (Bertani 2005, 2010; Liou 2010).

However, it was not until the early 2010s that significant new develop-ments in Taiwan's geothermal sector took shape. The government set a 200 MW installed geothermal capacity target by 2030 under its New Energy Policy strategy launched in 2011. Earlier that year, the Fukushima disaster in Japan further heightened Taiwanese public opposition to nuclear power and turned attention to renewable energy sources, such as geothermal, where considerable indigenous potential for exploitation existed. With state support, construction of Taiwan's first new geothermal plant for decades commenced in 2012, a small 1 MW installation devel-oped by the Kavalan Geothermal Power company. Additionally, a 50 kW demonstration plant built by ITRI for local community use was in devel-opment at this time. Both test installations were operational by 2014. In April 2013, the MOEA signed a memorandum of understanding with Idaho state government in the United States on green energy industry and technology collaboration, with a strong emphasis on geothermal.[1] The future development of Taiwan's geothermal sector, though, will be constrained by various locational issues, especially where the best prospec-tive sites lie in national parks or near recreational hot springs and residen-tial areas. In high-level volcanic zones, there is also the aforementioned problem with water acidity.

7.7. Southeast Asia

7.7.1. Introduction

Geothermal is the second most important renewables sector in Southeast Asia on power generation. The history of the sector's development in the sub-region dates back to the early 20th century, largely concentrated in the Philippines and Indonesia. By the early 2010s, Thailand and Vietnam were also exploring their geothermal energy potential, and many nations from the sub-region have developed direct-use applications, but not on a significant scale. Growth in Southeast Asia's geothermal industry has been generally slow, where development has often occurred in periodic phases followed by consecutive years of inertia and neglect. However, this corresponds with industry trends in most countries globally and the sector's pattern of development in East Asia generally. Notwithstanding these points, recent Southeast Asian government strategies on geothermal energy have introduced new policy measures intended to realise ambitious targets for future growth over the next decade and beyond, especially in Indonesia. Yet, these should be viewed circumspectly given the clear failures of realising previous strategies introduced in the mid-2000s.

7.7.2. Indonesia

Its position along a number of tectonic faultlines confers Indonesia with enormous geothermal energy potential. Most of its volcanic-based resources lie to the south and east (the islands of Sumatra, Java, Bali, and Flores) and northeast (northern Sulawesi and Maluku), while its non-volcanic resources are mostly located in the middle of the country's archipelago territory of Kalimantan and Sulawesi. Geothermal surveys were first conducted in 1919 during Dutch colonial rule. Based on their results, drilling operations were conducted in the Kamojang area in West Java from 1926 to 1928 overseen by the Dutch East Indian Volcanological Society, but no plant development followed. It was not until 1964 that further work on geothermal development occurred some years after the country had become independent. New field surveys were commissioned that year and also during 1968–1969 with the assistance of the United Nations Educational, Scientific and Cultural Organization and the French government, leading to drilling operations at Dieng in Central Java from 1970 to 1972, but that did not reveal a geothermal reservoir. Additional surveys were undertaken in the early and mid-1970s by the country's main state-owned energy company, Pertamina, with assistance from Japanese, New Zealand, and United States aid programmes, and in 1978 Indonesia's first test 250 kW geothermal plant was built at Kawah Kamojang (Shulman 1981). At this time, Pertamina estimated the national geothermal power generation potential at 8 to 10 GW, but more recent estimates suggest this to be around 28.6 GW based on more accurate survey information.

Working closely with the government, Pertamina continued to lead geothermal exploration activities in Indonesia. By the early 1980s, it was conducting nationwide surveys of potential plant sites, either independently or in partnership with foreign energy companies. Collaboration between Pertamina and Indonesia's state-owned grid company Perusahaan Listrik Negara (PLN) converted the initial test plant at Kawah Kamojang into the country's first commercially operating geothermal power station (30 MW capacity) by 1982. The sector continued to expand up in accordance the government's 5-year macro-development plans (REPELITA framework) until the 1997–1998 East Asian financial crisis, with national geothermal power capacity reaching 142 MW by 1990 and then 310 MW by 1997 across five developed fields–four in Java and one in north Sulawesi (Radja 1990, 1997). By then, a 1991 government regulation allowed Pertamina and other energy companies to develop geothermal plants themselves rather than only PLN. Indonesia was severely hit by the 1997–1998 financial crisis, suffering a major economic recession and four-fold depreciation of its currency. Then-planned geothermal projects (11 in total, with a combined capacity of around 1,500 MW) were contracted in US dollars, realistically leaving PLN and Pertamina with no real chance of further financing plant development, despite a policy regulation announced in 2000 that sought to attract new investment into the industry (Brophy et al. 2011).

As the economy began to recover, in 2003 the government introduced its Geothermal Law, providing an ambitious new policy and legislative framework to reinvigorating the sector, with the centrepiece being business permits granted to plant developers via a competitive bidding process. This helped complete many of the plant projects that had been put on hold after the 1997–1998 financial crisis. The following year, a multi-sector Green Energy Policy was introduced, which included a separate Blueprint for Geothermal Development up to 2020 (Saptadji 2006). These new initiatives helped lay the ground for the National Energy Blueprint 2005–2025, in which geothermal was assigned as the priority RE sector for strategic development. Its target of achieving 9.5 GW of installed geothermal capacity by 2025 was by far the highest set across the renewables spectrum. The 2004 Blueprint for Geothermal Development to 2020 was merged into the 2005 National Energy Blueprint to 2025 and their sector targets made compatible. However, these have proven to be very optimistic. Based on the 2004 starting point of 800 MW, installed geothermal capacity was supposed to reach 2,000 MW by 2008, 3,442 MW by 2012, 4,600 MW by 2016, and finally 6,000 MW by 2020. This strategy was based on the assumption of the industry attracting significant new investments that would expand existing plants and establish new ones. By 2012, Indonesia's installed geothermal capacity stood at 1,341 MW, less than half the target level for that year (Table 7.1). This may be instructive for the country's prospects of reaching its future set targets. Meanwhile, running parallel to this was the first Fast Track or 'Crash' programme (2004–2008), which launched in 2004 and aimed at quickly expanding grid capacity by

10 GW overall to cope with the demands of post-crisis economic recovery. While primarily focused on coal-fired power generation, the second Fast Track programme (2009–2014, an additional 10 GW) afforded preference to renewables, thereby principally to geothermal given its aforementioned strategic priority status (Munandar and Widodo 2013).

In the late 2000s and early 2010s, the Indonesian government enacted a number of policies and legislation to support geothermal sector development, continuing to set bold future goals. In 2010, its 2025 target was raised to 12.6 GW, and the Ministry of Energy and Mineral Resources sought to leverage new large-scale investments for the country's geothermal sector from various sources, including a far higher percentage from private investors and around US$2.7 billion via the Climate Investment Funds programme with ADB assistance (IEA 2011c). By September 2013, a total of 14 of the country's geothermal projects had also received CDM investment funding, over half the East Asian total in this area.

Although Indonesia may not have reached its targets set over recent years, the sector's installed power capacity had expanded at well over twice the average global rate during 2004 to 2012 (67 percent to 29 percent, respectively). Geothermal currently accounts for 3.5 percent of Indonesia's electricity generation, and the sector's further expansion is seen as critical to meeting national targets on carbon emissions and increasing the government's electrification rate from around 67 percent presently to 90 percent by 2020.

7.7.3. Philippines

The Philippines is East Asia's leading geothermal energy producer and has around 60 plants distributed across the resource fields of Bacman, Leyte, Mindanao, Negros, Mak-Ban, and Tiwi. In 2012, geothermal accounted for 11.0 percent of the country's total installed power capacity, but this share has actually fallen over recent years due to sector development not keeping pace with electricity grid expansion. The country began to develop its geothermal sector in the early 1960s, hence some decades after Indonesia. In 1962, the Philippine Commission on Volcanology conducted the first surveys on locating the optimum geothermal resources, and 5 years later in 1967 the government enacted legislation that permitted commercial operations in the industry to commence. An experimental 2.5 kW test installation was built at Tiwi near the midwest Luzon island coast in 1969; the following year, US firm Union Oil, under its newly formed foreign subsidiary (Philippine Geothermal Inc.), as well as the state-owned Philippine National Oil Company (PNOC), started exploration activities in the Tiwi region and other parts of Luzon (Finn 1980). The government's 1978 Act to Promote the Exploration and Development of Geothermal Resources provided tax incentives on various forms of business investment that spurred the first wave of industry expansion. The nation's first commercial plant (3 MW capacity) was installed at Tongonan on Leyte Island in the same year developed by PNOC subsidiary, the Energy Development Corporation (EDC), in joint venture with New Zealand firm Kingston, Reynolds, Thom, and Allardice. In this first wave of development from 1978 to 1984, total national capacity rose to 894 MW by the end of the period, making the Philippines the world's second largest geothermal energy producer at this time. The industry then entered a plateau phase for a few years, with no additional net capacity added.

In the immediate years leading up and after the downfall of the Marcos regime in 1986, government energy and economic policy-making was in state of flux. It was not until political stability was more firmly established by the early 1990s that the state's capacity for strategic economic and energy policy-making returned in some form. This broadly explains the second phase of significant geothermal development occurring from 1992 to 1999, when around a gigawatt of new installed capacity was added to reach 1,909 MW in total by the end of the decade (Karunungan and Requejo 2000). The most important policy measure spurring this capacity growth was the government's 1991 Build-Operate-Transfer (BOT) scheme, which opened up the country's power generation industry generally to private investors under state contract terms and conditions. This essentially marked the beginning of a privatisation process that would fully blossom a decade or so later. In the meantime, the Philippines, like Indonesia, was badly affected by the 1997–1998 financial crisis and another period of inertia in the nation's geothermal sector ensued. The Renewable Energy Policy Framework 2003–2013 included only relatively weak measures to stimulate new investment from the now-dominant private geothermal firms. With no other policy stimulus offered over this time, geothermal industry activity remained more or less flat over this decade. By 2009, total installed capacity had increased to just 1,953 MW, a net increase of only 44 MW from the 1999 level. The only notable developments over this period concerned new company takeovers and the full privatisation of EDC in 2007, which currently operates the majority of the country's geothermal plants. In addition, state-owned power generation company, National Power Corporation, sold more of its plants to private firms from 2008 to 2010, including EDC (Ogena 2011).

The 2009 Renewable Energy Act proved more effective at stimulating new private investment in the country's geothermal industry than the 2003 Renewable Energy Policy Framework, but only marginally so. The IEA (2011c) reported the policy incentives introduced by the 2009 Act induced new foreign companies to enter the Philippines geothermal industry, yet this led to no discernible expansion of it. Interestingly, the government decided to exclude the sector from its 2010 FiT scheme, most likely because of its already well-established and dominant position in the country's RE portfolio. The National Renewable Energy Programme 2011–2030 set a relatively modest target for installed geothermal capacity to reach 3,467 MW by 2030, which represented less than double the national level at the time. In contrast, we saw earlier that the Indonesian government was planning an over ninefold capacity increase by 2025.

It is too early to assess whether the National Renewable Energy Programme 2011–2030 will lead to a 'third wave' in geothermal sector development in the Philippines, but certainly there has been little progress to report in recent years. By June 2013, Ogena et al. (2013) noted that national capacity has declined slightly to 1,902 MW (1,967 MW in 2012; Table 7.1) due to plant decommissioning in Negros. They also observed that the main barriers to the industry's recent development have been complex permit and licensing procedures, difficulties over inter-connection rights, complications arising in power purchase agreement negotiations, and high exploration, operation, and grid-connection costs in remote area plant projects. In sum, the Philippines geothermal sector has experienced periodic spurts of development and, despite being dominated increasingly by private business, this development remains highly dependent on robust government policy support, when this is forthcoming.

7.7.4. Thailand and Vietnam

Thailand and Vietnam are the only two other Southeast Asian nations to have embarked on developing geothermal potential for power generation purposes. Since 1989, Thailand has operated a 300 kW singular binary-cycle test plant at Fang near Chiang Mai in the far north, which is also where the country's best geothermal resources are located and principally for small binary cycle plants (Bertani 2010, Raksaskulwong 2011). Regular surveys were undertaken from the 1970s to 1990s, although this activity dropped off somewhat during the 2000s but was reactivated in the early 2010s. In 2011, the government formed a new geothermal partnership agreement involving national oil company Petroleum Authority of Thailand, the Department of Alternative Energy Development and Efficiency, and the Department of Groundwater Resources to collaborate in developing a 5 to 10 MW capacity plant sometime in the future (Charuratna et al. 2013). In the same year, the Alternative Energy Development Plan set a very modest target of 1 MW installed capacity by 2021. We should therefore not expect Thailand to become a major geothermal producer anytime soon given the government's low ambitions for the sector.

Vietnam, meanwhile, has a relatively large recreational hot springs industry, but it was not until the mid-2000s with German government assistance that systematic surveys of the country's geothermal energy potential were commissioned (Mathews et al. 2008). Although there were found to be commercially viable resources to exploit, especially in the north, no plant development plans have yet followed (Thang et al. 2013).

7.8. Conclusion

Geothermal energy has a long history in East Asia but its development has been slow, often periodic, and notably uneven across the region. Japan and

Indonesia began to conduct geothermal field surveys in the 1910s and 1920s, but it was not until the 1960s and 1970s that the first utility-scale geothermal power plants became operational. The Philippines first emerged as East Asia's sector leader by the 1980s, around the time that South Korea and Taiwan began to also establish their own geothermal energy industries; the former concentrated more on GHP applications and direct-use technology rather than power generation. In the densely populated, mountainous Northeast Asian territories (Japan, South Korea, Taiwan), there have been significant spatial constraints to overcome; this has been particularly acute when plant developers compete for space with incumbent recreational hot spring businesses in the same geothermal fields. In China and Southeast Asia, the remote location of their best geothermal resources, often far from existing infrastructural networks, has been a problem.

Arguably, however, the most important constraint on geothermal energy development in East Asia concerns how the sector is perceived from a strategic industry perspective. Despite even the region's largest producer states having only still tapped into a fraction of their exploitable potentials, geothermal has failed to attract substantial investment and prioritisation in the RE strategies of most East Asian governments. Put simply, it is not viewed as possessing much techno-innovatory dynamism or significant growth potential. Yet, it should be remembered that geothermal possesses many virtues, including having relatively low generation and operational costs, providing constant and reliable energy streams, offering a wide-range scale of different technical applications, and having minimal adverse environmental and socio-economic impacts. High initial capital costs incurred from exploratory surveys and utility-scale plant development help to explain why many investors and policy-makers have thus far eschewed the sector. However, state capacity failure is also partly to blame. This chapter noted a number of policy-related barriers that continue to impede East Asia's huge geothermal potential, including those of a regulatory, financial incentive mechanism, agency coordination, and infrastructural nature.

As for other RE sectors in the region, geothermal energy's fate has been closely tied to state capacity and economic development cycles. In this chapter, we saw that the sector started to flourish in the Philippines when a more robust state capacity for strategic policy-making was established in the early 1990s, and it became a major component in its national energy mix. Both the Philippines and Indonesia were severely affected by the 1997–1998 financial crisis, and in the latter's case geothermal played an important role in expanding power generation capacity to meet the needs of the country's post-crisis economic resurgence. Indeed, Indonesia was the only real driver of geothermal industry expansion in East Asia during the latter half of the 2000s (Table 7.1). The Philippines' recovery from the 1997–1998 crisis has not been so impressive, largely explaining why investment opportunities for expanding geothermal and the nation's other major energy sectors have been more limited by comparison.

This chapter further revealed more generally how state policies and strategies introduced by East Asian governments in the 2000s often lacked substance and sufficient incentivisation for catalysing new investment. Even in the region's recent key driver country, Indonesia, government targets on raising installed capacity levels proved to be overly optimistic during this time, with less than half the assumed new strategic investments failing to materialise. Nevertheless, in the early 2010s, governments in the region gave fresh impetus to the sector through the introduction of new plans and policy incentives for geothermal development. In post-Fukushima Japan, the government's 2012 Innovative Strategy for Energy and Environment included a bold long-term strategy for a seven-fold increase in geothermal by 2030, and Indonesia still maintains a plan for a nine-fold expansion by 2025 with foreign investment assistance. Meanwhile, the South Korean government's New Renewable Energy Medium-Term Plan has a technical roadmap to establish a more substantive geothermal power generation industry using start-of-the-art EGS technology, as well as to further expand its already well-established GHP sector up to 2030. Taiwan, another relative latecomer to the sector, announced in 2011 the government's intention to realise similar objectives in its New Energy Policy to 2030, also based on strengthened international cooperation on geothermal technology development. Thailand and Vietnam are furthermore on the possible verge of becoming more active players in the sector's development in East Asia. However, given the historic evidence studied in this chapter, the chances of such ambitious new plans coming to fruition could be slim. Moreover, even if a new growth spurt is achieved, the geothermal energy sector in absolute terms will remain relatively small in East Asia and globally compared with many other renewables for some years to come.

Note

1   https://www.itri.org.tw/eng/econtent/news/news01_01.aspx?sid=34 (accessed January 24, 2014).

8   Ocean energy

8.1. Introduction

Ocean energy is the smallest mainstream renewables sector—the only one with a global installed capacity less than 1 GW. It comprises a cluster of mainly tidal and wave energy technologies. Most of these are still in the experimental stages of development, although tidal range barrages are now a well-established power generation method where East Asian states—and especially South Korea—are emerging world leaders. Tidal barrage projects also account for the vast majority of ocean energy's generation output. Although ocean energy may be considered a micro-sector, the combination of high rates of techno-innovation, close links with incumbent engineering science industries, and huge exploitable power potential levels have made many East Asian states view it as a promising strategic industry of the future. However, even though ocean energy could be on the verge of rapid expansion over forthcoming years in East Asia and elsewhere, it is likely to remain a micro-sector compared to other renewables for some time yet.

8.2. Ocean energy: an overview

8.2.1. Ocean energy explained

Ocean energy exploits the kinetic, thermal, or chemical energy of seawater, converting this into electricity generation or other forms of usable power. It thus has some close technological links to conventional hydropower, and it is sometimes viewed as a sub-category of it. Modern ocean energy technologies have developed over many decades. By the early 2010s, there were over a hundred different technological applications under development in over 30 countries.

The sector can be divided into the following five conversion technology types (Brown et al. 2011, IPCC 2011):

- *Tidal range*: This type is based on tidal barrages situated normally in river estuaries, capturing the kinetic energy of rising and falling tides, and thus essentially vertical movements of water. Tidal barrage projects using offshore or coastal basins away from estuaries (tidal lagoons), and usually in bay areas, are also in development. Many coastal areas experience two high and two low tides every day (semi-diurnal zones), whereas others experience just one tide (diurnal zones); the former tend to be the most effective plant locations.

- *Wave energy*: Wave energy is generated from the kinetic energy of wind on ocean surfaces. Key variables for determining energy potential include the magnitude of different wave motions (heaving, surging, pitching), water depth (deep, intermediate, shallow), and distance from shore (shoreline, near-shore, offshore). Three main types of devices have been thus far developed: oscillating water columns (shore-based, floating), oscillating bodies (surface-buoyant, submerged), and overtopping devices (shore-based, floating).
- *Tidal and marine currents*: This technology harnesses both tidal range and horizontal movements of seawater. Underwater planted turbines, akin to wind turbine designs, are the most common technology thus far developed in this category. Those capturing tidal currents must work on a bidirectional basis, while devices located further offshore capturing oceanic currents work unidirectionally. Minimum current velocities of around 2 metres per second (4 knots) are needed for commercially viable operations using technologies thus far developed.
- *Thermal gradients*: Here, the temperature (thermal) differences between warmer upper layers of seawater and cooler deeper layers are exploited using ocean thermal energy conversion processes. Southeast Asia, China, and south Japan are situated in or near the world's most best maritime zone (midwest Pacific) to exploit this energy source.
- *Salinity gradients*: At river mouths, the mix of freshwater and seawater creates energy sources from salinity gradients, which can be harnessed using pressure-retarded reverse osmosis power techniques.

Tidal range is by far the most developed of the above technologies. It is a predictable source of power, but its scope for application is limited to where tidal range and marine current variations make it commercial viable to locate plants. Furthermore, installations can only produce power for each part of the tidal cycle and are therefore a variable energy source, with average capacity factors varying between 20 and 35 percent (Denny 2009, IPCC 2011, Johnstone et al. 2013). Estuary-based tidal barrages have the potential for GW-scale electricity generation, but their construction is extremely expensive, they can cause notable environmental damage in fragile estuarine ecosystems, and there are relatively few sites worldwide where they would prove commercially viable. However, advances in both installation design (e.g. multi-basin tidal lagoons) and turbine technology (axial, cross-flow) are providing new opportunities for the sub-sector to flourish (Johnstone et al. 2013). China and South Korea are thought to have the best potential for developing tidal barrages in East Asia.

Wave energy is the next most developed technology type. The oscillating water column (OWC) is the most common design where wave-induced rises and falls of water inside the device's chamber produces airflows that spin 'air turbines' located at the top, which is the mechanical movement used for generating electricity. While only a small handful of grid-connected installations were in operation by the early 2010s, the potential of wave energy is huge. Being subject to similar intermittency problems of wind energy, it is an unpredictable power source. Moreover, there are significant challenges concerning its utilisation, mainly concerning equipment and infrastructural durability given the immense natural forces wave energy plants are subjected to (Falcao 2010). In addition, certain geographic areas are much better suited to developing wave farms than others, with the best broadly lying in the south and north temperate zones, which possess strong regular winds and oceanic currents close to coastal areas. As with wind energy, most wave energy potential in East Asia is concentrated in Northeast Asia, and especially off eastern Japan (Mork et al. 2010). The low-wind 'doldrums' meteorological phenomenon, first described in Chapter 4, explains why Southeast Asia has rather limited potential to exploit wave energy. *Tidal and marine current* devices are still largely at the experimental stage of development, mainly in Europe and Canada. Thermal and salient gradients technologies are mainly at the conceptual or very experimental stage of development.

Ocean energy is the smallest renewables sector under study in this book, with only 535.5 MW of installed power generation capacity worldwide by 2012 (Ernst & Young 2013, OES 2013). Moreover, just the six tidal range barrage installations (combined capacity 519.6 MW) operational world-wide at this time accounted for virtually all of this total (Table 8.1). For many decades, the 240 MW Rance tidal range barrage in France, which became operational in 1966, was the world's only significant tidal energy installation. In 2011, though, South Korea's 254 MW Sihwa Lake barrage became the world's largest, consequently doubling the sector's global installed capacity. South Korea has plans to expand its tidal energy power capacity to up 3 GW by 2017, based mainly on the construction of three other large-scale tidal range barrages at Garorim Bay (520 MW), Ganghwa (420 MW or 840 MW), and Incheon (1,320 MW). Meanwhile, China started an experimental tidal power programme in the late 1970s and has operated a small-scale 3.9 MW tidal barrage at Jiangxia at the Yula River estuary since 1980.

By 2012, there were only six nations with grid-connected tidal current devices, led by the United Kingdom (total capacity 6.7 MW), with the remaining 2.1 MW of aggregated capacity in China, South Korea, Italy, the Netherlands, and Canada. As Table 8.1 also indicates, there are only a small handful of grid-connected wave energy installations worldwide that together produce a combined 7.1 MW of power, the approximate equivalent of one of the largest offshore wind turbines in operation by the early 2010s. China was the only East Asian nation contributing to this wave energy total by 2012, with just 190 kW capacity. It is later examined how Japan was the world's first pioneer of wave energy technology but had no grid-connected devices in operation by the early 2010s. Only the Netherlands and Norway had grid-connected experimental salinity gradient devices in operation, both with very small power output ratings (Table 8.1). Taking all ocean energy technologies together, East Asia's total installed capacity by 2012 was 259.7 MW—more than any other region and just under half the global total.

8.2.2. Techno-innovation, production, and investment

Ocean energy can be thought of as a micro-sector with a small industry base. Most of its firms are relatively large organisations that have a division specifically tasked to ocean energy-related activities. These are typically companies working in marine engineering (primarily offshore oil and gas exploration) or infrastructure engineering (e.g. Alstom, ABB, Hyundai, Siemens, Voith) businesses that can apply their relevant expertise and technological competences to this renewable sector. Ocean energy development has significantly benefitted from techno-innovatory advances in marine engineering, such as in material science, construction, submarine cables and communications, and anti-corrosion technologies (Johnstone et al. 2013). The expansion of the offshore wind industry could have particularly useful spinoff effects for the tidal and marine currents sub-sector in terms of technology and infrastructure development given the similar dynamics and challenges of turbine design, and for establishing offshore transmission cable systems. As earlier noted, there is a wide range of largely experimental devices in development, and technological convergence has only really occurred thus far in tidal range barrages (IPCC 2011). Analysts tend to concur that the ocean energy industry's development is around 15–25 years behind wind energy, but it has the potential to follow a similar path to commercialisation if it receives sufficiently strong policy support for attracting investment (OES 2013, REN21 2014).

In 2013, the sector had reportedly attracted investment of only US$100 million in total worldwide (Table 1.3, p. 16). Governments have hitherto been the principal investors (IPCC 2011). Research institutes are also vital partners in the development, testing, and certification of new ocean energy prototype technologies. Institutions from Japan, China, South Korea, Taiwan, and Singapore have, for example, signed agreements with the world-leading European Marine Energy Centre (EMEC) on a joint research collaboration (REN21 2013). In 2001, the OES Implementing Agreement was established as an initiative by the International Energy Agency to advance ocean energy technology collaboration and industry development globally. The OES has played a key role in promoting the sector since its creation. For ocean energy to advance further, industry development road maps and supply chain studies will need to be commissioned, and many developed countries have already started this process (Brown et al. 2011).

8.2.3. Obstacles to the expansion of ocean energy

Obstacles primarily concern typical technological, financial, and corporate factors confronting any emerging energy sector at the demonstration development phase. To move to the cost-reduction phase, substantially more investments, technical resources, and policy support measures will be required over a sustained period. Tidal and wave energy installations do not emit carbon dioxide or any other greenhouse gases when operational, but CO_2 emissions do arise over plant lifecycles from the equipment manufacturing and plant construction processes to future decommissioning. However, compared to other energy technologies, the net greenhouse gas impact of ocean energy is very low (IPCC 2011). The main environmental concerns have centred on tidal range barrage projects located or proposed in river estuaries with fragile marine ecosystems. Environmental non-government organisations (NGOs) have been vehement opponents of such projects, citing the destruction caused to avian, fish, and other wildlife habitats. Large-scale tidal barrage development could potentially lead to displacements of local people, although nowhere near to the same degree as hydropower, where impact zones are far greater in size (Chapter 6). Depending on their design and location, there may be public opposition to their visual unsightliness and corresponding negative impacts on local tourism. Compared to most other energy sectors, however, the socio-economic and environmental 'hazard' obstacles to expansion are of a low magnitude. The most significant and fundamental obstacle to ocean energy development concerns the distribution of its resource potentials. Many countries lack access to those resources—or where they possess it, it may not yet be feasible for commercial exploitation.

8.3. China

8.3.1. Introduction

China has the most broadly developed ocean energy sector in East Asia. With 18,000 km of coastline of a large maritime territory, the country has the potential to substantially exploit all ocean energy technology types. As in many other countries, Chinese government policies on ocean energy have historically been singular–project oriented. It has only been since the late 2000s that a broader and more coherent policy on ocean energy development has emerged in China.

8.3.2. Policy development

China's first ocean energy development endeavours date back to the mid-1950s, when the government proposed a number of tidal range projects (Charlier 2001), leading to an initial test installation at Shun De County, Guangdong province in 1958. A total of 44 similar devices were constructed, mostly across the coastal provinces of Zhejiang, Shandong, Jiangsu, Shanghai, Fujian, and Liaoning over the following decade or so, ranging between 5 kW to 200 kW (Lund 2010, Qin et al. 2013). However, most of these test plants failed due to poor site selection, technical faults, and mismanagement. By the 1980s, only eight were still operational (Wang et al. 2011).

In the meantime, the government introduced plans in the early 1970s to build China's largest and most important tidal range project at Jiangxia in Zhejiang. Construction started in 1974 and was completed in 1985, with an initial generation capacity of 3.2 MW. Further upgrading in the late 2000s boosted this to 3.9 MW, making it at this time the world's third largest tidal range installation. Operational tests and upgrades undertaken at Jiangxia advanced the country's technological and environmental impact learning on tidal range energy. However, apart from commissioning new oceanographic surveys and experiments with a few new test devices under the '863' State High-Tech Development programme (Chapter 3), the government reduced its funding for the sector until the introduction in 2007 of its Medium to Long-Term Development Plan for Renewable Energy, which set a goal of developing 100 MW of installed tidal energy capacity by 2020. New plans and feasibility studies for a number of tidal range plants were set in motion soon thereafter to meet this objective: in Zhejiang, Jiantiao Port (21 MW); in Fujian, plants at Daguanban Port (10 MW), Bachimen (10 MW), and Maluan Bay (20 MW); and in Shandong, a 10 MW installation or possibly larger at Rushan (OES 2013, Qin et al. 2013).

Chinese government funding of tidal current technology experiments began in the late 1970s, when a 6 kW test device was installed in Dinghai, Zhejiang. Further kW-scale test prototypes and oceanographic surveys followed from the 1980s onwards (Wang et al. 2011). By the early 2010s, there were two MW-scale tidal current projects under development, at Daishan, Zhejiang (1 MW) and Longxudao, Shandong (1.2 MW), as well as three 0.5 MW devices at Zhaitang (Shandong), Qidong, and Rudong (both Jiangsu province), all of which would be later grid-connected (Xia 2012). As with a number of other smaller test installations, these were under long-term experimental development and had received strategic funding under the 863 programme (Qin et al. 2013). By 2012, though, China had established only 0.11 MW of installed tidal current capacity supplying power to the grid (Table 8.1).

Chinese government surveys on wave energy potential started in the late 1960s, and in 1972 the first test 1 kW device was positioned offshore at Shengshan Island, Zhejiang. This was a navigation light buoy similar to what the Japanese government were simultaneously developing. China was one of the first countries to test OWC devices, starting in the mid-1980s and drawing inspiration from Masuda's initial experiments in Japan some time earlier.

Table 8.1 Ocean energy development, East Asia and global by 2012

	Installed capacity (MW)				
	Tidal range	Tidal current	Wave energy	Salinity gradient	Total
China	3.9	0.11	0.19		4.2
South Korea	254.0	1.5			255.5
East Asia total	**257.9**	**1.61**	**0.19**		**259.7**
Denmark			0.25		0.25
France	240.0				240.0
Italy	0.006	0.1			0.106
Netherlands		0.1		0.01	0.11
Norway				0.004	0.004
Portugal			0.7		0.7
Russia	1.7				1.7
Spain			0.296		0.296
Sweden			0.15		0.15
United Kingdom		6.7	4.3		11.0
Europe total	**241.706**	**6.9**	**5.696**	**0.014**	**254.316**
Australia			1.15		1.15
Canada	20.0	0.25			20.25
New Zealand			0.02		0.02
United States			0.04		0.04
Asia-Pacific (developed) total	**20**	**0.25**	**1.21**		**21.46**
World	**519.606**	**8.670**	**7.096**	**0.014**	**535.476**

Sources: Ernst & Young (2013), OES (2013)

Note: Figures are for grid-connected installations.

Around this time, the 7th Five-Year Plan (FYP; 1986–1990) and its associated 863 programme introduced a more ambitious and systematic approach to wave energy development that brought ad hoc experimeannts around the country into a more organised and better funded research and development (R&D) policy framework for the sub-sector. By the 10th FYP (2001–2005), OWC devices up to 100 kW ratings were being developed, more th 600 navigation light buoys had been deployed, and other wave energy conversion technologies were in development, such as pendulum, point absorber, and backward bent duct mechanisms (Charlier 2001, Qin et al. 2013, Wang et al. 2011, Zhang et al. 2009). An experimental hybrid renewables plant installed at Daguan Island in 2011 incorporated a 30 kW wave energy system along with 60 kW wind and 15 kW solar energy capacity. In addition, a 500 kW hybrid wave energy installation was being tested at Danwanshan Island, Guangdong (Xia 2012). The government also has set a target of developing MW-scale grid-connected wave energy plants by 2020 (Qin et al. 2013).

Meanwhile, China began experimenting with both thermal gradient and salinity gradient technologies from the early 1980s onwards, but no scaled-up test installations had been developed after 30 years of effort (OES 2013, Qin et al. 2013). In sum, China's ocean energy policy has a deep history and has undergone a decades-long stage of research development and demonstration, where techno-innovatory advancement has occurred but at a slow gradual pace. From the late 2000s onwards, though, the government has shown more strategic intent towards realising the country's significant ocean energy potential and set ambitious future targets for 2015 and 2020.

8.3.3. National market and industry development

China has only developed an industry for navigation light buoys in the wave energy sector. These are low kW devices that Chinese firms are exporting in relatively small numbers to a niche international market, and it is the only element of China's ocean energy sector to have attained commercialisation status (Qin et al. 2013). Firms involved in developing the country's previously discussed MW-scale tidal range projects are the same state-owned power companies that currently hold dominant positions in many other renewable energy sectors. The most important is Longyuan Group (Box 3.1, p. 84), China's largest operator of wind farms that also manages the long-established Jiangxia tidal energy plant and is developing the 21 MW Jiantiao Port installation as well as being involved in other projects in Fujian (Lund 2010). China Huaneng Group, an active player in the solar park industry, is developing the two tidal current installations in Jiangsu, while fellow state-owned power company Datang Corporation is responsible for developing the larger tidal current plant in Shandong. National oil company China National Offshore Oil Corporation is the developer of the other MW-scale project in the same province.

8.3.4. Main challenges ahead

China has not yet scaled up its existing ocean energy installations to the same level as South Korea, nor does the Chinese government have the same degree of ambition to do so in its strategic plans. Correspondingly, the adverse environmental impacts of developing its tidal and wave energy plants will be much lower. The main future challenge facing China's ocean energy sector concerns progressing beyond its long phase of experimental development to paths of commercialisation. This may only be achieved gradually with commitment from both state and business to invest in the inter-related endeavours of plant-scale expansion, improving operational efficiency, and generating cost reductions. These are challenges that face the sector worldwide.

As a nation with long-standing experience in ocean energy R&D and a relatively ambitious plan for development up to 2020, China could make a significant contribution to future global efforts on commercialising wave and tidal energy technologies. However, moving from a position of only 4.2 MW grid-connected installed capacity for the sector overall by 2012–where there has only been a 1 MW increase in this total since the mid-1980s–to 50 MW by 2015 and 100 MW alone for tidal energy by 2020 will prove to be a real challenge for the country, and it will largely depend on whether the afore-mentioned planned tidal range projects can be completed on schedule.

8.4. Japan

Japan's coastline is around 30,000 km–one of the world's longest–and is endowed with very significant wave energy resources. It was also the world's first pioneer of wave energy technology. The experiments of Yoshio Masuda conducted in the 1940s helped to pave the way for the development of OWC devices that were later copied in other countries. In the early 1970s, Japan started to develop small kW-scale navigation light buoys using wave energy as the power source (Charlier 2001). Later that decade, Masuda's OWC designs were refined and scaled-up for the Kaimei project (1978–1986), where a floating barge housed many different types of test machines that generated grid-connected electricity, the world's first wave energy device to do so. This was followed by the floating, 50-metre long Mighty Whale (1998–2003), an OWC installation developed by the Japan Agency for Marine-Earth Science and Technology with a total rated power of 110 kW, deployed at Gokasho Bay, Mie Prefecture (Falcao 2010, Ushiyama 1999b). Both projects were sponsored by the government, but in the early 2000s a decision was taken to curtail funding for its wave energy programme. Meanwhile, tests undertaken up to the 1990s in tidal energy at Ariake Bay in the country's south were terminated due to unfavourable economic pros-pects for the technology in the country (Lund 2010).

However, renewed interest in developing Japan's ocean energy sector arose in the late 2000s based on collaboration between policy-makers, busi-nesses (e.g. companies such as Mitsui Engineering and Shipbuilding), and research institutions. The government funded a new national research programme for 2011–2015, coordinated by the New Energy and Industrial Technology Development Organization to conduct feasibility studies (for mainly wave, tidal current, and thermal gradient technologies) and install new test devices, with the aim of reducing power generation costs by 2020 (OES 2013). This helped Japan to restore its position as a technology leader in wave energy in particular, although its test installations were still small scale, very experimental, and not generating grid-connected electricity.

8.5. South Korea

8.5.1. Introduction

Over time, South Korea has emerged as a global leader on ocean energy, especially on tidal range projects. Endowed with favourable coastal and oceanographic resources, the country has good prospects for ocean energy development. The west (tidal range) and south (tidal current) coasts of the Korean peninsula are especially suited to tidal energy installation due to significant range and current flow factors (Lee 2006, Lund 2010). Since 2011, South Korea has been home to the world's largest ocean energy installation, the 254 MW Sihwa Lake tidal barrage. Although the country also has considerable potential for wave energy development, this has not yet materialised (Ernst and Young 2013).

8.5.2. Policy development

The South Korean government's first policy actions on ocean energy date back to the mid-1970s. In 1974, it announced proposals to explore the country's potential for tidal range energy. A feasibility study was undertaken in 1975 at Cheonsu Bay, then at Garorim Bay in 1976, leading to the publication of the general Korea Tidal Power Study of 1978 (Charlier 2001). Feasibility studies on the Garorim Bay project then followed in the 1980s and 1990s. Another proposal to build a tidal range barrage at Sihwa Lake was announced in 1997. The emergent policy pattern in this early phase was then very much project-specific rather than generally sector-oriented. Thus, there was no overall policy as such on ocean energy development, but rather separate plans for constructing two tidal barrages along the west coast.

A more coherent policy approach emerged by the early 2000s under the government's Study for Ocean Energy Development in Korea initiative launched in 2000, which came in the lead up to its 2001 Basic Plan for New and Renewable Energy Development, Utilisation and Promotion (Chapter 3). The study was divided into two consecutive stages over 11 years. In the first stage (2000–2005), a US$13.5 million budget was allocated to further developing the Garorim Bay and Sihwa Lake tidal range projects, constructing a tidal current installation at Uldolmok, and a few small-scale wave energy design projects. The study's second stage (2006–2010) enjoyed a much larger US$34.5 million budget to start developing a third tidal range barrage at Incheon Bay, further advance the Uldolmok project, conduct other tidal current device experiments, and install a 0.5 MW wave energy test device near Jeju Island, also on the south coast (Lee 2006). At the same time as the study launch, a 'basic plan' for the Sihwa Lake tidal barrage's development was initiated in 2000 and coordinated by the Korea Ocean Research Development Institute. A fourth large-scale tidal range barrage project, at Ganghwa, was announced during the second study phase, further consolidating South Korea's position as the emerging world leader in tidal energy development. This came around the time that the new Lee Myun-bak government came to power and launched its Green Growth Strategy (GGS).

Therefore, a basic strategy was already forming in South Korea's ocean energy policy before the GGS, but more substantive strategy-making became evident with the GGS's introduction, providing significant new impetus to the sector's overall development. In the GGS's new National Energy Plan 2008–2030 was a separate national strategy and roadmap for ocean energy. In its first phase, as well as receiving increased R&D funds for core technology development,[1] the sector would benefit from various general policy measures aimed at developing renewables generally, such 2012 renewable portfolio standard legislation mandating power companies to increasingly source their electricity from renewable energy sectors (Chapter 3). Consequently, these state-owned power companies ramped up investment levels in their ongoing ocean energy projects, especially in the four tidal barrage installations (Ernst & Young 2013). In the second phase (2013–2020), policy measures to strengthen domestic industry capacity and indigenous techno-innovation were to be introduced, followed in the third phase (2021–2030) by South Korea's private enterprises leading the broader commercial development of ocean energy with a vision of 'the hybrid utilisation of technologies combining multiple ocean energy resources' (OES 2013: 92). The exact nature of these policy measures to be deployed in these second and third plan phases is not yet clear. The National Energy Plan 2008–2030 has set a target of generating 6,159 GWh of electricity output from ocean energy by the plan period end, and around two-thirds of this could come from the country's four large tidal range barrage projects (Table 8.2).

After the introduction of the GGS, in sum, South Korea's ocean energy sector has benefitted from stronger general policy measures to support renewable energy development overall, continued and more generous funds for R&D activity, and a national strategy roadmap to develop the industry over the long-term. Largely in response to the GGS, local government authorities, such as Incheon city, and the state-owned, regionally based power companies have afforded greater strategic priority to ocean energy development.

8.5.3. National market and industry development

National market and industry development have always centred on the construction of large-scale tidal range barrage projects located on South Korea's west coast, with all four thus far developed being in close geographic proximity to each other. As Table 8.2 indicates, the already completed 254 MW Sihwa Lake barrage is the smallest of the four in terms of installed capacity and power output, and it is expected to help reduce carbon emissions annually by 315,440 metric tonnes (Lund 2010). The installation was also originally conceived back in the 1990s as serving another environmental function: to help improve flow circulation in what had become a rather stagnant body of water. The construction of the oldest project in development–the 520 MW Garorim Bay barrage–is due for completion by 2015, while the 1,320 MW Incheon Bay barrage is the largest by far, with a 2017 target date for completion and the potential to provide around 60 percent of household electricity consumption in Incheon, South Korea's

Table 8.2 South Korea's tidal range barrage projects

	First proposed	Completion date	Installed capacity (MW)	Estimated annual output (GWh)	Cost (US$ million)	Technical details	Power company developers (and main contractors)
Garorim Bay	1978	2015	520	880	1,000	Mean tidal range: 4.7 m Spring tidal range: 6.6 m Barrage length: 2.0 km Basin area: 45.5 km²	Korea Western Power (POSCO)
Sihwa Lake	1997	2011	254	553	350	Mean tidal range: 5.6 m Spring tidal range: 7.8 m Barrage length: 12.7 km Basin area: 43 km²	Korea Water Resource Corporation (Daewoo, VA Tech Hydro – Austria)
Incheon Bay	2006	2017	1,320	1,800	2,500	Mean tidal range: 5.3 m Spring tidal range: 7.3 m Barrage length: 20 km Basin area: 106 km²	Korea Hydro & Nuclear Power (GS Construction)
Ganghwa	2009	2017	840 (or 420, tbc)	tbc	1,900	Mean tidal range: 5.5 m Spring tidal range: 7.7 m Barrage length: 7.8 km Basin area: tbc	Korea Midland Power (Daewoo)

Sources: Ernst & Young (2013), Lund (2010), OES (2013).

Note: tbc, to be confirmed.

third largest city with a population of 3 million (Lund 2010). The future of the most recently developed project at Ganghwa–which would also supply power to Incheon and other nearby densely populated metropolitan areas– is less certain, mainly due to strong public opposition, which has lead to planned installed capacity being possibly halved from 840 MW to 420 MW.

South Korea's flagship tidal current energy project has been the Uldolmok installation, located in the southwest tip of the Korean peninsula where tidal currents can reach speeds up to 11 knots. It was initially rated at 1 MW when first constructed in 2008, then increased to 1.5 MW grid-connected capacity by 2011, with the original target of raising this substantially to 90 MW by the end of 2013. However, this was not achieved because the proposed plant extension failed to pass its commercial feasibility test that year and its development was consequently put on hold.[2] Another 110 kW test tidal turbine device was installed nearby in 2011 with an ambitious plan devised by a consortia of companies (Korea Hydro & Nuclear Power, POSCO, Renetec and Voith) in conjunction with local government to develop the site into a world-leading 400 MW plant by 2018 (Lund 2010, OES 2013). A third 100 kW tidal current device was under experiment on the mid-south coast at Daebang from 2009, with plans to possibly develop up to a 20 MW installation here in the future (Cho et al. 2010). Neither of these smaller devices was grid-connected. Although there have been a number of wave energy device designs under development in South Korea since the early 2000s, only one device has progressed to actual installation–the aforementioned 0.5 MW oscillating water column project developed by the Korea Institute of Ocean Science and Technology at Yongsoo, off Jeju Island. Tests were hampered, though, by exceptionally frequent typhoons during 2012 (OES 2013).

A number of Korean companies (e.g. Daewoo, Hyundai, Iljin Electric), universities, and government ministries (particularly the Ministry of Land, Transport and Maritime Affairs and Ministry of Knowledge Economy[3]), and a number of state-related research institutes[4] have collaborated in different combinations on advancing all types of ocean energy technology. This has comprised new turbine designs for tidal range and tidal current installations, various wave energy test devices, and experiments with thermal and salinity gradient technological applications offshore. R&D projects in these areas are currently scheduled to run up to 2018 (OES 2013).

8.5.4. Main challenges ahead

Opposition from both environmental lobby groups and the general public to tidal range barrage development in South Korea has been strong. It has proved by far the most important challenge to installation developers to date, and it is likely to remain so for the foreseeable future. Voiced concerns over the adverse impacts on local ecosystems, wildlife, and the livelihoods of local fishing communities have led to delays and, in the case of the Ganghwa project, a scaling back of ambitions, as earlier noted. Environmental

NGOs have mobilised local opposition to great political effect and used legal channels to hinder the path of barrage development. They have, for example, argued that building the Incheon Bay and Ganghwa installations would contravene legislation introduced in the early 2000s aimed at protecting unique wetland habitats in the area (Ko and Schubert 2011). It is possible that societal opposition could lead to the further development of the large barrage projects currently in development being put on hold in the future.

8.6. Taiwan

Under its New Energy Policy introduced in 2011, the Taiwanese government set a target of establishing 600 MW of installed capacity of ocean energy devices by 2030. As an island territory with around 1,500 km of coastline, it has significant potential for developing this sector. The 2005 National Energy Conference formally decreed that the ocean energy sector should become a priority; new research initiatives on wave, tidal, and thermal gradient technologies followed, coordinated by the Institute for Industrial Technology Research Institute (ITRI).[5]

The northeast offshore region of Taiwan was identified as having wave energy potential of several hundred megawatts, while the east coast and the Pescadores Channel (off the Penghu Islands) were believed to have GW-scale tidal current energy potential (Chen et al. 2013b). With ITRI funding, the National Taiwan Ocean University installed a very small test wave energy device near Keelung Harbour to the island's northeast corner. However, by the early 2010s, only a few oceanographic data surveys on possible tidal current and wave energy installations and small machine test experiments had been conducted under the government's ocean energy policy.

According to Chiu et al. (2013), like Southeast Asia, the wave energy potential of Taiwan is mostly dependent on monsoon winds in northern hemispheric autumn and winter seasons, with some dependence on typhoon winds–although these carry relatively high risks of installation damage. Taiwan has limited potential for tidal range energy because western coast is mostly flat and sandy, whereas the eastern coast is rocky and straight. Tidal lagoon development options are limited too, with the best being situated in the offshore Penghu, Kinmen, and Matsu Islands (Chen et al. 2010b). In December 2012, the European Marine Energy Centre signed a collaboration agreement on ocean energy technology development with ITRI and the National Taiwan Ocean University.[6]

8.7. Southeast Asia

Ocean energy development in Southeast Asia has been limited to testing experimental devices across just a few countries from the sub-region. Most of its potential in this sector lies in the more difficult and expensive areas of thermal gradient and salient gradient technologies. National governments in the sub-region have lacked the financial and technical resources to facilitate substantive research and development programmes. As earlier noted, most of Southeast Asia has relatively limited wave energy potential due to it lying in the low-wind 'doldrums' zone. While there are some good locations for capturing tidal current energy, the prospects for tidal range energy are generally slim owing to the sub-region's relatively narrow tidal range bands.

Indonesia has tested tidal current turbine devices in both the Malacca Strait and the Alas Strait, where current flows of up to 2 metres per second (4 knots) just make them viable commercial prospects for development if adequate investment funding were to follow (Blunden et al. 2013). Of all countries from the sub-region, Malaysia appears the most active, having installed a number of test installations over the years. The Malaysian government, like its Indonesian counterpart, has also helped to fund a number of tidal current device experiments in the Malacca Strait, the world's second busiest sealane and principal trade route between East Asia and other regions to its west. The Marine Renewable Energy research group at University of Malaysia, which undertook these tests, is the country's main centre for ocean energy research. Other identified suitable sites for tidal current energy development are Pulau Jambongan and Kota Belud in northern Sabah and Sibu in south Sarawak (Hassan et al. 2012). The consensus of Malaysia's scientist community is that in addition to tidal current energy there exists considerable potential for thermal gradient energy given seawater temperature differentials in its maritime zones (Chong and Lam 2013, Sakmani et al. 2013). The best possible areas for wave energy development thus far identified are along the east coast of Peninsular Malaysia based on test data collected between 1998 to 2009, yet even here the energy yield was deemed relatively low and highly seasonally dependent (Chiang et al. 2003, Muzathik et al. 2010).

The Philippines government had funded thermal gradient survey work in the 1980s in 14 different offshore areas in the country's northern islands, and low MW rated conceptual designs were proposed but no further work followed (Uehara et al. 1988). It is only one of two Southeast Asia countries to set a target on ocean energy development, which is to establish 70.5 MW of installed capacity by 2030 under the National Renewable Energy Programme 2011–2030. However, there has been no notable evidence of any substantive progress being made toward realising this goal. Thailand is the only other nation from the sub-region to set a target for ocean energy, but this is extremely modest at 2 MW by 2021, and no discernible sector development was evident by the mid-2010s. Finally, Singapore conducted tests on tidal range and current energy technologies in the 1990s but did not proceed to beyond the demonstration phase due to low-level commercial prospects and what were deemed prohibitive costs of further technology development. One installation site option was inside a causeway road link construction between Singapore and Malaysia (McGregor and DeSouza 1997). The Energy Research Institute (ERI) at Nanyang Technological University in

Singapore is the city-state's principal centre for ocean energy research, and which by the early 2010s had installed a test wave energy device off Sentosa Island to the south, as well as developing three different low-flow tidal current turbines and other aspects of advanced turbine design.[7] In December 2013, ERI became the latest East Asian research institute to sign an ocean energy collaboration agreement with the EMEC. In sum, the ocean energy sector offers rather limited prospects for clean energy generation in Southeast Asia due to a combination of significant natural resource, technical, and financial constraints, which explains why it has not been afforded much priority in government renewable energy policies and strategies in the sub-region.

8.8. Conclusion

Ocean energy is a dynamic, new-frontier renewables sector with enormous potential for future development that continues to make significant techno-innovatory advances. However, it is likely to remain a micro-sector for some years yet due to a combination of natural resource, technical, and financial constraints. Many ocean energy applications (e.g. tidal range barrages) involve huge initial capital costs of installation. Most of the sector's technologies are still at the experimental and demonstration stages. State support on further R&D activity, including pushing towards technology convergence and scaling up operations to achieve greater production efficiencies, is crucial to the sector's future development. Furthermore, many of the sector's resources can be scarce (e.g. tidal range), and many parts of East Asia simply do not possess them—or if they do, the commercial prospects of exploiting them are limited. This has especially applied to a number of Southeast Asian nations and Taiwan, and here too extreme weather patterns pose a risk to certain offshore installations.

Ocean energy's future looks far more promising in South Korea, China, and Japan. All three Northeast Asian countries have over the years sought to become technology leaders in this field. Japan was the world's first pioneer of wave energy technology in the 1940s, and China began to develop experimental tidal range devices in the 1950s, since then progressing to install a wide range of grid-connected ocean energy plants. Sometime later, South Korea emerged as a global player in the tidal range sub-sector, whereas Japan has yet to convert its long-standing technology-oriented approach to ocean energy development into substantiated installed capacity. Similarities can be drawn here with the country's solar photovoltaics sector (Chapter 5). The Japanese government's partial retreat from industrial policy in the early 2000s (Chapter 2) led to curbed state support of its wave and other ocean energy programmes. However, revitalised state capacity, evident towards the end of the decade, created new impetus for the sector. Japan's policy-makers, leading companies in the industry, and research institutions renewed their collaborative efforts within the framework of a

2011–2015 national research programme–connected with the government's New Growth Strategy and Industrial Structure Vision initiative–that focused on multiple technological fronts and was based on realising various strategic development targets.

Meanwhile, the Chinese government was steadily supporting ocean energy's development through its 863 new technology programme from the mid-1980s onwards, in which state agencies worked closely with industry. For some time, though, the pace of techno-innovatory advancement achieved was slow and gradual. Like Japan, it was not until the late 2000s that the Chinese government showed more strategic intent towards realising the country's significant ocean energy potential, setting ambitious development targets for 2015 and 2020 within the 12th FYP and Medium to Long-Term Development Plan for Renewable Energy.

Up to now, China clearly has had the opportunity to dominate ocean energy development in East Asia but chose not to exploit it, where instead South Korea has taken the lead primarily through its tidal range barrage programme. China could have, for example, applied its considerable resources and expertise from the hydropower sector to develop large-scale tidal range installations that exceeded South Korea's own ambitious plans here. Nonetheless, China could still make important contributions to future global efforts on commercialising wave and tidal energy technologies through substantial investments in new R&D and mass production of ocean energy equipment.

South Korea's sector development programmes, first initiated in the mid-1970s came to fruition also by the late 2000s, when the government's GGS provided a more substantive strategy-making basis for further sectoral development. The GGS's new National Energy Plan 2008–2030 introduced a multi-phase ocean energy strategy and roadmap with ambitious techno-innovatory and installation objectives set over the two-decade plan period. At the strategy's core is the world's most ambitious tidal range development programme. The 254 MW Sihwa Lake barrage is already operational. If the three other even more powerful barrages at Garorim Bay (520 MW), Incheon Bay (1,320 MW), and Ganghwa (420 MW or 840 MW) are completed on schedule, then South Korea will become the global leader on ocean energy power generation by some considerable distance. There is, nevertheless, the chance that opposition from environmental NGOs and the general public may halt or delay the country from realising this position.

Overall, we can see that ecological modernisation influenced exercises of state capacity in Japan, China, and South Korea from the late 2000s onwards, spurred ocean energy development in Northeast Asia, and assigned the sector as a 'new frontier' energy source and potentially important future strategic industry. At the same time, it is likely to make only limited contributions to delivering cleaner energy and lower carbon development in East Asia, and globally into the foreseeable future, due to its relatively small size.

Notes

1 At this time, these came from the Ministry of Land, Transport and Maritime Affairs, and Ministry of Knowledge Economy under their respective plans, Development of Activity Plan on Ocean Energy R&D Programme and R&D Strategy 2030 for New and Renewable Energy–Ocean, both launched in 2009.
2 Research interview, Korea University, January 14, 2014, Seoul.
3 The Ministry of Knowledge Economy was superseded by the newly formed the Ministry of Trade, Industry and Energy in 2013 (Chapter 3).
4 These include the Korea Institute of Ocean Science and Technology, the Korea Ocean Research Development Institute, the Korea Institute of Marine Science and Technology Promotion, Korea Electric Power Research Institute, the Korea Marine Equipment Research Institute, and the Korea Advanced Institute of Technology.
5 http://www.renewableenergyworld.com/rea/news/article/2011/02/country-profile-taiwan (accessed January 14, 2014).
6 http://www.emec.org.uk/press-release-emecs-4th-collaboration-agreement-in-asia (accessed January 14, 2014).
7 Research interview with Energy Research Institute (ERI) at Nanyang Technological University representatives, January 2013, Singapore.

9   Bioenergy

9.1. Introduction

Bioenergy is a very widely used renewable energy (RE) in East Asia and globally, and it is one of the most complex and controversial. It is unique in the renewables spectrum for its dependence on and production of tangible energy fuels. Various biomatter feedstocks are now deployed to generate outputs for power, transportation, and heating purposes. No other RE sector is so comprehensive in its energy applications. Bioenergy is the world's fourth largest renewables power generator, while the biofuels sub-sector is making important contributions to greening global transportation. Furthermore, biogas technology is helping to reduce dependency on natural gas for heating and cooking and improve energy efficiency by replacing traditional biomass for the same purposes, particularly in rural areas and low-income households.

Most parts of East Asia have significant bioenergy potential in at least one or two sub-sectors. This especially applies to the region's developing economies, while its developed economies have demonstrated strong techno-innovatory capabilities in various bioenergy fields. Notwithstanding the complexities and adverse impacts that can arise from bioenergy development on other policy domains, such as agriculture, rural development, and food security, East Asian governments have over time afforded greater priority to the sector when formulating their renewable energy policies and strategies. However, serious challenges lie ahead regarding the future scaling-up of bioenergy in the region.

9.2. Bioenergy: an overview

9.2.1. Bioenergy explained

Bioenergy concerns unlocking the latent chemical energy found in biological matter. This can take the form of various renewable biomass feedstocks, which can be generally categorised into natural or agricultural oil crops (e.g. rape seed, sunflower, soya, animal fats), lignocellulosic material (e.g. wood, straw), sugar and starch crops, biodegradable municipal solid waste (MSW; e.g. food wastes, sewerage sludge, manure), and photosynthetic micro-organisms (e.g. micro-algae, bacteria; Intergovernmental Panel on Climate Change 2011). Utilising bioenergy fuels normally involves a combustion process. However, these fuels take carbon out of the atmosphere while growing and then release this carbon back again when burned: hence, there is no net addition to carbon emissions, unlike the combustion of fossil fuels.

Bioenergy can take many solid, liquid, and gaseous forms. It comprises a cluster of conversion technology sub-sectors, which may be broadly categorised as follows:

- *Biomass power and heat generation*: Biomass feedstocks are combusted in combined heat and power (CHP) stations, primarily to generate electricity. The simultaneous capturing of heat energy outputs significantly improves energy efficiency ratings (75–90 percent achievable). This enables such co-generation plants to compete effectively against conventional forms of energy in power generation. Industry and buildings are the main recipients of this energy, which is often produced by onsite plants working outside national grid systems. In simple power generation plants, biomass feedstock (e.g. wood pellets) can be either direct-fired as the sole energy input source (up to 40 percent energy efficiency ratings) or co-fired, where it is mixed in certain proportions (currently up to around 20 percent of the total) with coal in conventional power stations. Biomass solids account for around four-fifths of the feedstock type globally, with MSW (which may be considered as recycled energy), biogases, and bioliquids making up the remainder.
- *Biofuels (liquid)*: Biofuels are used for transportation purposes. These primarily concern biodiesel and bioethanol blended with conventional fuels (gasoline and diesel) as the minority element of the mix (typically 5–15 percent). More specifically, bioethanol (or fuel ethanol) is derived from fermenting sugar components of certain plants, such as corn, grasses, and trees. Biodiesel is derived from material based on vegetable oils, recycled grease, and animal fats.
- *Biogas*: Biogas involves various thermochemical gasification processes that transform biological matter (e.g. at landfill sites by using anaerobic digestion technology) into fuel gases used for power generation, transportation, or heating purposes. For example, biomethane can be with co-fired with natural gas in co-generation plants, in a similar manner as biomass power.
- *Traditional biomass*: Biomatter can be combusted for domestic cooking and heat purposes, which is relevant to the livelihoods of billions of people in low-income nations. In terms of primary energy consumption, this sub-sector is the main contributor to the bioenergy sector overall and the fourth largest source of primary energy supply (around 10 percent of the world total) after the three fossil fuel sectors.

Notable challenges persist in bioenergy feedstock processing and production. For instance, MSW material is very heterogeneous and often highly contaminated, requiring the use of relatively costly control technologies (IEA 2012j). Generally speaking, MSW plants are particularly common in urbanised high-income economies, such as Japan, South Korea, Taiwan, and Singapore, because these produce higher levels of disposable waste and moreover lack agricultural land to grow biomass energy crops. The production of these crops, even in countries with large agricultural zones, can be a problem where acute land-use competition arises. This is especially relevant to East Asia's burgeoning middle-class societies with fast-growing food demand levels. However, farmers in these countries are often given strong financial incentives from government policies to grow biomass crops instead of food crops. Expanded biomass crop production during the mid-2000s was reportedly blamed for food price hikes worldwide (Brown et al. 2011). Here, renewable energy policy clearly intersected with another policy domain and—where policy choices on reconciling energy security and food security interests must be considered—caused complexity in development strategy-making.

9.2.2. Recent growth trends and developments

National and global bioenergy data gathering can often prove difficult compared to other RE sectors. From the verifiable information available, we can nevertheless discern that bioenergy sub-sector growth has ranged between the relatively slow and the reasonably fast, and that uneven growth patterns are apparent over time. For example, Table 9.1 shows that biomass power generation globally did not expand significantly during the mid-2000s; however, since the late 2000s, it has grown at annual average rates of between 12 and 14 percent—about half the rate achieved by wind energy. By 2013, total world capacity stood at 88 GW.

In recent years, the European Union has emerged as the key global player in the sub-sector, overtaking the United States, with Europe's share of the global total capacity more than doubling from 18.2 percent in 2005 to 39.2 percent in 2013. Meanwhile, East Asia's relative position has remained relatively stable over this period, at between roughly 15–17 percent, and is still behind the US share of 18.0 percent (Table 9.1). On a national basis, China and Japan are globally ranked fourth and fifth, respectfully. During the same period, biomass heat output has increased from 220 GW_{th} to 296 GW_{th}, with the European Union as the leading consumer worldwide (REN21 2006, 2014).

Liquid biofuels accounted for 2.5 percent of global transport fuels in 2012—more specifically, 3.4 percent of road transport fuels and a very small but fast-growing share of aviation fuels. Estimates from the IEA (2011d) suggest that by 2050, biofuels could provide up to 27 percent of total transport fuel and achieve around 2.1 gigatonnes of carbon emission savings annually. Table 9.2 reveals that year-over-year output levels across countries

Table 9.1 Biomass power generation, East Asia and global, 2004–2013

	Installed capacity (GW)									
	2004	**2005**	**2006**	**2007**	**2008**	**2009**	**2010**	**2011**	**2012**	**2013**
China	0.0	2.0	2.5	3.0	3.3	4.6	5.5	7.0	8.0	8.5
Japan	3.0	3.0	3.0	3.0	3.0	3.0	3.3	3.3	3.3	3.4
Thailand	0.0	0.6	0.7	0.7	0.8	0.8	0.8	0.8	0.8	0.8
Taiwan	0.6	0.7	0.7	0.7	0.7	0.7	0.7	0.7	0.7	0.7
South Korea	0.0	0.1	0.1	0.1	0.1	0.1	0.2	0.2	0.3	0.3
East Asia	**3.6**	**6.4**	**7.0**	**7.5**	**7.9**	**9.2**	**10.5**	**12.0**	**13.1**	**13.7**
% share of world total	*9.2*	*15.4*	*15.6*	*14.9*	*15.2*	*17.0*	*16.9*	*16.7*	*15.8*	*15.6*
European Union	n/a	8.0	10.0	n/a	15.0	16.0	20.0	26.0	31.0	34.5
% share of world total	*–*	*18.2*	*22.2*	*–*	*28.8*	*26.9*	*32.3*	*36.1*	*37.4*	*39.2*
Brazil	n/a	6.1	6.3	6.4	7.0	6.1	7.8	10.9	10.8	11.4
% share of world total	*–*	*13.9*	*14.0*	*12.7*	*13.5*	*11.3*	*12.6*	*15.1*	*13.0*	*13.0*
United States	10.9	11.0	11.0	11.6	12.0	12.1	12.3	13.5	14.3	15.8
% share of world total	*27.9*	*25.0*	*24.4*	*23.1*	*23.1*	*22.4*	*19.8*	*18.8*	*17.2*	*18.0*
World	**39.0**	**44.0**	**45.0**	**50.2**	**52.0**	**54.0**	**62.0**	**72.0**	**83.0**	**88.0**

Sources: EIA (2014), REN21 (Global Status Reports from 2005 to 2014).

can fluctuate quite significantly, but the global trend has been steadily upward. World bioethanol production has almost tripled, from 510,900 barrels per day on average in 2004 to 1,493,500 barrels; it has been duopolised for some time by the United States and Brazil, who together have accounted for at least 85 percent of the total since 2004 (Table 9.2). East Asia's share of the global total has remained low, at around 3–4 percent. China is by far the region's largest bioethanol producer (39,000 barrels per day) and is ranked third globally, followed by Thailand (8,900 barrels), which is globally ranked seventh.

Biodiesel production has increased nine-fold worldwide from 2004 to 2011–three times faster than bioethanol–but it remains the much smaller sub-sector. Here, the European Union has historically dominated, although as Table 9.2 shows, its global share has more than halved in recent years, to 44.1 percent by 2011. East Asia has a more notable presence in this biofuels industry, with its share rising gradually, but only to 12.4 percent (50,000 barrels per day) by 2011. Indonesia is the region's highest globally ranked producer (sixth, with 20,000 barrels) and Thailand ranked seventh with 10,200 barrels. The United States, Argentina, and Brazil are the other major biodiesel producers, with the Latin American countries being

Table 9.2 Biofuels development, East Asia and global, 2004–2012

	2004	2005	2006	2007	2008	2009	2010	2011
Bioethanol production						Thousands barrels per day		
China	17.2	20.7	28.0	27.8	24.4	37.0	37.0	39.0
Thailand	0.1	1.2	2.3	3.0	5.7	6.9	7.5	8.9
Japan	0.0	0.0	0.0	0.0	0.0	0.0	1.0	1.0
Vietnam	0.0	0.0	0.0	0.0	0.0	0.0	0.2	1.0
Philippines	0.0	0.0	0.0	0.0	0.0	0.5	1.0	0.5
Indonesia	0.0	0.0	0.1	0.2	0.2	0.2	0.1	0.1
East Asia	**17.3**	**21.9**	**30.4**	**31.0**	**30.3**	**44.6**	**46.8**	**50.5**
% share of world total	*3.4*	*3.7*	*4.2*	*3.4*	*2.5*	*3.4*	*3.1*	*3.4*
European Union	8.0	14.5	27.1	31.3	47.1	59.2	72.3	72.3
% share of world total	*1.6*	*2.5*	*3.8*	*3.4*	*3.9*	*4.5*	*4.7*	*4.8*
United States	221.5	254.7	318.6	425.4	605.6	713.5	867.4	908.6
% share of world total	*43.4*	*43.5*	*44.5*	*46.0*	*49.8*	*53.8*	*56.8*	*60.8*
Brazil	251.7	276.4	306.1	388.7	466.3	449.8	486.0	392.0
% share of world total	*49.2*	*47.2*	*42.8*	*42.0*	*38.4*	*33.9*	*31.8*	*26.2*
World	**510.9**	**585.0**	**715.9**	**924.5**	**1,215.2**	**1,326.3**	**1,527.6**	**1,493.5**
Biodiesel production						Thousands barrels per day		
Indonesia	0.0	0.2	0.4	1.0	2.0	6.0	8.0	20.0
Thailand	0.0	0.4	0.4	1.2	7.7	10.5	11.0	10.2
China	0.1	0.8	4.0	2.0	5.0	6.0	6.0	7.8
South Korea	0.1	0.2	0.9	1.7	3.2	5.0	6.5	6.3
Philippines	0.0	0.2	0.4	0.6	1.1	2.0	2.4	2.5
Taiwan	0.0	0.0	0.0	0.1	0.4	0.6	0.4	1.0
Malaysia	0.0	0.0	1.1	2.5	4.5	4.5	2.0	1.0
Singapore	0.0	0.2	0.3	0.7	2.0	1.0	0.5	0.5
Vietnam	0.0	0.0	0.0	0.0	0.0	0.0	0.4	0.4
Japan	0.0	0.0	0.1	0.1	0.1	0.2	0.3	0.3
East Asia	**0.2**	**2.0**	**7.6**	**9.9**	**26.0**	**35.8**	**37.5**	**50.0**
% share of world total	*0.5*	*2.8*	*6.1*	*5.5*	*9.9*	*11.6*	*11.1*	*12.4*
European Union	41.0	62.2	96.1	121.9	150.5	174.4	183.8	178.2
% share of world total	*93.6*	*87.4*	*77.1*	*68.2*	*57.4*	*56.4*	*54.4*	*44.1*
United States	1.8	5.9	16.3	32.0	44.1	33.6	22.4	63.1
% share of world total	*4.1*	*8.3*	*13.1*	*17.9*	*16.8*	*10.9*	*6.6*	*15.6*
Argentina	0.2	0.2	0.6	3.6	13.9	23.1	36.0	47.3
% share of world total	*0.5*	*0.3*	*0.5*	*2.0*	*5.3*	*7.5*	*10.7*	*11.7*
Brazil	0.0	0.0	1.2	7.0	20.1	27.7	41.1	46.1
% share of world total	*0.0*	*0.0*	*1.0*	*3.9*	*7.7*	*9.0*	*12.2*	*11.4*
World	**43.8**	**71.2**	**124.6**	**178.8**	**262.1**	**309.1**	**337.8**	**403.7**

Source: EIA (2014).

large exporters. Southeast Asia's major biodiesel producers also export in significant quantities, with Europe being an important market. However, biofuels are still mostly consumed within the country of production. Although there is currently no international comparative data on biogas, developed nations primarily use it in CHP plants and developing nations use it for household heating and cooking.

9.2.3. Techno-innovation, production, and investment

All mainstream bioenergy sub-sector technologies have attained reasonably well-established levels of commercialisation. Current frontier areas of new techno-innovation are focused on advancements in gasifier and energy conversion techniques. By the early 2010s, biomethanol had entered the early commercial or cost-reduction stage of development, while biogas-reforming hydrogen-based technologies had made good progress at the demonstration stage (IEA 2012j). Other key areas of techno-innovation include biosynthetic gases, hydrotreated vegetable oils, non-food biomass materials and feedstocks (e.g. jatropha, very relevant in East Asia's case), 'fungible fuels' that can be blended in any proportion with fossil fuel energy sources, and scaled-up power generation plants (Brown et al. 2011).

However, bioenergy has failed to attract anywhere near the same levels of investment as wind and solar energy. In 2013, only US$8 billion of investment globally was directed at biomass power generation and just US$4.9 billion went toward biofuels (see Table 1.3, p. 16). Indeed, investment in the sector has recently fallen in many countries due to lower profit margins, commodity price spikes, intensifying competition for biomass feedstocks, drought-induced reductions in crop productivity, concern over competing land and other resource use (e.g. water), sustainability of production concerns generally, and policy uncertainty. Many bioenergy sub-sectors now compete very favourably against other technologies in the market worldwide, but more substantial investments will be required if bioenergy as a whole is to establish itself as a truly dynamic renewables sector. By September 2013, bioenergy had attracted the third highest number of Clean Development Mechanism projects, although East Asia's representation (287 out of a global 848) was just 33.8 percent of the total, its lowest for any RE sector (Box 1.2, p. 32).

The bioenergy industry comprises a broad range of business actors, including power generation plant constructors and operators; bioenergy plant equipment manufacturers; biomass feedstock processors and suppliers; biofuel and biogas processors and distributors; and producers of specialist biomass harvesting, handling, and storage equipment (REN21 2013). Direct-fired and CHP biomass power plants tend to be relatively small by sectoral comparison, with the larger ones rated at 50 MW or above and the vast majority rated less than 100 MW of installed capacity. However, there is a trend now toward building significantly larger plants (between 300 MW and 800 MW), and many are in development in Europe (especially in Britain),

East Asia, and North America.[1] These installations will depend on many million tonnes of feedstock annually, presenting certain supply security and storage challenges. Because biomass feedstocks tend to be sourced locally or at least nationally, countries are not so dependent on foreign suppliers compared to conventional fossil fuel sectors; thus, like renewables, biomass generally is viewed as an indigenous energy source. However, as we later discuss, East Asia's resource-poor developed economies of Japan, South Korea, and Taiwan will have to import significant levels of feedstocks to scale up their bioenergy sectors. Meanwhile, the micro-scale biomass heating industry–which includes domestic wood burners and biogas stoves (<100 kW), pellet burners (<500 kW), and small boilers (<1 MW)–has continued to grow strongly over recent years, helping to expand bioenergy's distributed or prosumer base (Chapter 1).

Material input costs in biofuel production are typically between 45 and 70 percent. Bioethanol and biodiesel can only compete directly with conventional gasoline and diesel in countries such as Brazil, where materials and hence production costs are low. In most other countries, the biofuels market is policy-driven (Brown et al. 2011). Many East Asian governments strongly supported the sub-sector's development. Moreover, in China and most Southeast Asian nations, it is additionally conceived as a socio-economic policy of safeguarding farmer incomes. Governments in East Asia and elsewhere also use regulatory mandates on fuel blending and fuel duty rebates to stimulate demand for biofuels and help meet policy targets.

A potentially important emerging biomass feedstock sub-sector is wood pellets, which are produced from compacted wood waste materials that include sawdust, shavings, and chips, which are byproducts from certain timber-based industries. Most pellet production supplies small biomass heat plants; the rest is used for power generation. Although a fast-growing industry, it still only accounts for 1–2 percent of total solid biomass demand globally. In 2012, global wood pellet production was 22.0 million tonnes. Over a third of this (8.2 million tonnes) is internationally traded, with the largest share (40 percent) being exported from the United States to Europe. The monetary value of this production and international trade is still currently low. However, in the near future, the sub-sector could become significant worldwide and perhaps in many parts of East Asia–a region with an abundance of waste wood materials and a fast-growing market for biomass power and heat generation.

9.2.4. Obstacles to the expansion of bioenergy

Bioenergy is a relatively contentious renewables sector; like hydropower, most obstacles to future development are of an environmental or socioeconomic character. When it was beginning to substantially develop in the 1990s, advocates argued that bioenergy fuel combustion was carbon-neutral. However, recent scientific studies have indicated that this process can make net contributions to other greenhouse gas emissions and climate change, although significantly less than fossil fuel combustion. Concern over the impact of expanding bioenergy crop production on food security was noted earlier. Currently, a tenth of the world's corn production is currently used as feedstock for the US bioethanol industry. In Brazil, more than half of the country's sugar cane crop is used for the same purpose, and the European Union's biodiesel producers account for more than two-thirds of home-grown rapeseed oil consumption (Koizumi 2013).

There are other obstacles of more technical and market-related matters. For liquid biofuels, arguably the most important technical challenge concerns increasing the blending proportions by which bioethanol and biodiesel can be mixed with conventional fossil fuels in mainstream combustion engine vehicles. Promoting investment in flex-fuel automotive technology (i.e. engines running on a wider range of blend mixes) is core to this particular challenge. The sensitivity of biofuel production costs to feedstock supply prices is likely to remain an important market-related challenge, especially if commodity prices are prone to significant and unpredictable fluctuations (Brown et al. 2011).

9.3. China

Bio-sustainability has become increasingly important to China. The country possesses 9 percent of the world's arable land, 7 percent of its freshwater, and 3 percent of its forests, from which it must sustain 20 percent of the global population. Its per-capita area of arable land is just 0.1 hectare, about a quarter of the world average (Sang and Zhu 2011). However, China possesses substantial bioenergy resources, and the government has devised a comprehensive strategy on the future development of this RE sector. Its endeavours on advancing modern bioenergy technology date back some decades. Household-scale biogas digesters were first introduced in the 1950s, and China has remained a world leader in this field. From the 1960s onwards, government-sponsored research began on utility-scale biomass power generation (Wu et al. 2010). The 6th Five-Year Plan (FYP) was the first to take a more systemic approach to bioenergy development, with the 863 State High-Tech Development programme focusing on the sector's techno-innovation (Chapter 3). MSW incineration technology was first introduced in the late 1980s, and incinerator plants have since proliferated nationwide. In the same decade, the government initiated new research and development (R&D) programmes on biofuels and their derivative energy crops, laying a foundation for the sub-sector's industrial production and commercialisation by the 1990s. Subsequent FYPs afforded increasing resources to bioenergy development, although not to the same level as hydropower, wind, and solar energy (Hu and Phillips 2011, Zhang et al. 2010).

During the 2000s, most bioenergy sub-sectors in China experienced significant growth (Table 9.1 and Table 3.3, p. 82), supported by more substantive and regular policy initiatives. This included various direct financial support measures (investment subsidies, price subsidies, public-funded R&D), market-based instruments (tax incentives and feed-in tariffs) and regulatory mandates, such as biofuel blending requirements (IEA 2014). Bioenergy development also benefitted from the government's deepening commitment to strategising on renewables generally. No other country has as comprehensive a set of targets, with goals set for more or less every bioenergy sub-sector. The 12th FYP comprises a series of 2015 objectives for biomass power generation (and within this, specific feedstock source targets for agricultural and forest residuals, biogas, and MSW), households using biogas, biogas treatment plants, solid biomass production, and bioethanol and biodiesel output. Additionally, the 2007 Medium to Long-Term Plan on Renewable Energy established ambitious 2020 goals across the whole bioenergy spectrum (Table 3.3, p. 82). Sub-sectoral strategies were created under the 12th FYP to help realise these targets, such as the National Development Plan of Forest-based Bioenergy (2011–2020), which included a programme to expand China's forests by an additional 19 million hectares for feedstock supply purposes; establish a coordinated production structure, new tax incentives, and subsidies; and provide extra R&D support on advanced energy conversion technologies regarding forestry-based biomass power and biofuel production applications (Kahrl et al. 2013, Yang et al. 2013). Notwithstanding the Chinese government's ambitious plans, growing policy support, and stronger strategic priority afforded to bioenergy development over recent years, there remain significant challenges in developing every sub-sector, as discussed here.

Biomass power generation in China has expanded steadily since 2005, rising from an installed capacity level of 2 GW to 8.5 GW by 2013 (Table 9.1) based on a combination of substantive investments, mostly from state-owned power companies but also from private companies and foreign multinationals. Agriculture (mainly straw) and animal residues still account for around three-quarters of the feedstock source in biomass power plants; the next most important, in order, are forestry residues, biogas, and MSW, which account for a small but growing share (Zhao et al. 2012). However, due to feedstock supply bottlenecks, most of China's biomass power plants operate only 30–40 percent of the time. Even with state subsidy support, only a third of plants make a profit. Furthermore, the country is still far from realising its forestry-based biomass potential, with only two power plants using woody matter as their primary feedstock source (Yang et al. 2013). According to Zhao et al. (2012), by the early 2010s the sub-sector's market remained generally immature, and notable legislative and regulatory gaps still existed in the industry. Another challenge concerns the expansion of China's MSW-based power generation capacity. The country's MSW levels have increased seven-fold since 1980 and by 2010 accounted for 29 percent of global waste levels (Zhang et al. 2010). Landfill sites are filling rapidly and the composition of China's waste is becoming less organic and more toxic.

Policy-makers are under pressure to simultaneously expand the number of incinerator plants and introduce the best emission-mitigation technologies (Zhuang et al. 2010).

Biofuels have demonstrated strong growth in China since the early 2000s, although production was cut back in 2007 after the expansion of biocrops caused food shortages at home and abroad. Food security concerns remain a significant policy issue for the Chinese government (Huang et al. 2012, Koizumi 2013). By 2011, national bioethanol production was more than double the 2004 level, while growth rates in the much smaller biodiesel sub-sector were even higher (Table 9.2). As transportation-related greenhouse gas emissions rise rapidly in China, government and business will come under increasing pressure to invest more heavily in biofuels; however, this must be reconciled with growing domestic food demands, especially for high-protein foodstuffs (e.g. meat) that are both land- and resource-intensive. At present, biodiesel is used mainly for industry power generation rather than vehicle fuel in China, and moreover it depends mainly on used cooking oil feedstock—most of which it must already import. Meanwhile, the country's bioethanol import levels are expected to rise substantially in the future due to competition with food agriculture (Chang et al. 2012). Like for other countries, developing and using non-food 'second-generation' feedstock technologies (e.g. forestry-based) offers perhaps the best way forward for biofuels, and it could become a future strategic priority for the Chinese government. The country's two largest national oil companies—Sinopec and China National Petroleum Corporation—were reported to be already making substantial investments in second-generation biofuels by the early 2010s (Qiu et al. 2012, Solidiance 2013).

Biogas consumption in China is highly distributed across mainly rural areas. Many millions of households operate their own biogas digesters, and government initiatives for establishing utility-scale plants are underway to better capture scale economies; however, this development would require considerable investment in new transmission infrastructure. Moreover, only 19 percent of rural households had digesters installed by the early 2010s, indicating there is much scope for expanding distributed generation (Yang et al. 2013). More generally, China still lags behind many other leading RE producer nations in bioenergy technology development. Hence, in addition to promoting greater international cooperation, the government's strategic planners may look to strengthen indigenous techno-innovatory capacity in this sector to better exploit the national bioenergy resource potential.

9.4. Japan

It took until the late 1990s before the Japanese government looked seriously at developing its bioenergy sector. The country faced predicaments that were similar to its developed Northeast Asian economy counterparts of South Korea and Taiwan: all three have limited space to grow bioenergy

crops due to being mountainous territories with high-level population densities. They are also high-income, material-intensive societies that generate
substantial amounts of industrial and domestic biowaste.

The first goals on Japan's bioenergy development were set under the 1999
New Renewable Energy Target, when the government set the goal of
increasing installed power generation to 4,500 MW by 2010, specifically
4,170 MW from MSW and 330 MW from non-waste biomass. This target
was 50 percent higher than what was set for wind energy and just below that
for solar PV, but none were actually realised by the decade's end. The
government's inaugural Biomass Nippon Strategy (BNS), approved in
December 2002, was based on four main objectives: prevention of global
warming; creation of a recycling-oriented society; fostering new competitive
strategic industries; and activation of agriculture, forestry, and fishery, as
well as rural development. Two new institutions–the Biomass Information
Headquarters and Biomass Japan Comprehensive Strategy Promotion
Council–were created to manage the BNS and design the necessary legislation and regulations to incentivise business development and new techno-
innovation (Kuzuhara 2005, Matsumura and Yokoyama 2005). Soon after,
the 'Green Power' renewable portfolio standards (RPS) scheme was
launched in April 2003, requiring Japan's power companies to source
increasing levels of wood-based biomass in co-fired power plants, providing
significant impetus to bioenergy supplying firms (Goto 2011).

No further new significant policy initiatives on bioenergy arose until the
late 2000s, when the Japanese government embarked on its second-phase
RE strategy-making (Chapter 3). The 2008 Biofuel Technology Innovation
Plan–part of the Cool Earth 50 Energy Innovative Technology Plan–aimed
to advance efficiency in bioethanol processing from lignocellulosic feedstocks, thus addressing food security predicaments. An accompanying
Biofuel Law introduced the same year provided tax incentives on both
consumption and production. Launched in the context of the New Growth
Strategy, the 2010 Basic Plan on Biomass was a revision of the BNS with
various targets set on bioenergy utilisation (e.g. MSW, forestry) residues by
2020. The upgraded 2012 feed-in tariff (FiT) scheme provided new incentivisation for higher levels of biomass power generation, although Japan's
installed capacity of has only risen from 3.0 GW in 2005 to 3.4 GW by 2013
(Table 9.1).

Most electricity and heat produced from biowaste in Japan is derived from
MSW feedstocks–by the early 2010s, around 90 percent of it (Institute for
Sustainable Energy Policies 2013). There are now hundreds of waste-to-
energy incinerators spread around the country and around three-quarters of
its MSW is incinerated (Sasaki et al. 2011). Over time, the government has
promoted the greater use of forest residue feedstocks, mainly from brush-
wood and tree thinning material, yet this has proved difficult. Around two-
thirds of Japanese territory is forested, but the nation's forestry industry fell
into decline after the government pursued an imported timber policy from

the early 1960s onwards. Mountainous terrain conditions, low timber prices, and a highly distributed pattern of forestry ownership have hitherto posed challenges for promoting Japan's wood residue harvesting business (Minami and Saka 2005, Kinoshita et al. 2010). Most of the country's consumption of wood pellets for biomass power generation is sourced from overseas, approximately 80 percent from Canada (Goto 2011).

Japan's levels of biofuel production have remained surprisingly low given the international stature of its economy and automotive industry. The country only processes an average of thousand barrels of bioethanol daily and is East Asia's lowest ranked biodiesel producer, with just 300 barrels daily in 2011 (Table 9.2). Government measures to promote both fuels started in 2003 under the BNS, followed up by support from the aforementioned 2008 Biofuel Law and Biofuel Technology Innovation Plan. In addition, 2007 and 2010 legislation on blending standards helped further consolidate the industry (Goto 2011). As with biomass power generation, Japan is severely resource constrained concerning indigenous supplies of biofuel feedstocks. Switching relatively scarce farm-land use from food production to growing energy crops is an issue mostly pertinent to bioethanol, and it raises important questions of food security in Japan and its Northeast Asian counterparts of South Korea and Taiwan. All three have comparatively small, inefficient agricultural sectors and have to import between 60 and 70 percent of their foodstuffs. Raising demand for imported energy crops from lower income countries whose average nourishment levels may be already low, such as lesser developed Southeast Asian countries, may exacerbate their own existing food security problems. However, by the early 2010s, approximately 97 percent of Japan's bioethanol consumption was imported (mainly from Brazil), and only very small amounts of energy crops for domestic production were sourced overseas (Koizumi 2013). As in other RE sectors, Japan is looking to address these issues and challenges through advances in techno-innovation (New Energy and Industrial Technology Development Organization 2013).

9.5. South Korea

Over time, the South Korean government's priorities on bioenergy development have settled into the following order: using organic wastes (mainly MSW), liquid biofuels production and consumption, and lastly biomass production from forest residues (Lee et al. 2011). South Korea has limited agricultural land space to grow substantial quantities of bioenergy crops. It was also a relative latestarter in promoting the sector generally; the first substantive state support measures were introduced in the 2003 Second Basic Plan for New and Renewable Energy, mainly focused on R&D activity and new technology dissemination. After financial incentives on power generation and biofuel production were launched in the mid-2000s, the Green Growth Strategy and associated 2008 National Energy Plan gave significant impetus to the sector's development overall by establishing a comprehensive policy framework of support and setting future output targets (2,628 GWh for biomass power, 161 GWh for biogas) to be realised by 2030 (Ryu 2010). This included sub-sectoral strategies, such as the Waste to Energy Strategic Plan, which aimed to increase energy recovery and utilisation for all available non-recyclable wastes from 32 percent in 2008 to 57 percent by 2012, and to 100 percent by 2020. New MSW-based power generation complexes founded on inter-province cooperation on clustering waste management facilities were integral to this plan. A FiT scheme provided generalised support for bioenergy development over 2008 to 2011, later superseded by a 2012 RPS scheme mandating power companies to increasingly cogenerate from biomass feedstocks. An anticipated Renewable Fuel Standards scheme for biodiesel will oblige transportation fuel producers to do the same from 2015 onwards (IEA 2013c).

Biowaste from MSW feedstock forms the predominant basis of biomass power and heat generation in South Korea. Most of the industrial economy's waste derives from non-combustible construction and demolition building materials, and usable biomatter constituted just over 20 percent of this total by the early 2010s.[2] By this time, more than 50 incinerators of 50–900 MSW tonnes daily capacity were operating in the country and well over 100 smaller plants (<50 tonnes per day capacity) were installed, mainly in rural areas (Ryu 2010, Ryu and Shin 2013). However, landfill biogas was South Korea's first commercialised bioenergy sub-sector to emerge, based on the anaerobic digestion of food wastes (Lee et al. 2011). Early test plants using this technology were established in 1996 as a joint venture between Halla Engineering and Heavy Industries and the Korea Institute of Energy Research public think-tank (Park et al. 2011).

As in Japan, forestry and other wood residues are the main source of non-waste biomass feedstocks used in South Korea for power and heat generation. Tree replanting policies undertaken from the early 1970s boosted the forestation rate to around two-thirds of national territory; however, like its Northeast Asian neighbour, this is mainly concentrated in mountainous zones, which makes residue harvesting costly. Nevertheless, state support for sustainable forest management and forestry biomass has helped to develop an expanding domestic wood-chip and wood-pellet industry in South Korea since the mid-2000s (Lee et al. 2011, Kraxner et al. 2014). REN21 (2013) reported that there were eight new wood-pellet plants under construction by the country's power companies to help them meet the government's RPS-mandated targets. In addition, they were importing growing quantities of pellets, with around 90 percent of these from Vietnam, China, and Malaysia. Meanwhile, Korean firm Eco Frontier was the lead developer of the world's largest biomass power plant in Teesside, northern England, which commenced operations in 2012.

South Korea also has a fast growing biofuels sector, especially the biodiesel industry, where the country is the fourth-ranked producer in East Asia, with production levels rising especially quickly from the late 2000s onwards (Table 9.2). State support for transportation biofuels was introduced in 2006 after small test demonstration projects in Seoul and Chonbuk province launched in 2002 proved successful. The scaling up of bioethanol production is more problematic due to South Korea's dependence on agricultural energy crop feedstocks. The country already has to import more than 80 percent of its feedstock requirements (vegetable oil) for biodiesel production. Alternative indigenous feedstock options, such as rapeseed and coastal algae, are being investigated by Korean public research institutes and companies as next-generation biofuel technologies (Lee et al. 2011). At present, the country only possesses between 23 and 30 percent of indigenous bioenergy resource potential to meet the government's ambitious 2030 targets for the sector (IEA 2013c). Given its limited available land for energy crop production, this will make South Korea ever more dependent on imported feedstocks.

9.6. Taiwan

Although it has relatively high levels of co-generation output, direct-fired biomass power generation in Taiwan is almost wholly based on MSW and biogas feedstocks. Thus, like South Korea, it relies on own biowaste rather than imported energy crops for this sub-sector. Installed capacity was 740.4 MW in 2013, and only small additions to new capacity have been added over recent years (Table 9.1): in 2006, the level stood at 726.3 MW. Municipal solid waste has accounted for around 85 percent of the total since the early 2000s, and landfill biogas was by far the greatest part of the remainder. The Taiwanese government commenced its MSW incinerator construction programme in the early 1990s, with the first built in 1991 and over 20 more installed over the next two decades. A more systemic approach to promoting bioenergy development emerged by the late 1990s. In 1999, the government's Waste Energy Application Technology Development provided subsidies and grants for R&D activity on landfill gas, gasification, liquefaction, and refuse-derived fuel conversion technologies. The 2003 New Renewable Energy Development Targets scheme set a biomass power goal of 1.35 GW by 2020, and the 2009 Renewable Energy Development Act gave new financial support to the development of Taiwan's biofuels industry. Further boost to the sector came from the 2011 FiT scheme based on 20-year contracts for biomass power producers. Most recently, the 2013 Promotion of Biogas-based Power Generation Systems programme gave subsidies for small-scale (up to 500 kW) biogas producers.

The government's current target for biomass power generation capacity, set by the 2011 New Energy Policy, is 1.4 GW by 2030. Since the early 1990s, Taiwan has gradually increased its MSW incineration rate. By the early 2010s, it had reached annual rates of between 50 and 60 percent of total waste, while waste recycling rates had risen to around 30 percent (up from just 2–3 percent in the late 1990s), and landfill's share had diminished to a very small percentage. Taiwan is a densely populated and developed economic society that produces high levels of biowaste, hence conferring a strong tendency toward MSW power generation (Kuo et al. 2011, Tsai and Chou 2006). However, concerns over incinerator dioxins emissions led to growing public opposition to expanding MSW-based power and heat generation. Consequently, in the mid-2000s, the government was compelled to cancel nine of the 15 incinerator projects planned for construction at the time (Hsu 2006, Tsai 2007), and total installed capacity has remained more or less static since then. Development of landfill biogas facilities has proved even more inert, which means that Taiwan may need to import energy crops for direct-fired biomass power generation to realise the aforementioned 2030 target of 1.4 GW.

Meanwhile, on the biofuels front, Taiwan has focused most attention to date on biodiesel production based on domestic waste oil sources (Huang and Wu 2008). State support for this industry began to cohere in the mid-2000s. Financial incentives for biodiesel producers were first introduced in 2004, soon followed by 'green bus' schemes utilising the fuel in major cities (Su and Lee 2009). By 2012, Taiwanese companies were producing around a thousand barrels of biodiesel daily, the sixth highest level in East Asia (Table 9.2). The government has not set any future target on biofuels production, and Taiwan would have to drastically increase its energy crop imports to develop a substantial bioethanol industry given its constrained agricultural land space. The use of indigenous coastal algae may provide an alternative feedstock option, and the government's continued support for new bioenergy techno-innovation generally should help improve efficiency performance across the sector overall (Kuo et al. 2011, Liou 2010).

9.7. Southeast Asia

Southeast Asia's experience with bioenergy utilisation has been both complex and contentious. The sub-region is rich in bioenergy resources and is an important global hub for agriculture, agroprocessing, and forestry. Byproduct wastes from these industries–such as rice husks, sugarcane bagasse, oil palm residues, and woody residues–are produced in very large quantities. Traditional biomass is also still widely used across the sub-region's rural and low-income households. As Bush (2008) contended, though, Southeast Asia has had a troubled history of 'bio-revolution' technologies and agrarian change, characterised by pervasive environmental and socioeconomic calamities–the general root causes of which have been corrupt politics in hand with the commercial exploitation of natural resources. In the context of the discussions in Chapter 2 on new developmentalism, the sub-region's bioenergy sector has often suffered from weak state capacity with respect to sustainable development principles and outcomes. For example, the unregulated expansion of plantation industries (e.g. palm oil) and illegal logging has led to substantial biomass burning practices, causing chronic, regular episodes of transboundary 'haze' pollution affecting numerous Southeast Asian nations, and moreover it has marginalised the livelihoods of rural communities.

Government support on R&D activity on modern bioenergy technologies and their applications in Southeast Asia can be traced back to the 1970s, with a focus on the biofuel and biogas sub-sectors. However, this was based on limited financing for relatively few projects. It was not until the early 2000s that more substantial policy frameworks and state financial assistance on bioenergy development arose, with an initial emphasis on the biofuel industry. Thailand led the way with its National Bioethanol Programme and Gasohol Strategic Plan in 2003, which included phased production targets up to 2011, excise tax incentives, and investment support to manufactures (Kumar et al. 2013). In 2005, Malaysia rolled out its National Biofuel Policy in 2005. Soon after, the Indonesian government launched its National Biofuel Roadmap (2006–2025) based on similar strategic plans and objectives, where the state outlined its intent to work closely with private/civil sector stakeholders on increasing biofuel utilisation, setting 2025 deadline targets (e.g. biodiesel 20 percent of total diesel consumption, bioethanol 15 percent of gasoline, biofuels total 5 percent of the total national energy mix) and created Special Biofuel Zones (SBZs) for biofuel crops plantation and transformation. This was followed in 2009 by the Biofuel Decree legislation, which set new mandatory requirements in biofuel production and consumption (e.g. 15 percent for transportation by 2025) and increased the number of SBZs across Indonesia.

Meanwhile, the Philippines followed in a similar vein by legislating its 2007 Biofuels Act, which incorporated tax incentives and subsidy support to help firms realise medium-term mandatory targets on minimum utilisation levels of local-sourced biofuels, as well as created the National Biofuels Board, with institutional responsibility to oversee this policy programme. FiT schemes and further regulatory mandates subsequently followed in all four countries to provide impetus to reach earlier-set strategic goals (IEA 2014). New revised objectives on bioenergy development arose from the second-phase RE strategies of Southeast Asian governments, namely Malaysia's Renewable Energy Policy and Action Plan (to 2050), the Philippines National Renewable Energy Programme (2011–2030), Thailand's Alternative Energy Development Plan (AEDP, 2012–2021), and Vietnam's National Power Development Plan (2011–2030). These have taken this development in new directions. For example, Thailand's AEDP strategy includes tax incentive support for the manufacture and purchase of high-ratio biofuel blend eco-cars, with the country being Southeast Asia's principal automobile producer.

Thailand is also the stand-out biomass power generation producer in the sub-region. Its grid-connected installed capacity level is the third highest in East Asia but has remained on a plateau level of 800 MW for some years

(Table 9.1). Under the AEDP, the Thai government is targeting an expansion to 3,630 MW (3,790 MW if MSW is included) by 2021, and almost doubling biomass heat generation to 8,200 ktoe over the same period. Sources of comparative international statistics currently indicate that no other Southeast Asian nation has significant levels of grid-connected biomass power, although surveys of the literature reveal that most biomass-generated electricity in the sub-region originates from agro-processing mills that use their own residue waste as feedstock, usually in a co-generation mix. Sugar cane bagasse is especially important to Thailand and the Philippines here, whereas in Malaysia and Indonesia palm oil waste-to-energy is common practice (Hansen and Nygaard 2014, Kumar et al. 2013, Ng et al. 2012, Singh and Setiawan 2013, Umar et al. 2013).

Chapter 3 outlined how many Southeast Asian governments have introduced small power producer schemes that promote independent power generation capacity due to gaps in national grid infrastructures. Distributed biomass power generation is thus the norm in the sub-region. Some of its countries (e.g. Indonesia, Malaysia, Vietnam) also export biomass feedstocks, such as wood pellets, in substantial quantities, with East Asia's developed economies being key importing markets. This has become increasingly relevant to Singapore, with the city-state having a narrower range of renewable energy options due to size and resource constraints; like Japan, South Korea, and Taiwan, it uses MSW as the principal feedstock for biomass power generation. In 1979, the Singapore government introduced its Energy Recovery from Biomass in Municipal Waste scheme, which mandated the use of municipal waste for CHP plants. Over time, it has sought to utilise as much indigenous woody biomass source material as possible to complement MSW feedstock but has inevitably had to import from neighbouring Southeast Asian nations, not just for power generation purposes but also for its expanding biofuel industry (Khoo et al. 2008).

The biofuels industry is arguably Southeast Asia's most important and also controversial bioenergy sub-sector. Indonesia, Malaysia, the Philippines, and Thailand all started to invest in the industry from the 1970s onwards. Some early derivative fuel types (e.g. bagasse-based alcogas and cocodiesel in the Philippines) were tested and developed but were ditched in favour of bioethanol and biodiesel (Montefrio and Sonnenfeld 2011). Malaysia was a relatively early mass producer of palm oil-based biodiesel after the government had help laying the foundation of the industry in the early 1980s, working in close collaboration with the state-owned national oil company Petronas on establishing both production and test laboratory facilities. Table 9.2 shows that four of the top six East Asian bioethanol producers are from Southeast Asia—in order of importance being Thailand, Vietnam, the Philippines, and Indonesia. The sub-region's nations are even more dominant in biodiesel production: Indonesia and Thailand are ranked first and second, respectively, in East Asia overall, followed by the Philippines (fifth), Malaysia (seventh), Singapore (eighth), and Vietnam (ninth).

While there have been notable fluctuations in output levels, the previously discussed biofuel strategies and plans of the mid-2000s played their part in establishing a general upward trend in production. Southeast Asian governments actually intervened to curb biofuel activities when food security concerns arose regionally and globally in 2006 and 2007. Industry producers have also at times cut production due to rising feedstock prices, and this is a particular problem for countries that depend heavily on one feedstock source–for example, Malaysia and palm oil, Thailand and bagasse (Goh and Lee 2010, Jayed et al. 2009). Consequently, certain blending mandates have not been met by firms and other policy goals not realised. For example, Thailand failed to reach the 2011 targets set in the aforementioned National Bioethanol Programme and Gasohol Strategic Plan but, as Table 9.2 shows, production growth has been impressive nevertheless (Kumar et al. 2013).

However, increased biofuel production in Southeast Asia has come at a cost, not only in many cases compromising food security interests of society but also associated with the worst excesses of 'cash crop' industry development. In Myanmar, Goh and Lee (2010) reported that the government allegedly forced local communities to cultivate jatropha plants bound by strictly enforced production quotas. The programme was poorly managed, caused severe local ecological damage, and undermined the livelihoods of many rural communities. The same authors also contended that political corruption remains prevalent, where state-sponsored biofuel projects are awarded to firms without due proper public tendering processes being followed, and where companies often exploit legislative and regulatory gaps to make speculative short-term profit with no commitment made to long-term industry development. In addition, Bush (2008) observed that illegal land acquisitions and ensuing violence have become frequent occurrences in the palm oil business. These are areas where state capacity could be considerably strengthened.

Biogas is a widely used technology across Southeast Asia, but in much smaller quantities than China. Thailand was an early leader from the sub-region, starting R&D and test demonstration projects on biogas digesters in the late 1970s (Gosling 1982). In 1995, the Thai government launched a 15-year strategy for scaling up the sub-sector, which proved reasonably successful at expanding production capacity. By the early 2010s, there were more than 2,000 farm-based biogas production units, but no data were available on how many biogas digesters were installed in Thai households (Aggarangsi et al. 2013). Indonesia is the sub-region's other substantial biogas producer; as in Thailand, the priority of government policy for installing household-scale digesters is mainly in rural areas (Singh and Setiawan 2013).

Bioenergy is also an RE sector where there is considerable scope in the sub-region for improving indigenous techno-innovatory capacity. While Southeast Asian states have engaged in bioenergy R&D activity for many decades, this has frequently relied on foreign assistance, such as from the EC–ASEAN COGEN Programme (Carlos and Khang 2008, Hansen and

Nygaard 2014). More developed states, such as Malaysia, Singapore, and Thailand are making notable progress on strengthening indigenous innovation through government support programmes (e.g. Malaysia's Green Technology Financing Scheme); deepening technology partnerships between government research agencies, universities, and companies; and supporting R&D on second-generation biofuels, such as jatropha-based biodiesel (Goh and Lee 2010, Le et al. 2013, Ng et al. 2012).

To conclude, we have seen how bioenergy development in Southeast Asia intersects with many other policy domains and socio-economic objectives, perhaps more than any other RE sector in the sub-region. In addition to the issues of food security, land use, and sustainable agriculture, many governments have used bioenergy policies as a means to safeguard small-scale farmer incomes and livelihoods through price support measures for bio-crop cultivation. Reconciling this renewable energy policy with other economic, social, and environmental policies will remain an important future challenge for Southeast Asian governments. However, many have weak state capacity in technocratic and institutional terms when confronting these complexities of issue linkage and associated policy linkages (Montefrio and Sonnenfeld 2011, Prasertsan and Sajjakulnukit 2006, Singh and Setiawan 2013).

9.8. Conclusion

Bioenergy is a widely used, multiple-purpose energy source but also a complex and contentious one. East Asian governments actively promoted bioenergy development for some decades, yet it has only been since the early 2000s that a more strategic, comprehensive, and programmatic approach has been adopted. It was not long after the introduction of first-phase RE strategies around this time that governments from the region encountered various issue-linkage and policy-linkage challenges. These especially related to food security, forestation, rural livelihoods, local ecosystems, political corruption, land governance, and commercial exploitation of natural resources, as most comprehensively evident in Southeast Asia. By the late 2000s and early 2010s, second-phase RE strategies on bioenergy have endeavoured to take these challenges more effectively into account, with a generally stronger emphasis placed accordingly on advancing techno-innovation (e.g. next-generation non-food biocrops), improved policy coordination, and new institution creation.

It was shown that China has devised the most substantive and wide-ranging strategy on future bioenergy development, but some crucial bottlenecks exist regarding installation efficiency, feedstock supply, competing land use with agriculture, and expanding waste-to-energy incineration capacity. However, increased bioenergy utilisation in the country could make major contributions to mitigating dependences on fossil fuels for power generation, transportation, heating, and domestic cooking, and hence significantly reduce China's greenhouse gas emissions. Key common challenges for the region's developed economies of Japan, South Korea, Taiwan, and Singapore have meanwhile focused on high rates of MSW-based biomass power generation and biogas production, as well as growing import dependencies for other sources of biomass feedstock. All four are densely populated, high-income, and material-intensive societies with limited bioenergy resources, and therefore scaling up this RE sector would inevitably depend on raising their biomass feedstock import levels still further, with potential deleterious impacts on the food security of lower-income biomass exporting countries from Southeast Asia and elsewhere. State support in Japan, South Korea, Taiwan, and Singapore for helping to find new techno-innovatory alternatives to traditional biocrops has been particularly strong in this part of East Asia. Improved forestry management techniques also are being investigated by these developed economies as a means of raising their indigenous bioenergy resource potential.

This chapter also discussed how various issues of conflict have marred Southeast Asia's troubled history of bioenergy development. As with hydropower, there have been numerous adverse socio-economic and environmental impacts arising from this sector's expansion in the sub-region, as well as strong associations with corrupt political and commercial practices. As was contended earlier, this is clearly an area of renewables development in East Asia where weak state capacity has led to sub-optimal outcomes with respect to sustainable development and social justice. This experience furthermore suggests that East Asia's new developmentalism, when both formulating and pursuing low-carbon development strategies, must take increasingly into consideration the need to coordinate the promotion of renewable energy with related policy domain processes, demonstrate a compatibility with other socio-economic and environmental objectives, and be consistent with the principles of good governance generally.

Notes

1 http://www.greenworldinvestor.com/2011/03/09/list-of-major-biomass-power-plants-in-the-world-scale-increasing (accessed June 2, 2014).
2 This was a three-fold increase from the mid-1990s level (Ryu 2010).

10 Conclusion

10.1. Renewable energy and East Asia's new developmentalism

This book has made a comprehensive study of renewable energy (RE) in East Asia in the context of the region's new developmentalism. In this concluding chapter, we summarise its main empirical findings and arguments using the analytical framework of new developmentalism outlined in Chapter 2. To initially recap its conceptual approach, new developmentalism may be defined as revitalised and refocused forms of state capacity aimed at realising the transformative economic objectives associated with low carbon development. Its key premise is that renewable energy development is embedded within East Asia's broader endeavours on achieving low-carbon economy transformation over time. Furthermore, new developmentalism is founded on the synthesis of state capacity theory (SCT) and ecological modernisation theory (EMT). The former is essentially concerned with how the state can play an effective role in transforming the economy towards more advanced development ends, whereas the latter prescribes the reform, modification, and improvement of existing economic, business, and social structures to realise environmental objectives through the application of new technologies, policies, and practices. In this theoretical synthesis, it is posited that EMT stresses the importance of state policies and institutions in helping to shape markets and paths of economic development towards more environmentally friendly outcomes. Thus, the core common ground between SCT and EMT lies in three closely inter-related areas: transformative economic development, the key role of the state, and the state's partnership with business and society.

10.2. Empirical and functional

10.2.1. Main drivers

It has been shown that the *environmental and sustainable development imperatives* driving the advance of renewable energy in East Asia have strengthened over time, but they have not always been the most important drivers. For example, the main purpose of hydropower in its early development period was simply to generate power for industrialisation; moreover large-hydro dams were then viewed as symbols of 20th century economic modernity. More recently, hydropower has become the principal renewables sector for producing cleaner, lower-carbon electricity, albeit causing significant adverse environmental and socio-economic impacts. Other increasingly scaled up sectors, such as wind, solar, and bioenergy, also are largely seen as helping to reduce air pollution, address climate change, and lay a viable foundation on which to establish cleaner energy systems.

The inherently indigenous nature of renewables will make *energy security*, as drivers for development, increasingly relevant as East Asia's own fossil fuel reserves deplete and the region becomes more import dependent on foreign sources of oil, gas, and coal. However, studies in previous chapters suggest that energy security drivers behind East Asia's renewable energy development are not yet that important. Although fossil fuel supply security risks have become a growing concern in the region, they are not yet perceived as critical, and East Asian governments continue to ardently promote the further development of carbon-intensive industrial activities.

A more relevant driver has been the framing of renewables as *emerging strategic industries*. This has depended both on the sector generally and sector selection of East Asian governments. Naturally, those sectors which have exhibited greatest scope for commercial expansion and techno-innovation have tended to be strategically prioritised. This especially applies to wind energy and solar photovoltaic (PV), and to a lesser extent bioenergy. These sectors are either approaching or already at grid parity regarding competition with fossil fuels in power generation markets. The Chinese and Japanese governments made decisions in the 1990s to strategically prioritise wind energy and solar PV industry development, respectively. Meanwhile, South Korea opted to specialise in tidal range energy. In Southeast Asia, bioenergy industry development was deemed to be strategically important.

10.2.2. Governance of developmental process

State capacity is generally concerned with strategic plans, targets, and outcomes, which provide multi-layered structures for state actions and policies, whereas ecological modernisation focuses on reflexive responses and tactical adaption to new changing realities, arising along the path of economic transformation. This study has revealed that many Southeast Asian governments still lack sufficient state capacity (principally technocratic and institutional aspects) to implement coherent low carbon development strategies and renewable energy policies. Often, these have suffered from legislative and regulatory gaps, and poor coordination exists among relevant state agencies. These problems have also been very evident in China's wind energy sector, where the pattern and quality of state capacity intervention has been somewhat patchy. As Chapter 4 discussed, this was partly due to rapid growth, multiple stakeholders, and dynamic entrepreneurship factors, which presented difficult governance challenges to the Chinese national government. Even in the developed Northeast Asian economies of Japan, South Korea, and Taiwan, governments have often failed to implement strong-enough governance structures to advance wind, solar, hydropower, geothermal, and bioenergy to the development levels one would expect from states of their techno-industrial capabilities and renewable energy resource potentials. Chapter 9's study on bioenergy development furthermore showed the need to coordinate RE policy-making with other connected policies (e.g. agriculture, forestation, rural livelihoods, land management) and related socio-economic and environmental objectives of state development strategy, and to be generally consistent with the principles of good governance. Clearly, there remains much scope for 'smarter' state capacity and national-level governance across East Asia.

10.2.3. Policy, strategy, and objectives

Following on from discussions above, recent developments in the solar PV sub-sector show what a positive difference well-designed state support can make. Chapter 5 showed how new measures introduced in the early 2010s by Japan, Thailand, South Korea, Taiwan, and China helped to achieve remarkable growth in installed PV capacity. In fast-growing industries, governments frequently find themselves on a step-learning curve, which applies to many renewable energy sectors in East Asia. Strategic planning has become an increasingly prominent feature of East Asia's RE policy-making, and two phases of this were identified. After long periods of largely ad hoc measures promoting renewable energy installations and technologies, a first phase of RE strategic plans emerged in the region during the early 2000s. Second-phase plans then followed in the late 2000s, which generally coincided with the 'green growth' fiscal stimulus policies implemented in response to the 2008–2009 financial crisis. In this second phase, governments strengthened state capacity measures on renewables development and set more ambitious objectives towards this end.

Furthermore, by then East Asian governments had more firmly linked their RE strategies with those on climate change and broader macro-plans on low-carbon development, with ecological modernisation having greater due influence over development policy-making overall. The propensity of the region's governments to adhere to strategic plans put in motion obviously depends on domestic political systems and governance structures. In established democracies such as Japan, South Korea, and Taiwan, while energy-related state bureaucracies maintain strong strategy-making capacity and ensure some degree of strategic development continuity on renewables, they are subject to significant changes in policy priorities and interests that potentially come with newly elected governments. This is less relevant in mono-regime states, such as China and Vietnam, which are able to pursue much longer-range strategic plans. However, as consistent with the ecological modernisation approach, all East Asian governments have to adapt and revise their strategies in response to new trends and intervening variables arising.

The current record of East Asian governments realising their objectives on renewable energy development has been mixed. Japan and South Korea most notably failed to meet most targets set in early policy initiatives and first-phase strategies, and the same applied in certain Southeast Asian cases, such as Indonesia and geothermal energy. Although China has experienced failure on this front, it has also well exceeded plan targets in key areas, such as wind energy and solar PV, and is well on track to meet almost all goals on renewables set within the 12th Five-Year Plan (2011–2015). The most important factor determining the realisation of development targets is invariably the success of policy measures inducing new business investment to the sector, whether from private firms or state-owned enterprises (SOEs). Securing investment is especially critical as well as difficult in sectors that involve high initial capital costs, such as geothermal, tidal range, and large-hydro. Much will depend on the strategic sectoral choices made by governments regarding the degrees of state support devoted to helping develop a renewables sector. For example, South Korea decided to back tidal range energy whereas China did not, despite being in a strong position to do so. Likewise, Japan and South Korea could have chosen to become world leaders in offshore wind, but instead their governments afforded priority to other energy industries, such as nuclear. Overall, a predilection towards substantive strategic planning on renewables and its programmatic integration with broader development macro-plans makes East Asia somewhat unique compared to other parts of the world, and it is rooted in the region's historically strong state capacity tradition.

10.2.4. Relational aspects: state, business, society

This study has argued that renewable energy development and decarbonising economic activity per se requires deep socio-cultural and socio-technical transformations, and therefore the relational dimension of East Asia's new developmentalism is critically important. Given the growing palpable environmental and welfare costs arising from fossil fuel combustion, there is growing public support in the region for renewables and low carbon development. Business is becoming more active and engaged in East Asia's renewable energy development, not least because its industrial and market base is expanding fast in many sectors—most notably wind energy, solar PV, and bioenergy. As observed in previous chapters, SOEs dominate many aspects of this development, and in many instances may be considered an extension of government RE policy. This is most clearly and significantly evident in China (Box 3.1, p. 84). Where the private sector is the most critically prominent in renewable energy business—wind energy and solar PV manufacturing–close state-business developmental ties have formed, set in the context of strategic planning, where both sides have interests in realising state plan targets. In addition to China, the region's most successful developmental states of Japan, South Korea, Taiwan, Singapore, and Malaysia deserve special mention, with solar PV industry development providing the best illustrative example. It was also noted that plant developers (e.g. of wind farms and solar parks) are often the same SOE power companies operating coal-fired, gas-fired, and nuclear power stations. Meanwhile, state-society relations regarding East Asia's renewable energy development is at a more formative stage. In the region's more open democratic societies, civil society has become increasingly engaged in shaping and generally promoting RE policy. However, in sectors such as hydropower, wind, and ocean energy, there has been vocal public opposition to many installation projects. As certain RE technologies become more distributed and East Asia's 'prosumer' community grows, so we may expect this relational aspect of the region's new developmentalism to strengthen, as is later discussed.

10.3. Outcome impacts

10.3.1. Sustainable development paths

As elsewhere around the world, renewable energy will play a key role in establishing more sustainable paths of economic development in East Asia. However, as was discussed in Chapter 2, the region's new strategic macro-development plans continue to promote carbon-intensive industries and energy systems closely associated with 'old developmentalism.' This is evident region-wide in both developed and developing countries alike. China's ambitious programme for expanding its coal-fired power sector has been the subject of high-profile criticism, as have state industrial policies causing persistent and substantial ecological damage. The key question is essentially how fast will the transition from 'dirty' to 'clean' energy in East Asia be; thus, it centres on the pace of structural and systemic transformation. At present, it looks likely that the region's 'modernist' industrialised development will remain entrenched for some considerable time, and that the ecological modernisation approach of incremental rather than radical change will prevail. Even the relatively large RE sectors of wind and solar PV that have also experienced dynamic growth still only make comparatively small single-digit percentage contributions to East Asia's power generation. Scaling them up to much larger double-digit shares will present resource sustainability challenges of their own given the material masses required (with some of them being scarce rare earth minerals; see Chapter 1) to produce sufficient numbers of wind turbines and solar modules. Chapter 9 examined how these kinds of resource constraints are also relevant to future bioenergy development. Meanwhile, many East Asian states are already near exploiting their commercially feasible hydropower potential, and further expansion of large-hydro is problematic for previously discussed local environmental and socio-economic cost reasons (Chapter 6).

Additionally, although many governments have progressively raised their ambitions on renewables development, East Asia is still the world's fastest-growing regional economy, with a forecasted energy demand growth trajectory expected to be twice that of the global average. This has meant that the expansion of renewables in the region has existed in the context of fast-growing national energy sectors overall. China again provides perhaps the best and most significant example. Renewables, fossil fuels, and nuclear are all anticipated to expand rapidly in the country over the foreseeable future. Chapter 3 showed that despite China well over tripling its installed renewables capacity in absolute terms from 2004 to 2013, the share of renewables in national power generation capacity rose from just 23.6 percent to 30.5 percent, and in terms of actual power output, even less proportionately from 16.2 percent to 20.0 percent. Similar rates of slow progression made my renewables apply to other East Asian states. This at least constitutes a steady forward movement. It is still too early to judge how effective East Asian state second-phase RE strategies will prove to be at accelerating this pace of renewable energy's relative advance against conventional energy sectors.

10.3.2. Energy security risk management

It was noted earlier that energy security motives have not yet become primary drivers behind renewable energy development in East Asia. A partial exception to this concerns the role many RE sectors initially played in supplying power to remote areas of the region. Renewables still make an important contribution to rural and remote community energy security, but have over time shifted focus to now mainly serving the load centres of East Asia's burgeoning urbanised population. However, due to being intermittent suppliers of power, key sectors such as wind and solar make them essentially complementary to more constant energy power streams. This predicament has meant that many East Asian states—and most significantly the region's three largest economies of China, Japan, and South Korea—have looked to nuclear as an alternative option on long-term energy security risk management. Even after the 2011 Fukushima disaster, previous chapters discussed how Japan is likely to turn again to nuclear and prioritise its development over renewables. China and South Korea look in the meantime to continue implementing the world's most ambitious plans on expanding their respective nuclear power sectors. Furthermore, East Asia's most developed states' overt technology-oriented approach to RE policy will only perpetuate the perception of renewable energy development serving strategic industry interests rather than energy security interests. This is less the case in developing East Asia, where renewables make a much more important contribution to energy supply. Chapter 9's study on bioenergy further showed how energy security risk management concerning renewables can have a strong international dimension. For example, the high imported biomass feedstock demand of countries such as Japan and South Korea could have adverse impacts on the energy security (as well as food security) of lower-income biomass exporting countries from Southeast Asia and elsewhere. Chapter 6's study of hydropower development in the Mekong River basin illustrated other international dimensions to managing energy security risks.

10.3.3. Techno-economic development

Renewables are generally viewed as relatively high-tech industries and thus are at the frontier of any country's techno-economic development. As Chapter 2 discussed, green energy, with renewables at its core, has a strong claim for being one the new technology clusters on which the next 'long wave' of future world development will be based. This would entail a creative destruction, process where renewables eventually displace fossil fuel energy systems to establish a new clean-energy paradigm. However, as earlier discussed, this structural and systemic transformation in East Asia and worldwide could take a very long time to be realised. Moreover, the most dominant renewables sector, hydropower, is a mature energy technology with long historic origins as a mainstream energy source, dating back over a century. Geothermal has almost as long of a history in the same sense, and it has grown at a glacial pace over past decades, notwithstanding how recent advances in enhanced or engineered geothermal systems technology provides some hope for more impressive future development. The primary focus will remain on the most dynamic innovatory sectors (i.e. wind, solar, bioenergy, and ocean) with respect to renewable energy's position at new frontiers of techno-economic development.

10.3.4. Socio-cultural and socio-technical issues

Renewables not only offer cleaner and more sustainable methods of producing energy but also a much wider variety of energy technology options for society. Most importantly, many renewable energy applications can operate on a small enough scale to allow individuals, households, and communities take greater control of their own energy needs. In this sense, renewables have the potential to bring about a systemic revolution in energy production and consumption. Solar modules, very small wind turbines, pico-hydro machines, geothermal heat pumps, biogas digesters, and other micro-level devices are helping to expand 'distributed' energy in East Asia and elsewhere, and thereby increasing the number of energy prosumers (Chapter 1). If this trend continues, it will bring about a socio-cultural and socio-technical transformation of energy practice, fundamentally changing conventions in energy stakeholder relationships. In East Asia's case, it would mean a shift from a state-centric to a more society-centric approach to energy development, at least regarding power and heat generation. However, for scale economy reasons, utility-scale forms of generation (e.g. wind farms and hydropower dams) will remain crucially important if the broader structural and systemic transformations in energy practice noted earlier are to be realised.

10.3.5. Continuation or break from economic modernity?

Chapter 2 asked the question of whether East Asia's new developmentalism marked a departure from the carbon-intensive and economic modernisation oriented policies of the past, and more generally whether ecological modernisation ideas and thinking will succeed in placing humanity on more sustainable paths to economic development. The region's promotion of renewable energy can be understood at this meta-level of analysis. It was argued in Chapter 2 that the realistic best expectations of even the continued fast expansion of renewables in East Asia would be delivering *relatively lower* carbon development over the medium-term. It was furthermore contended that East Asia's new developmentalism has thus far been driven more by economistic objectives rather than environmentalism, to economic growth rather than sustainable development per se. This is consistent with the strong emerging strategic industries motive driving renewable energy development in the region, notwithstanding strong environmental and sustainable development imperatives similarly driving that development. There are, of course, a complex set of factors at play here, and East Asia has only just embarked on the very long transition to meaningful low carbon development. It is essentially an evolving process that is likely to take many decades–if indeed it will be achieved and humanity survives into the 22nd century and beyond.

10.4. Final comments

This author hopes that greater faith will be placed in renewables to help secure our common low carbon development future. East Asian countries have made a promising start on this front since the mid-2000s, but this is all that has been achieved thus far. Greater efforts are required in promoting the diffusion of renewables across the region's economy and society, as well as an evolving new developmentalism that is more effectively achieving decarbonised, sustainable development. By maintaining and improving their various forms of state capacity over time, East Asian states will be well positioned to sustain the significant recent advance of renewable energy and thereby further strengthen the low carbon development orientation of strategic industry policies and macro-development plans. Both business and society have important roles to play as developmental partners in the process. The continued growth of distributed renewable energy applications and ensuing expansion of the region's prosumer community should too lead to stronger societal engagement in East Asia's new developmentalism, making people more active stakeholders in energy sustainability. This broadening of societal utilisation of and support for renewables is likely to prove crucial to the prospects of renewable energy's own future.

References

Aanesan, K., Heck, S. and Pinner, D. (2012). *Solar Power: Darkness before Dawn*, New York: McKinsey and Company.

Agency for Natural Resources and Energy/ANRE. (2006). *Japan's New and Renewable Energy Policy*, Tokyo: ARNE.

Agency for Natural Resources and Energy/ANRE. (2010). *The Strategic Energy Plan of Japan: 2010 Revision*, Tokyo: ARNE.

Agency for Natural Resources and Energy/ANRE. (2010). *Energy in Japan: 2010*, Tokyo: ARNE.

Aggarangsi, P., Tippayawong, N., Moran, J.C. and Rerkkriangkrai, P. (2013). "Overview of Livestock Biogas Technology Development and Implementation in Thailand," *Energy for Sustainable Development*, 17, 371–377.

Aldrich, D.P. (2011). "Future fission: Why Japan won't Abandon Nuclear Power," *Global Asia*, 6, 62–67.

Alexander, J.C. (1990). "Between Progress and Apocalypse: Social Theory and the Dream of Reason in the Twentieth Century," in J. Alexander and P. Sztompka (eds). *Rethinking Progress: Movements, Forces, and Ideas at the End of the 20th Century*, Boston: Unwin Hyman.

Amsden, A. (1989). *Asia's Next Giant: South Korea and Late Industrialisation*, Oxford: Oxford University Press.

Andersen, M.S. and Massa, I. (2000). "Ecological Modernization: Origins, Dilemmas and Future Directions," *Journal of Environment and Policy Planning*, 2, 337–345.

APCO Worldwide. (2010). *China's 12th Five-Year Plan*, Shanghai: APCO Worldwide.

APCO Worldwide. (2010). *China's 2011 National People's Congress: Fine-tuning the Economy with an Eye on Social Stability*, Shanghai: APCO Worldwide.

Arat, Z.F. (1988). "Democracy and Economic Development: Modernisation Theory Revisited," *Comparative Politics*, 21(1), 21–36.

Aratani, F. (2005). "The Present Status and Future Direction of Technology Development for Photovoltaic Power Generation in Japan," *Progress in Photovoltaics: Research and Applications*, 13, 463–470.

Asian Development Bank/ADB. (2009). *The Economics of Climate Change in Southeast Asia: A Regional Review*, Manila: ADB.

Asian Development Bank/ADB. (2010). *Sector Assistance Programme Evaluation for the Energy Sector in Lao People's Democratic Republic*, Manila: ADB.

Asian Development Bank/ADB. (2012a). *Clean Energy Programme*, Manila: ADB.

Asian Development Bank/ADB. (2012b). *Green Growth, Resources, and Resilience: Environmental Sustainability in Asia and the Pacific*, Manila: ADB.

Association of Academies of Sciences in Asia/AASA. (2011a). *Towards a Sustainable Asia: Green Transition and Innovation*, Beijing: Science Press.

Association of Academies of Sciences in Asia/AASA. (2011b). *Towards a Sustainable Asia: Environment and Climate Change*, Beijing: Science Press.

Association of Southeast Asian Nations/ASEAN. (2011). *ASEAN Plan of Action for Energy Co-operation 2010–2015*, Jakarta: ASEAN Secretariat.

Avril, S., Mansilla, C. and Lemaire, T. (2012). "Solar Energy Support in the Asia–Pacific Region," *Progress in Photovoltaics: Research and Applications*, 20, 785–800.

Ayres, R.U. and Ayres, L.W. (1996). *Industrial Ecology: Towards Closing the Materials Cycle*, Cheltenham: Edward Elgar.

Bagliani, M., Dansero, E. and Puttilli, M. (2010). "Territory and Energy Sustainability: The Challenge of Renewable Energy Sources," *Journal of Environmental Planning and Management*, 53(4), 457–472.

Barbier, E.B. (2010a). *A Global Green New Deal: Rethinking the Economic Recovery*, Cambridge: Cambridge University Press.

Barbier, E.B. (2010b). "Green Stimulus, Green Recovery and Global Imbalances," *World Economics*, 11(2), 1–27.

Beeson, M. and Pham, H.H. (2012). "Developmentalism with Vietnamese Characteristics: The Persistence of State-led Development in East Asia," *Journal of Contemporary Asia*, 42(4), 539–559.

Beck, U., Giddens, A. and Lash, S. (1994). *Reflexive Modernization: Politics, Tradition and Aesthetics in the Modern Social Order*, Stanford: Stanford University Press.

Bernstein, H. (1971). "Modernisation Theory and Sociological Study of Development," *Journal of Development Studies*, 7(2), 141–160.

Bertani, R. (2005). "World Geothermal Power Generation in the Period 2001–2005," *Geothermics*, 34, 651–690.

Bertani, R. (2009). "Geothermal Energy: An Overview on Resources and Potential," paper presented the *International Geothermal Days Slovakia 2009 Conference and Summer School*, Casta Papiernicka, Slovakia, May 26–29, 2009.

Bertani, R. (2010). "Geothermal Power Generation in the World, 2005–2010 Update Report," paper presented at the *World Geothermal Congress 2010*, Bali, Indonesia, April 25–3, 2010.

Bloomberg New Energy Finance/BNEF. (2011a). *The Geopolitics of Clean Energy*, West Hartford: BNEF.

Bloomberg New Energy Finance/BNEF. (2011b). *Sun Sets on Oil for Gulf Power Generation*, West Hartford: BNEF.

Bloomberg New Energy Finance/BNEF. (2012a). *Re-considering the Economics of Photovoltaic Power*, West Hartford: BNEF.

Bloomberg New Energy Finance/BNEF. (2012b). *Solar will be Biggest Beneficiary of Japan's New Clean Energy Policy*, West Hartford: BNEF.

Blunden, L.S., Bahaj, A.S. and Aziz, N.S. (2013). "Tidal Current Power for Indonesia? An Initial Resource Estimation for the Alas Strait," *Renewable Energy*, 49, 137–142.

Bowen, A. and Fankhauser, S. (2011). "Editorial–The Green Growth Narrative: Paradigm Shift or Just Spin?" *Global Environmental Change*, 21, 1157–1159.

BP (2013). *Statistical Review of World Energy 2012*, London: BP.

Bradford, T. (2006). *Solar Revolution: The Economic Transformation of the Global Energy Industry*, Cambridge, MA: MIT Press.

Brand, U. (2010). "Sustainable Development and Ecological Modernization: The Limits to a Hegemonic Policy Knowledge," *Innovation–The European Journal of Social Science Research*, 23(2), 135–152.

Branker, K., Pathak, M.J.M. and Pearce, J.M. (2011). "A Review of Solar Photovoltaic Levelised Cost of Electricity," *Renewable and Sustainable Energy Reviews*, 15, 4470–4482.

Brennand, T.P. (2001). "Wind Energy in China: Policy Options for Development," *Energy for Sustainable Development*, 5(4), 84–91.

Bresser-Pereira, L.C. (2011). *From Old to New Developmentalism in Latin America*, Oxford: Oxford University Press.

Broadbent, J. (2002). "Japan's Environmental Regime," U. Desai (ed.). *Environmental Policy in Industrialized Countries*, Cambridge, MA: MIT Press.

Brophy, P., Nelson, G. and Majumdar, R. (2011). "The Emerging Geothermal Development Sector in Indonesia," *Geothermal Resources Council–Transactions*, 35, 1159–1163.

Brown, A., Müller, S. and Dobrotková, Z. (2011). *Renewable Energy Markets and Prospects by Technology*, Paris: International Energy Agency.

Brown, P.H., and Xu, K. (2010). "Hydropower Development and Resettlement Policy on China's Nu River," *Journal of Contemporary China*, 19(66), 777–797.

Brundtland Commission. (1987). *Our Common Future*, Oxford: Oxford University Press.

Bureau of Energy/BOE. (2007). "Strategies for Renewable Energies in Taiwan," in *2007 Strategic Review Board Meeting*, Taipei: BOE.

Bureau of Energy/BOE. (2012). *The Future Outlook of Renewable Energy Development Policy in Taiwan under Global Trends*, Taipei: BOE.

Bureau of Energy/BOE. (2013). *Energy Statistical Databook 2012*, Taipei: BOE.

Burgassi, P.D. (1999). "Historical Outline of Geothermal Technology in the Larderello Region to the Middle of the 20th Century," in R. Cataldi, S. Hodgson, J.W. Lund (eds.). *Stories from a Heated Earth*, Sacramento: Geothermal Resources Council.

Bush, S.R. (2008). "The Social Science of Sustainable Bioenergy Production in Southeast Asia," *Biofuels, Bioproducts and Biorefining*, 2, 126–132.

Buttel, F.H. (2000). "Ecological Modernization as Social Theory," *Geoforum* 31(1), 57–65.

Carbon Trust. (2011). *Integrating Renewables into Existing Buildings*, London: Carbon Trust.

Carlos, R.M. and Khang, D.B. (2009). "Characterization of Biomass Energy Projects in Southeast Asia," *Biomass and Bioenergy*, 32, 525–532.

Cataldi, R. (1999). "The Year Zero of Geothermics," in R. Cataldi, S. Hodgson, J.W. Lund (eds.). *Stories from a Heated Earth*, Sacramento: Geothermal Resources Council.

Chang, H.J. (2010). "Industrial Policy: Can we go Beyond an Unproductive Confrontation?" *Turkish Economic Association Discussion Paper*, No. 2010/1, Ankara: Turkish Economic Association.

Chang, K.C., Lee, T.S. and Chung, K.M. (2008). "Solar Water Heaters in Taiwan," *Renewable Energy*, 31, 1299–1308.

Chang, K.C., Lin, W.M. and Chung, K.M. (2013). "Solar Thermal Market in Taiwan," *Energy Policy*, 55, 477–482.

Chang, K.C., Lin, W.M., Lee, T.S. and Chung, K.M. (2008). "Outlook for Solar Water Heaters in Taiwan," *Energy Policy*, 36, 66–72.

Chang, K.C., Lin, W.M., Lee, T.S. and Chung, K.M. (2009). "Local Market of Solar Water Heaters in Taiwan: Review and Perspectives," *Renewable and Sustainable Energy Reviews*, 13, 2605–2612.

Chang, K.C., Lin, W.M., Lee, T.S. and Chung, K.M. (2011). "Subsidy Programs on Diffusion of Solar Water Heaters: Taiwan's Experience," *Energy Policy*, 39, 563–567.

Chang, P.L., Ho, S.P. and Hsu, C.H. (2013). "Dynamic Simulation of Government Subsidy Policy Effects on Solar Water Heaters Installation in Taiwan," *Renewable and Sustainable Energy Reviews*, 20, 385–396.

Chang, S., Zhao, L., Timilsina, G.R. and Zhang, X. (2012). "Biofuels Development in China: Technology Options and Policies Needed to Meet the 2020 Target," *Energy Policy*, 51, 64–79.

Charlier, R.H. (2001). "Ocean Alternative Energy: The View from China–Small is Beautiful," *Renewable and Sustainable Energy Reviews*, 5, 403–409.

Charuratna, A., Buarapa, P., Boongthong, T. and Pirarai, K. (2013). "New Phases of Recurrent Geothermal Exploration for Electricity Generation in Thailand," paper presented at the *Proceedings of the 10th Asian Geothermal Symposium*, Manila, September 22–24, 2013.

Chen, C.H. (2009). *Dawning Green-Energy Industry Programme*, Taipei: Bureau of Energy.

Chen F.L., Lu S.M., Lee S.C., Tseng K.T. and Wang E. (2010a). Assessment of Renewable Energy Reserves in Taiwan, *Renewable and Sustainable Energy Reviews*, 14, 2511–2528.

Chen F.L., Lu S.M., Wang E. and Tseng K.T. (2010b). "Renewable Energy in Taiwan," *Renewable and Sustainable Energy Reviews*, 14, 2029–2038.

Chen, J. (2012). "Development of Offshore Wind Power in China," *Renewable and Sustainable Energy Reviews*, 15, 5013–5020.

Chen, X., Wang, Z., He, S. and Li, F. (2013a). "Programme Management of World Bank Financed Small Hydropower Development in Zhejiang Province in China," *Renewable and Sustainable Energy Reviews*, 24, 21–31.

Chen, W.B., Liu, W.C., Hsu, M.H. (2013b). "Modelling Assessment of Tidal Current Energy at Kinmen Island, Taiwan," *Renewable Energy*, 50, 1073–1082.

Cheng, W.T. (1985). *Geothermal Update Report: Taiwan, Republic of China*, Taiwan: ITRI.

Chiang, H.D. (2004). "The Driving Role of the Government: Promotion of Renewable Energy in Taiwan," paper presented at the *Second World Renewable Energy Forum*, Bonn, Germany, May 29–31, 2004.

Chiang E.P., Zainal, Z.A., Aswatha Narayana P.A. and Seetharamu, K.N. (2003). "The Potential of Wave and Offshore Wind Energy in Around the Coastline of Malaysia that Face the South China Sea," paper presented at the *International Symposium on Renewable Energy: Environment Protection & Energy Solution for Sustainable Development*, September 14–17, 2003, Kuala Lumpur.

Chien, J.C.L. and Lior, N. (2011). "Concentrating Solar Thermal Power as a Viable Alternative in China's Electricity Supply," *Energy Policy*, 39, 7622–7636.

China Council for International Cooperation on Environment and Development/ CCICED. (2009). *China's Pathway towards a Low Carbon Economy*, Beijing: CCICED.

China Council for International Co-operation on Environment and Development/ CCICED. (2010). *China's Pathway towards a Low Carbon Economy*, Beijing: CCICED.

China Council for International Co-operation on Environment and Development/ CCICED. (2011a). *China's Low Carbon Industrialization Strategy*, Beijing: CCICED.

China Council for International Co-operation on Environment and Development/ CCICED. (2011b). *Li Keqiang Emphasizes the Importance of Ecological Improvement and Vigorous Development of the Energy-Conserving Environmental Protection Industry*, Beijing: CCICED.

China Council for International Cooperation on Environment and Development/ CCICED. (2012). *Annual Policy Report 2011*, Beijing: CCICED.

China National Renewable Energy Centre/CNREC. (2013a). *Renewable Energy in China Database*, Beijing: CNREC.

China National Renewable Energy Centre/CNREC. (2013b). *China Renewable Energy*, 2(1), Beijing: CNREC.

China Photovoltaic Industry Alliance/CPIA. (2013). *Annual Report of China PV Industry 2012*, Beijing: CPIA.

Chiu, A.S.F. and Yong, G. (2004). "On the Industrial Ecology Potential in Asian Developing Countries," *Journal of Cleaner Production*, 12(8–10), 1037–1045.

Chiu, F.C., Huang, W.Y. and Tiao, W.C. (2013). "The Spatial and Temporal Characteristics of the Wave Energy Resources around Taiwan," *Renewable Energy*, 52, 218–221.

Cho, Y.R. (2004). "Emergence and Evolution of Environmental Discourses in South Korea," *Korea Journal*, 44(3), 138–164.

Cho, C.H., Lee, K.H. and Rho, Y.H. (2010). Recent Tidal Current Power Projects in Korea, *Science China Technological*, 53(1), 57–61.

Chong, H.Y. and Lam, W.H. (2013). "Ocean Renewable Energy in Malaysia: The Potential of the Straits of Malacca," *Renewable and Sustainable Energy Reviews*, 23, 169–178.

Christoff, P. (1996). "Ecological Modernisation, Ecological Modernities," *Environmental Politics*, 5(3), 476–500.

Chu, Y.W. (2009). "Eclipse of Reconfigured? South Korea's Developmental State and Challenges of the Global Knowledge Economy," *Economy and Society*, 38(2): 278–303.

Chua, S.C. and Oh, T.H. (2012). "Solar Energy Outlook in Malaysia," *Renewable and Sustainable Energy Reviews*, 16, 564–574.

Cohen, G. (2005). *Solar Thermal Parabolic Trough Electric Power Plants for Electric Utilities in California*, Raleigh: Solargenix Energy.

Commission of the European Communities/CEC. (1993). *White Paper on Growth, Competitiveness and Employment: The Challenges and Ways Forward into the 21st Century*, Luxembourg: Office for Official Publications for the EC.

Council for Economic Planning and Development/CEPD. (2012). *Adaptation Strategy to Climate Change in Taiwan*, Taipei: CEPD.

Darma, S. (2010). "Geothermal Energy Development and Utilisation in Indonesia," paper presented at the *Proceedings World Geothermal Congress*, Bali, April, 25–30, 2010.

Dasuki, A.S., Djamin, M. and Lubis, A.Y. (2001). "The Strategy of Photovoltaic Technology Development in Indonesia," *Renewable Energy*, 22, 321–326.

Delman, J. and Odgaard, O. (2011). "From Worn to Green China Model: The Environment, Renewable Energy, Climate Change, and the 12th Five Year Plan," paper presented at the *2011 NTNU Japan Seminar: Renewables and Energy Security in Japan, East Asia and Norway*, Norwegian University for Science & Technology (NTNU), Oslo, November 9–10, 2011.

Denny, E. (2009). "The Economics of Tidal Energy," *Energy Policy*, 37, 1914–1924.

Dent, C.M. (2008). *East Asian Regionalism*, London: Routledge.

Dent, C.M. (2010). "Organising the Wider East Asia Region," *Asian Development Bank Series on Regional Economic Integration*, No. 62, Manila: Asian Development Bank.

Dent, C.M. (2012). "Renewable Energy and East Asia's New Developmentalism: Towards a Low Carbon Future?" *The Pacific Review*, 25(5), 561–587.

Dent, C.M. (2013a). "Understanding the Energy Diplomacies of East Asian States," *Modern Asian Studies*, 47(3), 935–967.

Dent, C.M. (2013b). "Paths Ahead for East Asia and Asia-Pacific Regionalism" (2013), *International Affairs*, 89(4), 963–985.

Department of Alternative Energy Development and Efficiency/DEDE. (2012). *Thailand's Solar Power Status*, Bangkok: DEDE.

Desker, B. (2013). "Southeast Asia Going Nuclear," *East Asia Forum*, 20 December 2013.

DeWit, A. and Iida, T. (2011). "The 'Power Elite' and Environmental-Energy Policy in Japan," *The Asia-Pacific Journal*, 9(4). Available at http://www.japanfocus.org/-Iida-Tetsunari/3479 (accessed August 10, 2014).

Dicken, P. (2011). *Global Shift*, London: Sage.

Dinmore, E. (2013). "Concrete Results?: The TVA and the Appeal of Large Dams in Occupation-Era Japan," *Journal of Japanese Studies*, 39(1), 1–38.

Dittmer, L. (2007). "The Asian Financial Crisis and the Asian Developmental State," *Asian Survey*, 47(6), 829–833.

Dryzek, J.S. (1997). *The Politics of the Earth: Environmental Discourses*, Oxford: Oxford University Press.

Economic Development Board/EDB. (2012). *Clean Energy*, Singapore: EDB.

Electric Power Research Institute/EPRI. (2010). *Electric Energy Storage Technology Options: A White Paper Primer on Applications, Costs, and Benefits*, Palo Alto: EPRI.

Energy Information Administration/EIA. (2011). *International Energy Outlook 2011*, Washington, DC: EIA.

Energy Information Administration/EIA. (2013). *International Energy Statistics Database*, Washington, DC: EIA.

Ernst & Young. (2013). *Rising Tide: Global Trends in the Emerging Ocean Energy Market*, London: Ernst & Young.

Esteban, M., Zhang, Q., Utama, A., Tezuka, T. and Ishihara, K.N. (2010). "Methodology to Estimate the Output of a Dual Solar–Wind Renewable Energy System in Japan," *Energy Policy*, 38, 7793–7802.

European Commission. (2010). *Europe 2020: A European Strategy for Smart, Sustainable and Inclusive Growth*, Brussels: European Commission.

European Photovoltaic Industry Association/EPIA. (2011a). *Global Market Outlook for Photovoltaics until 2015*, Brussels: EPIA.

European Photovoltaic Industry Association/EPIA. (2011b). *Solar Generation 6: Solar Photovoltaic Electricity Empowering the World*, Brussels: EPIA.

European Photovoltaic Industry Association/EPIA. (2011c). *Solar Photovoltaics Competing in the Energy Sector: On the Road to Competitiveness*, Brussels: EPIA

European Photovoltaic Industry Association/EPIA. (2012a). *Global Market Outlook for Photovoltaics until 2016*, Brussels: EPIA.

European Photovoltaic Industry Association/EPIA. (2012b). *Market Report 2011*, Brussels: EPIA.

European Photovoltaic Industry Association/EPIA. (2012c). *Connecting the Sun: Solar Photovoltaics on the Road to Large-Scale Grid Integration*, Brussels: EPIA.

European Photovoltaic Industry Association/EPIA. (2013). *Global Market Outlook for Photovoltaics 2013–2017*, Brussels: EPIA.

European Photovoltaic Industry Association/EPIA. (2014). *Global Market Outlook for Photovoltaics 2014–2018*, Brussels: EPIA.

European Small Hydropower Association/ESHA. (2005). *Small Hydropower for Developing Countries*, Brussels: ESHA.

Evans, P. (1995). *Embedded Autonomy: States and Industrial Transformation*, Princeton: Princeton University Press.

Falcao, A.F. (2010). "Wave Energy Utilization: A Review of the Technologies," *Renewable and Sustainable Energy Reviews*, 14, 899–918.

Fearnside, P.H. (1988). "China's Three Gorges Dam: 'Fatal' Project or Step Toward Modernization?" *World Development*, 16(5), 615–630.

Finn, D. (1979). "Geothermal Developments in the People's Republic of China," *Geothermal Resources Council–Transactions*, 3, 209–210.

Finn, D. (1980). "Geothermal Developments in the Philippines, 1980," *Geothermal Resources Council–Transactions*, 4, 771–773.

Fischer, C. and Preonas, L. (2010). "Combining Policies for Renewable Energy: Is the Whole Less than the Sum of its Parts?" *Discussion Paper Series*, No. 10–19, Washington, DC: Resources for the Future.

Fleming, P.D. and Probert, S.D. (1984). "The Evolution of Wind Turbines: An Historic Review," *Applied Energy*, 18(3), 163–177.

Fletcher, R. (2010). "When Environmental Issues Collide: Climate Change and the Shifting Political Ecology of Hydroelectric Power," *Peace and Conflict Review*, 5(1), 1–17.

Frosch, R.A. and Gallopoulos, N.E. (1989). "Strategies for Manufacturing," *Scientific American*, 261(3), 144–152.

Garcia, E. (2012). "Degrowth: The Past, the Future, and the Human Nature," *Futures* 44(6), 546–552.

Garnaut, R. (2013). "China's Contribution to the Global Mitigation Effort," *East Asia Forum*, 26(June).

Geothermal Energy Association/GEA. (2011). *Geothermal Energy: The Basics*, Washington, DC: GEA Secretariat.

Gerschenkron, A. (1962). *Economic Backwardness in Historical Perspective*, Cambridge, MA: Harvard University Press.

Gibbs, D. (2000). "Ecological Modernisation, Regional Economic Development and Regional Development Agencies," *Geoforum*, 31(1), 9–19.

Giddens, A. (1990). *The Consequences of Modernity*, Cambridge: Polity Press.

Gills, B. (ed.). (2010). *Globalisation in Crisis*, London: Routledge.

Gipe, P. (1991). "Wind Energy Comes of Age: California and Denmark," *Energy Policy*, 19(8), 756–767.

Global Wind Energy Council/GWEC. (2007a). *Global Wind 2006 Report*, Brussels: GWEC Secretariat.

Global Wind Energy Council/GWEC. (2007b). *China Wind Power Report, 2007*, Brussels: GWEC Secretariat.

Global Wind Energy Council/GWEC. (2008a). *Global Wind Energy Outlook 2008*, Brussels: GWEC Secretariat.

Global Wind Energy Council/GWEC. (2008b). *Global Wind 2007 Report*, Brussels: GWEC Secretariat.

Global Wind Energy Council/GWEC. (2009a). *Wind Power is Crucial for Combating Climate Change*, Brussels: GWEC Secretariat.

Global Wind Energy Council/GWEC. (2009b). *Global Wind 2008 Report*, Brussels: GWEC Secretariat.

Global Wind Energy Council/GWEC. (2010a). *Global Wind Energy Outlook 2010*, Brussels: GWEC Secretariat.

Global Wind Energy Council/GWEC. (2010b). *China Wind Power Outlook, 2010*, Brussels: GWEC Secretariat.

Global Wind Energy Council/GWEC. (2010c). *Global Wind 2009 Report*, Brussels: GWEC Secretariat.

Global Wind Energy Council/GWEC. (2011). *Global Wind Report: Annual Market Update 2010*, Brussels: GWEC Secretariat.

Global Wind Energy Council/GWEC. (2012). *Global Wind Report: Annual Market Update 2011*, Brussels: GWEC Secretariat.

Global Wind Energy Council/GWEC. (2013). *Global Wind Report: Annual Market Update 2012*, Brussels: GWEC Secretariat.

Global Wind Energy Council/GWEC. (2014). *Global Wind Report: Annual Market Update 2014*, Brussels: GWEC Secretariat.

Goh, C.S. and Lee, K.T. (2010). "Will Biofuel Projects in Southeast Asia Become White Elephants?" *Energy Policy*, 38, 3847–3848.

Gold, T.B. (1986). *State and Society in the Taiwan Miracle*, Armonk: M.E. Sharpe.

Goldstein, B.A., Hiriart, G., Tester, J., Bertani, R., Bromley, L., Guitierrez-Negrin, C.J., et al. (2011). "Great Expectations for Geothermal Energy to 2100," paper presented at the 36th Workshop on Geothermal Reservoir Engineering, Stanford: Stanford University, January 31–February 2, 2011.

Gosling, D. (1982). "Biogas for Thailand's Rural Development: Transferring the Technology," *Biomass*, 2, 309–316.

Goto, S. (2011). *IEA Bioenergy Task 40: Country Report 2011 for Japan*, Paris: IEA.

Gottesfeld, P. and Cherry, C.R. (2011). "Lead Emissions from Solar Photovoltaic Energy Systems in China and India," *Energy Policy*, 39, 4939–4946.

Grau, T., Huo, M. and Neuhoff, K. (2012). "Survey of Photovoltaic Industry and Policy in Germany and China," *Energy Policy*, 51, 20–37.

Green, D. (2004). "Thailand's Solar White Elephants: An Analysis of 15 yr of Solar Battery Charging Programmes in Northern Thailand," *Energy Policy*, 32, 747–760.

Green, M.A. (2005). "Silicon Photovoltaic Modules: A Brief History of the First 50 Years," *Progress in Photovoltaics: Research and Applications*, 13(5), 447–455.

Grumbine, R.E., Dore, J. and Xu, J. (2012). "Mekong Hydropower: Drivers of Change and Governance Challenges," *Frontiers in Ecology and the Environment*, 10(2), 91–98.

Gu, S. and Liu, W. (1996). "Development of Small Hydropower (SHP) in China," *Energy for Sustainable Development*, 3(3), 46–49.

Gunningham, N. (2011). "Energy Governance in Asia: Beyond the Market," *East Asia Forum Quarterly*, 3(1): 29–30.

Hajer, M.A. (1995). *The Politics of Environmental Discourse: Ecological Modernisation and the Policy Process*, Oxford: Oxford University Press.

Han, H. (2013). "China's Policymaking in Transition: A Hydropower Development Case," *Journal of Environment and Development*, 22(3), 313–336.

Han, S.Y., Kwak, S.J. and Yoo, S.H. (2008). "Valuing Environmental Impacts of Large Dam Construction in Korea: An Application of Choice Experiments," *Environmental Impact Assessment Review*, 28, 256–266.

Han, Jingyi, Mol, A.J., Lu, Y. and Zhang, L. (2009). "Onshore Wind Power Development in China: Challenges Behind a Successful Story," *Energy Policy*, 37, 2941–2951.

Hang, Q., Jun, Z., Xiao, Y., and Junkui, C. (2007). "Prospect of Concentrating Solar Power in China: The Sustainable Future," *Renewable and Sustainable Energy Review*, 12(9), 2505–2514.

Hansen, U.E. and Nygaard, I. (2014). "Sustainable Energy Transitions in Emerging Economies: The Formation of a Palm Oil Biomass Waste-To-Energy Niche in Malaysia 1990–2011," *Energy Policy*, 66, 666–676.

Harborne, P. and Hendry, C. (2009). "Pathways to Commercial Wind Power in the US, Europe and Japan: The Role of Demonstration Projects and Field Trials in the Innovation Process," *Energy Policy*, 37, 3580–3595.

Hassan, H.F., El-Shafie, A. and Karim, O.A. (2012). "Tidal Current Turbines: Glance at the Past and Look into Future Prospects in Malaysia," *Renewable and Sustainable Energy Reviews*, 16, 5707–5717.

Hayashi, S. (2010). "The Developmental State in the Era of Globalisation: Beyond the Northeast Asian Model of Political Economy," *The Pacific Review*, 23(1): 45–69.

Hayes, P. and von Hippel, D. (2006). "Energy Security in Northeast Asia," *Global Asia*, 1(1): 90–105.

Heinberg, R. (2004). *Powerdown: Options and Actions for a Post-Carbon World*, Gabriola: New Society.

Herberg, M.E. (2009). "Fuelling the Dragon: China's Energy Prospects and International Implications," in A. Wenger, R. Orttung and J. Perovic (eds). *Energy and the Transformation of International Relations*, Oxford: Oxford University Press.

Hirschl, B. (2009). "International Renewable Energy Policy: Between Marginalisation and Initial Approaches," *Energy Policy*, 37, 4407–4416.

Holden, K. and Derneritt, D. (2008). "Democratising Science? The Politics of Promoting Biomedicine in Singapore's Developmental State," *Environment and Planning D*, 26(1), 68–86.

Hong, L. and Möller, B. (2012). "Feasibility Study of China's Offshore Wind Target by 2020," *Energy*, 39, 1–10.

Hoselitz, B.F. (1952). *The Progress of Underdeveloped Areas*, Chicago: University of Chicago Press.

Hsu, S.H. (2006). "NIMBY Opposition and Solid Waste Incinerator Siting in Democratising Taiwan," *The Social Science Journal*, 43, 453–459.

Hu, M.C. and Phillips, F. (2011). "Technological Evolution and Interdependence in China's Emerging Biofuel Industry," *Technological Forecasting & Social Change*, 78, 1130–1146.

Hu, R., Sun, P. and Wang, Z. (2012). "An Overview of the Development of Solar Water Heater Industry in China," *Energy Policy*, 51, 46–51.

Huang, J., Yang, J. and Msangi, S., Rozelle, S. and Weersink, A. (2012). "Global Biofuel Production and Poverty in China," *Applied Energy*, 98, 246–255.

Huan, Q. (2007). "Ecological Modernisation: A Realistic Green Road for China?" *Environmental Politics*, 16(4), 683–687.

Huang, Y.H. and Wu, J.H. (2007). "Technological System and Renewable Energy Policy: A Case Study of Solar Photovoltaic in Taiwan," *Renewable and Sustainable Energy Reviews*, 11, 345–356.

Huang, Y.H. and Wu, J.H. (2008). "Analysis of Biodiesel Promotion in Taiwan," *Renewable and Sustainable Energy Reviews*, 12, 1176–1186.

Huang, Y.H. and Wu, J.H. (2009). "A Transition Toward a Market Expansion Phase: Policies for Promoting Wind Power in Taiwan," *Energy*, 34, 437–447.

Huang, Y.H., Bor Yunchang, J. and Peng, C.Y. (2011). "The Long-Term Forecast of Taiwan's Energy Supply and Demand: LEAP Model Application," *Energy Policy*, 39, 6790–6803.

Huber, J. (1982). *Die Verlorene Unschuld der Okologie*, Frankfurt am Main: Fischer Verlag.

Huber, J. (1985). *Die Regenbogengesellschaft: Okologie und Sozialpolitik*, Frankfurt am Main: S. Fischer.

Huber, J. (2000). "Towards Industrial Ecology: Sustainable Development as a Concept of Ecological Modernization," *Journal of Environmental Policy and Planning*, 2(4), 269–285.

Huenteler, J., Schmidt, T.S., and Kanie, N. (2012). "Japan's Post-Fukushima Challenge: Implications from the German Experience on Renewable Energy Policy," *Energy Policy*, 45(1), 6–11.

Hughes, L. (2013). "Japan's Radical Incrementalism in Energy," *East Asia Forum*, 25 June 2013.

Huo, M.L. and Zhang, D.W. (2012). "Lessons from Photovoltaic Policies in China for Future Development," *Energy Policy*, 51, 38–45.

Hwang, J.J. (2010). "Promotional Policy for Renewable Energy Development in Taiwan," *Renewable and Sustainable Energy Reviews*, 14(3), 1079–1087.

Industrial Development Bureau/IDB. (2005). *Renewable Energy Equipment Industry Promotional Plan*, Taipei: IDB.

Inoue, Y. and Miyazaki, K. (2008). "Technological Innovation and Diffusion of Wind Power in Japan," *Technological Forecasting and Social Change*, 75, 1303–1323.

Institute for Sustainable Energy Policies/ISEP. (2013). *Renewables 2013 Japan Report*, ISEP: Tokyo.

Intergovernmental Panel on Climate Change/IPCC. (2011). *Special Report on Renewable Energy Sources and Climate Change*, Geneva: IPCC Secretariat

International Atomic Energy Agency/IAEA. (2014). Available at http://www.iaea. org/pris/home.aspx (accessed June 1, 2014).

International Energy Agency/IEA. (2003a). *National Survey Report of PV Power Applications in Japan*, 2002, Paris: IEA

International Energy Agency/IEA. (2003b). *National Survey Report of PV Power Applications in Korea*, 2002, Paris: IEA

International Energy Agency/IEA. (2004). *Renewable Energy: Market and Policy Trends in IEA Countries*, Paris: IEA.

International Energy Agency/IEA. (2007a). *World Energy Outlook: China and India Insights*, Paris: IEA.

International Energy Agency/IEA. (2006). *IEA World Energy Outlook 2006*, Paris: IEA.

International Energy Agency/IEA. (2007b). *National Survey Report of PV Power Applications in Japan*, 2006, Paris: IEA.

International Energy Agency/IEA. (2007c). *National Survey Report of PV Power Applications in Korea*, 2006, Paris: IEA.

International Energy Agency/IEA. (2007d). *National Survey Report of PV Power Applications in Malaysia*, 2006, Paris: IEA.

International Energy Agency/IEA. (2008). *Japan 2008 Review*, Paris: IEA.

International Energy Agency/IEA. (2009a). *IEA World Energy Outlook 2009*, Paris: IEA.

International Energy Agency/IEA. (2009b). *Technology Roadmap: Wind Energy*, Paris: IEA.

International Energy Agency/IEA. (2009c). *Technology Roadmap: Solar Energy*, Paris: IEA.

International Energy Agency/IEA. (2009d). *National Survey Report of PV Power Applications in Korea, 2008*, Paris: IEA

International Energy Agency/IEA. (2010a). *National Survey Report of PV Power Applications in Japan, 2009*, Paris: IEA.

International Energy Agency/IEA. (2010b). *National Survey Report of PV Power Applications in Malaysia, 2009*, Paris: IEA.

International Energy Agency/IEA. (2011a). *Technology Roadmap: Smart Grids*, Paris: IEA.

International Energy Agency/IEA. (2011b). *China Wind Energy Development Roadmap 2050*, Paris: IEA.

International Energy Agency/IEA. (2011c). *Technology Roadmap: Geothermal Heat and Power*, Paris: IEA.

International Energy Agency/IEA. (2012a). *IEA World Energy Outlook 2012*, Paris: IEA.

International Energy Agency/IEA. (2012b). *Coal Power Database*, Paris: IEA.

International Energy Agency/IEA. (2012c). *National Survey Report of PV Power Applications in China, 2011*, Paris: IEA

International Energy Agency/IEA. (2012d). *Solar Energy Perspectives*, Paris: IEA.

International Energy Agency/IEA. (2012e). *National Survey Report of PV Power Applications in Japan, 2011*, Paris: IEA

International Energy Agency/IEA. (2012f). *National Survey Report of PV Power Applications in Korea, 2011*, Paris: IEA

International Energy Agency/IEA. (2012g). *Technology Roadmap: Hydropower*, Paris: IEA

International Energy Agency/IEA. (2012h). *IEA Geothermal Implementing Agreement: Republic of Korea Country Report 2012*, Paris: IEA

International Energy Agency/IEA. (2013a). *National Survey Report of PV Power Applications in China, 2012*, Paris: IEA.

International Energy Agency/IEA. (2013b). *National Survey Report of PV Power Applications in Japan, 2012*, Paris: IEA.

International Energy Agency/IEA. (2013c). *National Survey Report of PV Power Applications in Korea, 2012*, Paris: IEA.

International Energy Agency/IEA. (2013d). *National Survey Report of PV Power Applications in Malaysia, 2011*, Paris: IEA

International Energy Agency/IEA. (2013d). *Southeast Asia Energy Outlook*, Paris: IEA.

International Energy Agency/IEA. (2013e). *Bioenergy News*, 25(2), December, Paris: IEA.

International Energy Agency/IEA. (2014). *Global Renewable Energy: Policy and Measures Database*, Paris: IEA.

International Geothermal Association/IGA. (2014). *Geothermal Database*, Bochum: IGA.

International Hydropower Association/IHA. (2010). *2010 Activity Report*, London: IHA.

International Hydropower Association/IHA. (2011). *Advancing Sustainable Hydropower: 2011 Activity Report*, London: IHA.

International Hydropower Association/IHA. (2013). *IHA Hydropower Report 2013*, London: IHA.

International Renewable Energy Agency/IRENA. (2012a). *Renewable Energy Technologies: Cost Analysis Series–Hydropower*, Masdar City: IRENA.

International Renewable Energy Agency/IRENA. (2012b). *30 Years of Policies for Wind Energy*, Masdar City: IRENA.

International Renewable Energy Agency/IRENA. (2012c). *Power Sector Costing Study Update*, Masdar City: IRENA.

International Renewable Energy Agency/IRENA. (2012d). *Renewable Energy Technologies: Cost Analysis Series–Hydropower*, Masdar City: IRENA.

International Renewable Energy Agency/IRENA. (2013a). *Renewable Energy Country Profiles: Asia*, Masdar City: IRENA.

International Renewable Energy Agency/IRENA. (2013b). *Intellectual Property Rights: The Role of Patents in Renewable Energy Technology Innovation*, Masdar City: IRENA.

International Rivers. (2011). *Nam Theun 2 Hydropower Project: The Real Cost of a Controversial Dam*, Berkeley: International Rivers.

International Rivers. (2014). *International Rivers: Dam-Building Database*, available at: www.internationalrivers.org

Jackson, S. and Sleigh, A. (2000). "Resettlement for China's Three Gorges Dam: Socio-Economic Impact and Institutional Tensions," *Communist and Post-Communist Studies*, 33, 223–241.

Jackson, T. (2009). *Prosperity without Growth: Economics for a Finite Planet*, London: Earthscan.

Jacobs, D. and Sovacool, B.K. (2012). "Feed-In Tarrifs and Subsidy Schemes to Support the Implementation of PV Technology," in W.V. Sark and L. Kazmierski (eds). *Comprehensive Renewable Energy*, New York: Elsevier.

Jacobs, M. (2013). "Green Growth," in R. Falkner (ed.). *Handbook of Global Climate and Environmental Policy*, Oxford: Wiley Blackwell.

Jänicke, M. (1985). *Umweltpolitische Pravention als Okologische Modernisierung und Strukturpolitik*, Berlin: Wissenschaftszentrum.

Jänicke, M. (1988). "Okologische Modernisierung: Optionen und Restriktionen Praventiver Umweltpolitik," in U. Simonis (ed.). *Praventive Umweltpolitik*, Frankfurt am Main: Campus.

Jänicke, M. (1990). *State Failure: The Impotence of Politics in Industrial Society*, Cambridge: Polity.

Jänicke, M. (2004). "Industrial Transformation Between Ecological Modernisation and Structural Change," in K. Jacob, M. Binder and A. Wieczorek (eds). *Governance for Industrial Transformation: Proceedings of the 2003 Berlin Conference on the Human Dimensions of Global Environmental Change*, Berlin: Environmental Policy Research Centre.

Jänicke, M. (2012). "Green growth: From a growing eco-industry to economic sustainability, *Energy Policy*, 48(1), 13–21.

Janjai, S., Laksanaboonsong, J. and Seesaard, T. (2011). "Potential Application of Concentrating Solar Power Systems for the Generation of Electricity in Thailand," *Applied Energy*, 88, 4960–4967.

Japan Commission on Large Dams/JCOLD. (2010). *The Role of Dams and Hydropower*, Tokyo: JCOLD.

Japan Commission on Large Dams/JCOLD. (2011). *List of Major Dams in Japan*, Tokyo: JCOLD.

Japan Ministry of Environment. (2013). *Japan's Climate Change Policies*, Tokyo: Ministry of Environment.

Japan Renewable Energy Policy Platform/JREPP. (2010). *Renewables: Japan Status Report 2010*, Tokyo: JREPP.

Jayed, M.H., Masjuki, H.H., Saidur, R., Kalam, M.A. and Jahirul, M.I. (2009). "Environmental Aspects and Challenges of Oilseed Produced Biodiesel in Southeast Asia," *Renewable and Sustainable Energy Reviews*, 13, 2452–2462.

Jefferson, M. (2006). "Sustainable Energy Development: Performance and Prospects," *Renewable Energy*, 31(5), 571–582.

Jensen, P.D., Basson, L. and Leach, M. (2011). "Reinterpreting Industrial Ecology," *Journal of Industrial Ecology*, 15(5), 680–692.

Johnson, C. (1982). *MITI and the Japanese Miracle*, Stanford: Stanford University Press.

Johnstone, C.M., Pratt, D., Clarke, J.A. and Grant, A.D. (2013). "A Techno-Economic Analysis of Tidal Energy Technology," *Renewable Energy*, 49, 101–106.

Jones, R.S. and Yoo, B. (2011). "Japan's New Growth Strategy to Create Demand and Jobs," *OECD Economics Department Working Papers*, No. 890, Paris: OECD.

Jones, W. (2011). "How Much Water Does it Take to Make Electricity?" *IEEE Spectrum Online*. Available at http://spectrum.ieee.org/energy/environment/how-much-water-does-it-take-to-make-electricity (accessed December 12, 2012).

Jönsson, K. (2009). "Laos in 2008: Hydropower and Flooding (or Business as Usual)," *Asian Survey*, 49(1), 200–205.

Jung, T.Y. and Ahn, J.E. (2010). "Sowing the Seeds for Green Growth in Korea," *East Asia Forum*, 13 December.

Kadir, M.Z.A.A., Rafeeu, Y. and Adam, N.M. (2010). "Prospective Scenarios for the Full Solar Energy Development in Malaysia," *Renewable and Sustainable Energy Reviews*, 14, 3023–3031.

Kahrl, F., Williams, J., Ding, J. and Hu, J. (2011). "Challenges to China's Transition to a Low Carbon Electricity System," *Energy Policy*, 39, 4032–4041.

Kahrl, F., Su, Y., Tennigkeit, T., Yang, Y. and Xu, J. (2013). "Large or Small? Rethinking China's Forest Bioenergy Policies, *Biomass and Bioenergy*, 59, 84–91.

Kaldellis, J.K. and Zafirakis, D. (2011). "The Wind (R)evolution: A Short Review of a Long History," *Energy Policy*, 36, 1887–1901.

Kang, D. and Park, S.S. (2002). "Emergy Evaluation Perspectives of a Multipurpose Dam Proposal in Korea, *Journal of Environmental Management*, 66, 293–306.

Kang, J., Yuana, J., Hud, Z. and Xu, Y. (2012). "Review on Wind Power Development and Relevant Policies in China during the 11th Five-Year-Plan Period," *Renewable and Sustainable Energy Reviews*, 16, 1907–1915.

Karagiannis, N. and Madjd-Sadjadi, Z. (2007). *Modern State Intervention in the Era of Globalisation*, Cheltenham: Edward Elgar.

Karunungan, V.M. and Requejo, R. (2000). "Update on Geothermal Development in the Philippines," paper presented at the *Proceedings World Geothermal Congress 2000*, Kyushu and Tohoku, May 28–June 10, 2000.

Kerr, A. (2002). *Dogs and Demons: Tales from the Dark Side of Japan*, New York: Hill and Wang.

Khan, S.R. and Christiansen J. (eds.) (2010). *Towards New Developmentalism: Market as Means rather than Master*, London: Routledge.

Khoo, H.H., Tan, R.B.H. and Sagisaka, M. (2008). "Utilization of Woody Biomass in Singapore: Technological Options for Carbonization and Economic Comparison with Incineration," *International Journal of Life Cycle Assessment*, 13, 312–318.

Kibler, K.M. and Tullos, D.D. (2013). "Cumulative Biophysical Impact of Small and Large Hydropower Development in Nu River, China," *Water Resources Research*, 49, 3104–3118.

Kim, H. Shin, E.S. and Chung, W.J. (2011). "Energy Demand and Supply, Energy Policies, and Energy Security in the Republic of Korea," *Energy Policy*, 39, 6882–6897.

Kim, J., Park, J., Kim, J. and Heo, E. (2013). "Renewable Electricity as a Differentiated Good? The Case of the Republic of Korea," *Energy Policy*, 54, 327–334.

Kim, S. (2010). "Greening the Dam: The Case of the San Roque Multi-purpose Project in the Philippines," *Geoforum*, 41, 627–637.

Kinoshita, T., Ohki, T., and Yamagata, Y. (2010). "Woody Biomass Supply Potential for Thermal Power Plants in Japan," *Applied Energy*, 87, 2923–2927.

Kirkegaard, J.F., Hanemann, T. and Weischer, L. (2009). "It Should Be a Breeze: Harnessing the Potential of Open Trade and Investment Flows in the Wind Energy Industry," *Peterson Institute for International Economics Working Paper Series*, No. 09-14, December.

Kirkegaard, J.K., Hanemann, T., Weischer, L. and Miller, M. (2010). "Towards a Sunny Future? Global Integration in the Solar PV Industry," *Peterson Institute for International Economics Working Paper Series*, No. WP10-06, May.

Kirtikara, K. (1997). "Photovoltaic Applications in Thailand: Twenty Years of Planning and Experience," *Solar Energy and Materials and Solar Cells*, 47, 55–62.

Ko, Y. and Schubert, D. (2011). "South Korea's Plans for Tidal Power: When a 'Green' Solution Creates More Problems," *Nautilus Institute: NAPSNet Special Reports*, November 29, 2011.

Koizumi, T. (2013). "Biofuel and Food Security in China and Japan," *Renewable and Sustainable Energy Reviews*, 21, 102–109.

Kondratiev, N.D. (1935). "The Long Waves in Economic Life," *Review of Economic Statistics*, No. 17, 105–115.

Korhonen, J. (2004). "Industrial Ecology in the Strategic Sustainable Development Model: Strategic Applications of Industrial Ecology," *Journal of Cleaner Production*, 12, 809–823.

Korea Electric Power Corporation/KEPCO. (2013). *Statistics of Electric Power in Korea 2012*, Seoul: KEPCO.

Korea Environmental Industry and Technology Institute/KEITI. (2009). *Korea's Green Growth Vision and Eco-Innovation Policies*, Seoul: KEITI.

Korea Institute of Energy Technology Evaluation and Planning/KTEP. (2011). *Strategic Roadmap for Greenhouse Gas Reduction Technology: Geothermal*, Seoul: KETEP (Ministry of Knowledge Economy).

Kraxner, F., Aoki, K. and Leduc, S. (2014). "BECCS in South Korea: Analyzing the Negative Emissions Potential of Bioenergy as a Mitigation Tool," *Renewable Energy*, 61, 102–108.

Kuhn, T. (1962). *The Structure of Scientific Revolutions*, Chicago: University of Chicago Press.

Kumar, S., Salam, P.A., Shrestha, P. and Ackom, E.K. (2013). "An Assessment of Thailand's Biofuel Development," *Sustainability*, 5, 1577–1597.

Kuo, J.H., Lin, C.L., Chen, J.C., Tseng, H.H. and Wey, M.Y. (2011). "Emission of Carbon Dioxide in Municipal Solid Waste Incineration in Taiwan: A Comparison with Thermal Power Plants," *International Journal of Greenhouse Gas Control*, 5, 889–898.

Kuzuhara, Y. (2005). "Biomass Nippon Strategy: Why 'Biomass Nippon' Now?" *Biomass and Bioenergy*, 29, 331–335.

Kwon, H.S. (2010). *Research on the Way to Foster Leading New and Renewable Energy Industry on Greater Sphere Economic Area*, KEEI Research Paper, 31 December, Seoul: Korea Energy Economic Institute.

Lai, K.H., Wong, C.W.Y. and Cheng, T.C.E. (2012). "Ecological modernisation of Chinese export manufacturing via green logistics management and its regional implications," *Technological Forecasting and Social Change*, 79, 766–770.

Laird, J. (2011). "PV's Falling Costs." *Renewable Energy Focus*, 12, 52–56.

Lall, S. (2003). "Reinventing Industrial Strategy: The Role of Government Policy in Building Industrial Competitiveness," *QEH Working Paper Series*, No. 111, Oxford: Queen Elizabeth House.

Langhelle, O. (2000). "Why Ecological Modernization and Sustainable Development Should Not Be Conflated," *Journal of Environmental Policy and Planning* 2(4), 303–322.

Le, L.T., van Ierland, E.C., Zhu, X., Wesseler, J., and Ngo, G. (2013). "Comparing the Social Costs of Biofuels and Fossil Fuels: A Case Study of Vietnam," *Biomass and Bioenergy*, 54, 227–238.

Lee, B., Iliev, I., and Preston, F. (2009). *Who Owns Our Low Carbon Future? Intellectual Property and Energy Technologies*, London: Chatham House.

Lee, J.S., Lee, J.P., Park, J.Y., Lee, J.H. and Park, S.C. (2011). "Status and Perspectives on Bioenergy in Korea," *Renewable and Sustainable Energy Reviews*, 15, 4884–4890.

Lee, K.S. (2006). *Tidal and Tidal Current Study in Korea*, Ansan: Korea Ocean Research Development Institute.

Lee, S.C. and Shih, L.H. (2011). "Enhancing Renewable and Sustainable Energy Development Based on an Options-Based Policy Evaluation Framework: Case Study of Wind Energy Technology in Taiwan," *Renewable and Sustainable Energy Reviews*, 15, 2185–2198.

Lee, T.J. and Song, Y. (2013). "2013 Country Update on Geothermal Energy in Korea," paper presented at the *Proceedings of the 10th Asian Geothermal Symposium*, Manila, September 22–24, 2013.

Lee, Y.H. and Park, T.Y. (2009). "Civil Participation in the Making of a New Regulatory State in Korea: 1998–2008," *Korea Observer*, 40(3): 461–493.

Lema, A. and Ruby, K. (2007). "Between Fragmented Authoritarianism and Policy Co-ordination: Creating a Chinese Market for Wind Energy, *Energy Policy*, 35, 3879–3890.

Len, C. (2011). "Rethinking Nuclear Power After Fukushima," *East Asia Forum*, 25 March 2011.

Leroy, P. and van Tatenhove, J. (1999). "Political Modernization Theory and Environmental Politics," in F. Buttel, G. Spaargaren and A.P.J. Mol (eds.). *Environment and Global Modernities*, London: Sage Publishers.

Li, J. (2010). "Decarbonising power generation in China: Is the answer blowing in the wind? *Renewable and Sustainable Energy Reviews*, 14(4), 1154–1171.

Li, J. (2013). *China's Wind Energy Outlook 2012*, Brussels: Global Wind Energy Council.

Li, J. and Zhu, L. (1999). "Wind Power Commercialization Development in China," *Renewable Energy*, 16, 817–821.

Li, J., Gao, H., Shi, J., Ma, L., Qin, H. and Song, Y. (2007). *China Wind Power Report 2007*, Beijing: China Environmental Science Press.

Li, J. and Ma, L. (2009). *Background Paper: Chinese Renewables Status Report*, Paris: REN21.

Li, V. and Lang, G. (2010). "China's 'Green GDP' Experiment and the Struggle for Ecological Modernisation," *Journal of Contemporary Asia*, 40(1), 44–62.

Li, X., Hubacek, K. and Siu, Y.L. (2012). "Wind Power in China: Dream or Reality?" *Energy*, 37, 51–60.

Liao, C., Jochem, E., Zhang, Y. and Farid, N.R. (2010). "Wind Power Development and Policies in China, *Renewable Energy*, 35, 1879–1886.

Lim, H. (2010). "The Transformation of the Developmental State and Economic Reform in Korea," *Journal of Contemporary Asia*, 40(2): 188–210.

Lin, K.C. and Purra, M.M. (2012). "Transforming China's Electricity Sector: Institutional Change and Regulation in the Reform Era," *Centre for Rising Powers Working Paper Series*, No. 8, Cambridge: CRP.

Lin, C.J., Yu, O.S., Chang, C.L., Liu, Y.H., Chuang, Y.F. and Lin, Y.L. (2009). "Challenges of Wind Farms Connection to Future Power Systems in Taiwan," *Renewable Energy*, 34, 1926–1930.

Liou, H.M. (2010a). "Policies and Legislation Driving Taiwan's Development of Renewable Energy," *Renewable and Sustainable Energy Reviews*, 14, 1763–1781.

Liou, H.M. (2010b). "Overview of the Photovoltaic Technology Status and Perspective in Taiwan," *Renewable and Sustainable Energy Reviews*, 14, 1202–1215.

Liou, H.M. (2011). "Wind Power in Taiwan: Policy and Development Challenges," *Energy Policy*, 39, 3238–3251.

Liserre, M., Sauter, T. and Hung, J.Y. (2010). "Future Energy Systems: Integrating Renewable Energy Sources into the Smart Power Grid Through Industrial Electronics, *Industrial Electronics Magazine*, 4(1), 18–37.

Liu, Y. and Kokko, A. (2010). "Wind Power in China: Policy and Development Challenges," *Energy Policy*, 38, 5520–5529.

Liu, T.Y., Tavner, P.J., Feng, Y. and Qiu, Y.N. (2013a). "Review of Recent Offshore Wind Power Developments in China," *Wind Energy*, 16, 786–803.

Liu, J., Zuo, J., Sun, Z., Zillante, G. and Chen, X. (2013b). "Sustainability in Hydropower Development: A Case Study," *Renewable and Sustainable Energy Reviews*, 19, 230–237.

Lloyd, B. and Subbarao, S. (2008). "Development Challenges Under the Clean Development Mechanism (CDM). Can Renewable Energy Initiatives be put in Place Before Peak Oil?" *Energy Policy*, 37(1), 237–245.

Low, L. (2001). "The Singapore Developmental State in the New Economy and Polity," *The Pacific Review*, 14(3): 411–441.

Low, L.P. (2011). *Green Growth: Implications for Development Planning: A CDKN Guide*, London: Climate and Development Knowledge Network.

Lund, J. (2004). "100 Years of Geothermal Power Production," *Geo-Heat Centre Bulletin*, September, 11–19.

Lund, J., Freeston, D. and Boyd, T. (2010). "Direct Utilisation of Geothermal Energy 2010 Worldwide Review," paper presented at the *World Geothermal Congress 2010*, Bali, Indonesia, April 25–30, 2010.

Lund, P.C. (2010). *Energy from Wind and Ocean: A Northeast Asia Market Study*, Tokyo: Innovation Norway.

Luo, C., Huang, L., Gong, Y. and Ma, W. (2012b). "Thermodynamic Comparison of Different Types of Geothermal Power Plant Systems and Case Studies in China," *Renewable Energy*, 48, 155–160.

Luo, G.L., Zhi, F. and Zhang, X. (2012a). "Inconsistencies between China's Wind Power Development and Grid Planning: An Institutional Perspective," *Renewable Energy*, 48, 52–56.

Mao, M.L. and Chan, Y.K. (2006). "Geothermal Energy Potential in Taiwan," paper presented at the *31st Workshop on Geothermal Reservoir Engineering*, Stanford University, January 30–February 1, 2006.

Marinova, D. and Balaguer, A. (2009). "Transformation in the Photovoltaics Industry in Australia, Germany and Japan: Comparison of Actors, Knowledge, Institutions and Markets," *Renewable Energy*, 34, 461–464.

Martinelli, A. and Midttun, A. (2012). "Editorial–Introduction: Towards Green Growth and Multi-Level Governance," *Energy Policy*, 48(1), 1–4.

Martinot, E., and Li, J.F. (2010). "Renewable Energy Policy Update for China," *Renewable Energy World*, 21 July. Available at http://www.renewableenergyworld.com/rea/news/article/2010/07/renewable-energy-policy-update-for-china (accessed August 10, 2014).

Maruyama, Y., Nishikido, M. and Iida, T. (2007). "The rise of community wind power in Japan: Enhanced acceptance through social innovation," *Energy Policy*, 35, 2761–2769.

Massa, I. and Andersen, M.S. (2000). "Special Issue Introduction: Ecological Modernization," *Journal of Environment and Policy Planning*, 2: 265–267.

Massot, P. and Chen, Z.M. (2013). "China and the Global Uranium Market: Prospects for Peaceful Coexistence," *The Scientific World Journal*.

Masuda, T. and Komiyama, R. (2012). *Positioning of Nuclear in the Japanese Energy Mix*, Brussels: IFRI

Mathews, J.A. (2012). "Green Growth Strategies: Korean Initiatives," *Futures*, 44(8), 761–769.

Mathews, T., Kessels, K., Jolie, E., Witz, J. and Schmidt-Sercander, B. (2008). "Study on the Socio-Economic Framework for the Use of Geothermal Energy in Vietnam," paper presented at the *Proceedings of the 8th Asian Geothermal Symposium*, Hanoi, December 9–10, 2008.

Matsumura, Y. and Yokoyama, S. (2005). "Current Situation and Prospect of Biomass Utilization in Japan," *Biomass and Bioenergy*, 29, 304–309.

Matthews, N. (2012). "Water Grabbing in the Mekong Basin: An Analysis of the Winners and Losers of Thailand's Hydropower Development in Lao PDR," *Water Alternatives*, 5(2), 392–411.

McCully, P. (2001). *Silenced Rivers: The Politics and Ecology of Large Dams*, London: Zed Books.

McGregor, R.C. and DeSouza, W.E.R. (1997). "On the Analysis of Tidal Energy Schemes with Large Diurnal Variations with Application to Singapore," *Renewable Energy*, 10, 331–334.

McManus, P. and Gibbs, D. (2008). "Industrial Ecosystems? The Use of Tropes in the Literature of Industrial Ecology and Eco-Industrial Parks," *Progress in Human Geography*, 32(4), 525–540.

Meadows, D.H., Meadows, D.L., Randers, J. and Behrens, F. (1972). *The Limits to Growth*, London: Pan.

Mekhilef, S., Safari, A., Mustaffa, W.E.S., Saidur, R., Omar, R. and Younis, M.A.A. (2012). "Solar Energy in Malaysia: Current State and Prospects," *Renewable and Sustainable Energy Reviews*, 16, 386–396.

Mertha, A. (2008). *China's Water Warriors: Citizen Action and Policy Change*, Ithaca, NY: Cornell University Press.

Middleton, C. (2008). *Cambodia's Hydropower Development and China's Involvement*, Berkeley: International Rivers.

Miller, D. and Hope, C. (2000). "Learning to Lend for Off-Grid Solar Power: Policy Lessons from World Bank Loans to India, Indonesia, and Sri Lanka," *Energy Policy*, 28, 87–105.

Minami, E. and Saka, S. (2005). "Biomass Resources Present in Japan: Annual Quantities Grown, Unused and Wasted," *Biomass and Bioenergy*, 29, 310–320.

Ministry of Economy, Trade and Industry. (2010a). *The History of Japan's Industrial Policies*, Tokyo: METI.

Ministry of Economy, Trade and Industry/METI. (2010b). *The New Growth Strategy: Blueprint for Revitalising Japan*, Tokyo: METI.

Ministry of Economy, Trade and Industry/METI. (2010c). *The Industrial Structure Vision 2010 (Outline)*, Tokyo: METI.

Moe, E. (2012). "Vested Interests, Energy Efficiency and Renewables in Japan," *Energy Policy*, 40, 260–273.

Mok, K.H. and Yep, R. (2008). "Globalisation and State Capacity in Asia," *The Pacific Review*, 21(2), 109–120.

Mol, A.J.P. (1995). *The Refinement of Production*, Utrecht: Van Arkel.

Mol, A.J.P. and Sonnenfeld, D.A. (2000). "Ecological Modernisation Around the World: An Introduction," in A.J.P. Mol and D.A. Sonnenfeld (eds). *Ecological Modernisation Around the World: Perspectives and Critical Debates*, Oxford: Franks Cass.

Mol, A.J.P. and Spaargaren, G. (1993). "Environment, Modernity and the Risk Society: The Apocalyptic Horizon of Environmental Reform," *International Sociology*, 8(4), 431–459.

Montefrio, M.J.F. and Sonnenfeld, D.A. (2011). "Forests, Fuel, or Food? Competing Coalitions and Biofuels Policy-Making in the Philippines," *Journal of Environment and Development*, 20(1), 27–49.

Moon, T.H. (2010). "Green Growth Policy in the Republic of Korea: Its Promises and Pitfalls," *Korea Observer*, 41(3): 379–414.

Mork, G., Barstow, S., Kabuth, A. and Pontes, M.T. (2010). "Assessing the Global Wave Energy Potential," paper presented at the *29th International Conference on Ocean, Offshore Mechanics and Arctic Engineering*, Shanghai, June 6–11, 2010.

Morris, A.C., Nivola, P.S. and Schultze, C.L. (2012). "Clean Energy: Revisiting the Challenges of Industrial Policy," *Energy Economics*, 34, S34–S42.

Morthorst, P.E. (2009). *Wind Energy, the Facts: The Economics of Wind Power (Part III)*, Riso: Technical University of Denmark.

Muhammad-Sukki, F., Ramirez-Iniguez, R., Abu-Bakar, S.H., McMeekin, S.G. and Stewart, B.G. (2011). "An Evaluation of the Installation of Solar Photovoltaic in Residential Houses in Malaysia: Past, Present and Future," *Energy Policy*, 39, 7975–7987.

Muhammad-Sukki, F., Munir, A.B., Ramirez-Iniguez, R., Abu-Bakar, S.H., Yasin, H.M., McMeekin, S.G. and Stewart, B.G. (2012). "Solar photovoltaic in Malaysia: The Way Forward," *Renewable and Sustainable Energy Reviews*, 16, 5232–5244.

Munandar, A. and Widodo, S. (2013). "Geothermal Resources Development in Indonesia," paper presented at the *Proceedings of the 10th Asian Geothermal Symposium*, Manila, September 22–24, 2013.

Murni, S., Whale, J., Urmee, T., Davis, J.K. and Harries, D. (2013). "Learning from Experience: A Survey of Existing Micro-Hydropower Projects in Ba'Kelalan, Malaysia," *Renewable Energy*, 60, 88–97.

Murphy, J. and Gouldson, A. (2000). "Environmental Policy and Industrial Innovation: Integrating Environment and Economy through Ecological Modernisation," *Geoforum*, 31(1), 33–44.

Muzathik, A.M., Wan Nik, W.B., Ibrahim, M.Z. and Samo, K.B. (2010). "Wave Energy Potential of Peninsular Malaysia," *ARPN Journal of Engineering and Applied Sciences*, 5(7), 11–23.

Musgrove, P. (2010). *Wind Power*, Cambridge: Cambridge University Press.

Nakamura, H. (1981). "Development and Utilization of Geothermal Energy in Japan," *Geothermal Resources Council–Transactions*, 5, 33–35.

National Development and Reform Commission/NDRC. (2001). *China's 10th Five-Year Plan for Economic and Social Development*, Beijing: NDRC.

National Development and Reform Commission/NDRC. (2006). *China's 11th Five-Year Plan for Economic and Social Development*, Beijing: NDRC.

National Development and Reform Commission/NDRC. (2007a). *Medium and Long-Term Development Plan for Renewable Energy in China*, Beijing: NDRC.

National Development and Reform Commission/NDRC. (2007b). *China's National Climate Change Programme*, Beijing: NDRC.

National Development and Reform Commission/NDRC. (2011). *China's 12th Five-Year Plan for Economic and Social Development*, Beijing: NDRC.

National Energy Bureau/NEB. (2008). *China's Renewable Energy: Development Overview 2008*, Beijing: NEB.

Navigant Research. (2014). *World Market Update 2013: International Wind Energy Development*, Boulder: Navigant Research.

New Energy and Industrial Technology Development Organization/NEDO. (2004). *Overview of PV Roadmap Toward 2030 (PV2030)*, Tokyo: NEDO.

New Energy and Industrial Technology Development Organization/NEDO. (2013). *Strategic Development of Next-Generation Bioenergy Utilisation Technology*, Tokyo: NEDO.

New and Renewable Energy Centre/NREC. (2012). *Rules for Issuing Renewable Energy Certificates and Managing REC Markets*, NREC: Gyeonggi-do.

New Energy and Industrial Technology Development Organization. (2009). *PV Roadmap 2030+ (Plus)*, Tokyo: NEDO.

New Energy Foundation/NEF. (2007). *Status of Hydropower in Japan*, Tokyo: NEF.

Ng, W.P.Q., Lam, H.L., Ng, F.Y., Kamal, M. and Lim, J.H.E. (2012). "Waste-to-Wealth: Green Potential from Palm Biomass in Malaysia," *Journal of Cleaner Production*, 34, 57–65.

Nguyen, K.Q. (2007). "Wind Energy in Vietnam: Resource Assessment, Development Status and Future Implications," *Energy Policy*, 35, 1405–1413.

Noland, M. (2007). "From Industrial Policy to Innovation Policy: Japan's Pursuit of Competitive Advantage," *Asian Economic Policy Review*, 2(2): 251–268.

Ocean Energy Systems/OES. (2013). *Annual Report 2012*, Lisbon: OES.

Ogena, M.S. (2010). Philippine Country Update, 2005–2010 Geothermal Energy Development, paper presented at the *Proceedings World Geothermal Congress*, Bali, Indonesia, April 25–30, 2010.

Ogena, M.S. (2011). "Philippine Geothermal Industry Updates 2011," paper presented at the *Proceedings of the 9th Asian Geothermal Symposium*, Kagoshima, November 7–9, 2011.

Ogena, M.S. and Fronda, A.D. (2013). "Prolonged Geothermal Generation and Opportunity in the Philippines," paper presented at the *Geothermal Resources Council 2013 Annual Meeting*, Las Vegas, September 30,2013.

Ogena, M.S., Marasigan, M.C. and Fronda, A.D. (2013). "Philippine Country Update: Present and Future Geothermal Energy Development," paper presented at the *Proceedings of the 10th Asian Geothermal Symposium*, Manila, September 22–24, 2013.

Oh, K.Y., Kim, J.Y., Lee, J.S. and Ryu, K.W. (2012). "Wind Resource Assessment Around Korean Peninsula for Feasibility Study on 100 MW Class Offshore Wind Farm," *Renewable Energy*, 42, 217–226.

Olz, S. and Beerepoot, M. (2010). *Deploying Renewables in Southeast Asia*, Paris: IEA.

Organisation for Economic Co-operation and Development/OECD. (2009). *Declaration on Green Growth*, adopted at the Meeting of the Council at Ministerial Level, June 25, 2009, Paris: OECD.

Organisation for Economic Co-operation and Development/OECD. (2010). *Interim Report of the Green Growth Strategy: Implementing our Commitment for a Sustainable Future*, Paris: OECD.

Organisation for Economic Co-operation and Development/OECD. (2011). *Towards Green Growth: Monitoring Progress*, Paris: OECD.

Organisation for Economic Co-operation and Development/OECD. (2012). *OECD Work on Green Growth*, Paris: OECD.

Pacudan, R. (2005a). *Renewable Energy Policies in ASEAN*, Jakarta: ASEAN Energy Centre.

Pacudan, R. (2005b). "Green Power Markets Development in Southeast Asia: Electricity Industry Reforms and Policy Convergence," paper presented at the *2005 Risø International Energy Conference*, Risø National Laboratory, Denmark, May 23–25, 2005.

Park, S., Choi, J.H. and Park, J. (2011). "The Estimation of N_2O Emissions from Municipal Solid Waste Incineration Facilities: The Korea Case," *Waste Management*, 31, 1765–1771.

Pechak, O., Mavrotas, G. and Diakoulaki, D. (2011). "Role and Contribution of the Clean Development Mechanism to the Development of Wind Energy," *Renewable and Sustainable Energy Reviews*, 15, 3380–3387.

Pereira, A.A. (2008). "Whither the Developmental State? Explaining Singapore's Continued Developmentalism," *Third World Quarterly*, 29(6): 1189–1203.

Perez, C. (2002). *Technological Revolutions and Financial Capital: The Dynamics of Bubbles and Golden Ages*, Cheltenham: Edward Elgar.

Perez, C. (2010). "Technological Revolutions and Techno-Economic Paradigms," *Cambridge Journal of Economics*, 34, 185–202.

Petro Vietnam. (2009). *PV Power: Wind Energy Development*, Ho Chi Minh City: Petro Vietnam.

Pew Charitable Trusts. (2011). *Who's Winning the Green Energy Race?* Washington, DC: Pew Charitable Trusts.

Photovoltaic Science and Engineering Conference, Taiwan. (2013). "About Taiwan PV Industry." Available at: http://www.pvsec23.com/about_pv_taiwan.html (accessed August 11, 2014).

Phuangpornpitak, N. and Tia, S. (2011). "Feasibility Study of Wind Farms Under the Thai Very Small Scale Renewable Energy Power Producer (VSPP). Programme," *Energy Procedia*, 9, 159–170.

Pirie, I. (2008). *The Korean Developmental State: From Dirigisme to Neo-Liberalism*, London: Routledge.

Polanyi, K. (1944). *The Great Transformation*, Boston, MA: Beacon Press.

Prasertsan, S. and Sajjakulnukit, B. (2006). "Biomass and Biogas Energy in Thailand: Potential, Opportunity and Barriers," *Renewable Energy*, 31, 599–610.

Presidential Committee on Green Growth/PCGG. (2009). *Korea's Green Growth Vision and Five-Year Plan*, Seoul: PCGG.

Qin, G., Lou, P. and Wu, X. (2013). "Development of Ocean Energy Technologies: A Case Study of China," *Advances in Mechanical Engineering*.

Qiu, H., Sun, L., Huang, J. and Rozelle, S. (2012). "Liquid Biofuels in China: Current Status, Government Policies, and Future Opportunities and Challenges," *Renewable and Sustainable Energy Reviews*, 16, 3095–3104.

Radice, H. (2008). "The Developmental State under Global Neoliberalism," *Third World Quarterly*, 29(6), 1153–1174.

Radja, V.T. (1990). "Review of the Status of Geothermal Development and Operation in Indonesia 1985 to 1990," *Geothermal Resources Council–Transactions*, 14, 127–145.

Radja, V.T. (1997). "Strategy for Geothermal Energy Development in Indonesia Facing the Year 2020," paper presented at the *Japan International Geothermal Symposium*, Sendai, March 11–12, 1997.

Raksaskulworg, M. (2011). "Four Decades of Geothermal Research and Development in Thailand," paper presented at the *Proceedings of the 9th Asian Geothermal Symposium*, Kagoshima, November 7–9, 2011.

Redclift, M. (2005). "Sustainable Development (1987–2005): An Oxymoron Comes of Age," *Sustainable Development*, 13, 212–227.

Reiko, A. (2001). *Damu to Nihon*, Tokyo: Iwanami Shinsho.

REN21 (2006). *Renewables 2006 Global Status Report*, Paris: REN21 Secretariat.

REN21 (2009a). *Recommendations for Improving the Effectiveness of Renewable Energy Policies in China*, Paris: REN21 Secretariat.

REN21 (2009b). *Background Paper: Chinese Renewables Status Report, October 2009*, Paris: REN21 Secretariat.

REN21 (2010). *Renewables 2010 Global Status Report*, Paris: REN21 Secretariat.

REN21 (2011). *Renewables 2011 Global Status Report*, Paris: REN21 Secretariat.

REN21 (2012). *Renewables 2012 Global Status Report*, Paris: REN21 Secretariat.

REN21 (2013). *Renewables 2013 Global Status Report*, Paris: REN21 Secretariat.

Renewable Energy and Energy Efficiency Partnership/REEEP. (2010). "Japan 2010 Policy Database Report." Available at http://www.reeep.org/index.php?id=9353&special=viewitem&cid=150 (accessed August 12, 2014).

Renewable Energy and Energy Efficiency Partnership/REEEP. (2014). *Policy and Regulation Review Database*, Vienna: REEEP. Available at http://www.reegle.info/policy-and-regulatory-overviews (accessed August 12, 2014).

Righter, R.W. (1996). "Pioneering in Wind Energy: The California Experience," *Renewable Energy*, 9(3), 781–784.

Robins, N., Clover, R. and Saravanan, D. (2010). *Delivering the Green Stimulus*, London: HSBC Global Research.

Rock, M.T., Angel, D.P. and Feridhanusetyawan, T. (2000). "Industrial Ecology and Clean Development in East Asia," *Journal of Industrial Ecology*, 3(4), 29–42.

Rodan, G. (1989). *The Political Economy of Singapore's Industrialisation: National State and International Capital*, Basingstoke: MacMillan.

Rodrik, D. (2004). "Industrial Policy for the Twenty-First Century," *KSG Working Paper Series*. No. RWP04-047, Cambridge, MA: Kennedy School of Government.

Rodrik, D. (2007). "Normalizing Industrial Policy," paper prepared for the *Commission on Growth and Development*, Harvard University, September 2007.

Roland Berger. (2010). *From Pioneer to Mainstream: Evolution of Wind Energy Markets and Implications for Manufacturers and Suppliers*, Berlin: Roland Berger Consultants.

Rostow, W. (1960). *The Stages of Economic Growth: A Non-Communist Manifesto*, Cambridge: Cambridge University Press.

Ru, P., Zhi, Q., Zhang, F., Zhong, X., Li, J. and Su, J. (2012). "Behind the Development of Technology: The Transition of Innovation Modes in China's Wind Turbine Manufacturing Industry," *Energy Policy*, 43, 58–69.

Ryu, C. (2010). "Potential of Municipal Solid Waste for Renewable Energy Production and Reduction of Greenhouse Gas Emissions in South Korea," *Journal of the Air and Waste Management Association*, 60(2), 176–183.

Ryu, C. and Shin, D. (2013). "Combined Heat and Power from Municipal Solid Waste: Current Status and Issues in South Korea," *Energies*, 6, 45–57.

Sakaguchi, K. (2013). "Present Status of Geothermal Development in Japan," paper presented at the *Proceedings of the 10th Asian Geothermal Symposium*, Manila, September 22–24, 2013.

Sakmani, A.S., Lam, W.H., Hashim, R. and Chong, H.Y. (2013). "Site Selection for Tidal Turbine Installation in the Strait of Malacca," *Renewable and Sustainable Energy Reviews*, 21, 590–602.

Sang, T. and Zhu, W. (2011). "China's Bioenergy Potential," *GCB Bioenergy*, 3, 79–90.

Saptadji, N.M. (2006). "Update on Geothermal Development in Indonesia," paper presented at the *Proceedings of the 28th New Zealand Geothermal Workshop*, Auckland, November 10–12, 2006.

Sasaki, N., Owari, T. and Putz, F.E. (2011). "Time to Substitute Wood Bioenergy for Nuclear Power in Japan," *Energies*, 4, 1051–1057.

Sawangphol, N. and Pharino, C. (2011). "Status and Outlook for Thailand's Low Carbon Electricity Development," *Renewable and Sustainable Energy Reviews*, 15, 564–573.

Schmitz, H. (2007). "Reducing Complexity in the Industrial Policy Debate," *Development Policy Review*, 25(4), 417–428.

Schlosberg, D. and Rinfret, S. (2008). "Ecological Modernization, American Style," *Environmental Politics*, 17(2), 254–275.

Schneider, M. and Froggatt, A. (2013). *The World Nuclear Industry Status Report 2013*, Paris: Mycle Schneider Consulting

Schumpeter, J.A. (1911). *The Theory of Economic Development*, New York: Oxford University Press.

Schumpeter, J.A. (1939). *Business Cycles*, Philadelphia: Porcupine Press.

Segal, A. (2010). "China's Innovation Wall: Beijing's Push for Homegrown Technology," *Foreign Affairs*, September 28. Available at http://www.foreignaffairs.com/articles/66753/adam-segal/chinas-innovation-wall? (accessed August 12, 2010).

Seippel, Ø. (2000). "Ecological Modernization as a Theoretical Device: Strengths and Weaknesses," *Journal of Environment and Policy Planning*, 2: 287–302.

Shapiro, J. (2001). *Mao's War Against Nature: Politics and the Environment in Revolutionary China*. Cambridge: Cambridge University Press.

Shi, L. (2013). "Removing System Barriers, Ensuring the Large Scale Wind Energy Development," *China Renewable Energy*, 2(1), 4–9.

Sibayan, F.S. (2010). *Opportunities for Scaling Wind Energy in the Philippines*, Manila: Department of Energy.

Simpson, A. (2013). "Challenging Hydropower Development in Myanmar (Burma): Cross-Border Activism under a Regime in Transition," *The Pacific Review*, 26(2), 129–152.

Singapore International Energy Week/SIEW. (2012). *Securing Our Energy Future*, Singapore: SIEW.

Singapore National Climate Change Secretariat. (2011). *Solar Energy Technology Primer: A Summary*, Singapore: NCCS.

Singapore National Climate Change Secretariat. (2012). *Climate Change and Singapore: Challenges, Opportunities, Partnerships*, Singapore: NCCS.

Singh, R. and Setiawan, A.D. (2013). "Biomass Energy Policies and Strategies: Harvesting Potential in India and Indonesia," *Renewable and Sustainable Energy Reviews*, 22, 332–345.

Shulman, G. [1981). "Geothermal Power Development in Indonesia," *Geothermal Resources Council–Transactions*, 5, 37–40.

Smith, C. (1995). "Revisiting Solar Power's Past," *Technology Review*, July, 38–47.

Smits, M. and Bush, S.R. (2010). "A Light Left in the Dark: The Practice and Politics of Pico-Hydropower in the Lao PDR," *Energy Policy*, 38, 116–127.

Soares, P. F. (2009). "State of the Wind Industry in China: Patterns of Growth," paper presented at the *Fourth Annual Renewable Energy Finance Forum*, Beijing, May 12–13, 2009.

Solar Club (2012). *Thailand PV Status Report*, Bangkok: Solar Club

Solar Energy Research Institute of Singapore/SERIS. (2012). *Annual Report 2012*, Singapore: SERIS.

Solidiance. (2013). *China's Renewable Energy Sector: An Overview of Key Growth Sectors*, Shanghai: Solidiance.

Song, J. (1993). "The National Photovoltaic Project in Korea," *Energy Sources*, 15(1), 51–58.

Song, Y., Kim, H.C. and Lee, T.J. (2010). "Geothermal Development in Korea: Country Update 2005–2009," paper presented at the *Proceedings World Geothermal Congress 2010*, Bali, April 25–29, 2010.

Soule, M.E. and Lease, G. (eds). (1995). *Reinventing Nature: Responses to Postmodern Deconstruction*, Washington, DC: Island Press.

Sovacool, B.K. (2009). "The Importance of Comprehensiveness in Renewable Energy and Energy-Efficiency Policy," *Energy Policy*, 37(4), 1529–1541.

Sovacool, B.K. (2010). "The Political Economy of Oil and Gas in Southeast Asia: Heading Towards The Natural Resource Curse?" *The Pacific Review*, 23(2): 225–259.

Sovacool, B.K. and Bulan L.C. (2012). "Energy Security and Hydropower Development in Malaysia: The Drivers and Challenges Facing the Sarawak Corridor of Renewable Energy (SCORE)," *Renewable Energy*, 40, 113–129.

State Council (2012). *China's Energy Policy White Paper 2012*, Beijing: State Council.

State Electricity Regulatory Commission/SERC. (2011). *Report on the Regulation of Wind Power Safety*, Beijing: SERC.

Sternberg, R. (2010). "Hydropower's Future, the Environment, and Global Electricity Systems," *Renewable and Sustainable Energy Reviews*, 14, 713–723.

Stubbs, R. (2009). "What ever Happened to the East Asian Developmental State? The Unfolding Debate," *The Pacific Review*, 22(1): 1–22.

Su, Y.S. (2013). "Competing in the Global Solar Photovoltaic Industry: The Case of Taiwan," *International Journal of Photoenergy*, 2013, 1–11.

Su, C.L. and Lee, Y.M. (2009). "Development Status and Life Cycle Inventory Analysis of Biofuels in Taiwan," *Energy Policy*, 37, 754–758.

Sugandhavanija, P., Sukchai, S., Ketjoy, N. and Klongboonjit, S. (2011). "Determination of Effective University-Industry Joint Research for Photovoltaic Technology Transfer (UIJRPTT). in Thailand," *Renewable Energy*, 36, 600–607.

Sugino, H. and Akeno, T. (2010). "2010 Country Update for Japan," paper presented at the *Proceedings World Geothermal Congress 2010*, Bali, April 25–29, 2010.

Suliman, O. (ed.). (1998). *China's Transition to a Socialist Market Economy*, London: Praeger.

Taiwan Bureau of Energy. (2011). *Energy Statistics Handbook 2010*, Taipei: Bureau of Energy.

Takase, K. and Suzuki, T. (2011). "The Japanese Energy Sector: Current Situation, and Future Paths," *Energy Policy*, 39, 6731–6744.

Tan, Z., Zhang, H., Xu, J., Wang, J., Yu, C. and Zhang, J. (2012). "Photovoltaic Power Generation in China: Development Potential, Benefits of Energy Conservation and Emission Reduction," *Journal of Energy Engineering*, 138(2), 73–86.

Tang, W., Li, Z., Qiang, M., Wang, S. and Lu, Y. (2013). "Risk Management of Hydropower Development in China," *Energy*, 60, 316–324.

Thang, T.T., Giang C.V. and Dang, V.H. (2013). "Introduction to the Geothermal Potential of the North-Eastern Vietnam," paper presented at the *Proceedings of the 10th Asian Geothermal Symposium*, Manila, September 22–24, 2013.

Thavasi, V. and Ramakrishna, S. (2009). "Asia Energy Mixes from Socio-Economics and Environmental Perspectives," *Energy Policy*, 37, 4240–4250.

Timilsina, G.R., Kurdgelashvili, L. and Narbel, P.A. (2012). "Solar Energy: Markets, Economics and Policies," *Renewable and Sustainable Energy Reviews*, 16(1), 449–465.

Toichi, T. (2003). "Energy Security in Asia and Japanese Policy," *Asia-Pacific Review*, 10(1), 44–51.

Toke, D. (2011a). "Ecological Modernisation, Social Movements and Renewable Energy," *Environmental Politics*, 20(1), 60–77.

Toke, D. (2011b). *Ecological Modernisation and Renewable Energy*, Basingstoke: Palgrave Macmillan.

Tokyo Electric Power Company/TEPCO. (2003). *The First Electric Power Historical Museum Project in Japan*, Tokyo: TEPCO.

Tominaga, K. (1991). "The Historical Stages of the Development of Japan: Towards a Theory of Modernisation of Non-Western Countries," *Sociologie du Travail*, 33(1), 189–206.

Tsai, W.T. (2007). "Bioenergy from Landfill Gas (LFG) in Taiwan," *Renewable and Sustainable Energy Reviews*, 11, 331–344.

Tsai, W.T. and Chou, Y.H. (2006). "An Overview of Renewable Energy Utilization from Municipal Solid Waste (MSW) Incineration in Taiwan," *Renewable and Sustainable Energy Reviews*, 10, 491–502.

Tsuchiya, H. (2012). "Electricity Supply Largely from Solar and Wind Resources in Japan," *Renewable Energy*, 48, 318–325.

Tung, A.C. (1998). "Hydroelectricity and Industrialisation," in M. Elvin and C. Liu (eds.). *Sediments of Time: Environment and Society in Chinese History*, Cambridge: Cambridge University Press.

Uehara, H., Dilao, C.O. and Nakaoka, T. (1988). "Conceptual Design of Ocean Thermal Energy Conversion (OTEC) Power Plants in the Philippines," *Solar Energy*, 41(5), 431–441.

Umar, M.S., Jennings, P. and Urmee, T. (2013). "Strengthening the Palm Oil Biomass Renewable Energy Industry in Malaysia," *Renewable Energy*, 60, 107–115.

United Nations. (2011). *Energy Statistics Database*. Geneva: UN.

United Nations Environment Programme/UNEP. (2010). *Overview of the Republic of Korea's National Strategy for Green Growth*, Geneva: UNEP.

United Nations Environmental Programme/UNEP. (2011). *Towards a Green Economy: Pathways to Sustainable Development and Poverty Eradication*, Nairobi: UNEP.

United Nations Environmental Programme/UNEP. (2011). *Global Trends in Renewable Energy Investment 2011*, Nairobi: UNEP.

United Nations Economic and Social Commission for Asia and the Pacific/UNESCAP. (2005). *Green Growth at a Glance*, Bangkok: UNESCAP.

United Nations Economic and Social Commission for Asia and the Pacific/UNESCAP. (2008). *Greening Growth in Asia and the Pacific*, Bangkok: UNESCAP.

United Nations Economic and Social Commission for Asia and the Pacific/UNESCAP. (2012a). *Green Growth, Resources and Resilience: Environmental Sustainability in Asia and the Pacific*, Bangkok: UNESCAP.

United Nations Economic and Social Commission for Asia and the Pacific/UNESCAP. (2012b). *Low Carbon Green Growth Roadmap for Asia and the Pacific*, Bangkok: UNESCAP.

United Nations Framework Convention on Climate Change/UNFCCC. (2005). *Sixth Compilation and Synthesis of Initial National Communications From Parties Not Included in Annex I to the Convention*, Geneva: United Nations.

United Nations Framework Convention on Climate Change/UNFCCC. (2011). *Compilation and Synthesis of Fifth National Communications: Executive Summary*, Geneva: United Nations.

United Nations Framework Convention on Climate Change/UNFCCC. (2012). *Benefits of the Clean Development Mechanism 2012*, Geneva: United Nations.

Urban, F., Nordensvard, J., Khatri, D. and Wang, Y. (2013). "An analysis of China's investment in the hydropower sector in the Greater Mekong Sub-Region," *Environment, Development and Sustainability*, 15, 301–324.

US–China Economic and Security Review Commission/USCESRC. (2010). *2010 Report to Congress*, Washington DC: USCESRC.

Urashima, K. (2007). "Considering of Lightning Damage Protection and Risk Reduction for a Safe and Secure Society," *Science and Technology Trends Quarterly Review*, 25(October), 21–35, Tokyo: National Institute of Science and Technology Policy.

Urmee, T. and Harries, D. (2009). "A Survey of Solar PV Programme Implementers in Asia and the Pacific Regions," *Energy for Sustainable Development*, 13, 24–32.

Ushiyama, I. (1999a). "Wind Energy Activities in Japan," *Renewable Energy*, 16, 811–816.

Ushiyama, I. (1999b). "Renewable Energy Strategy in Japan," *Renewable Energy*, 16, 1174–1179.

Ushiyama, I. (2012). "Wind Power Development in Japan," presentation at the Japan Wind Energy Association, March 9, 2012.

Valentine, S.V. (2010). "Disputed Wind Directions: Reinvigorating Wind Power Development in Taiwan," *Energy for Sustainable Development*, 14, 22–34.

Valentine, S.V. (2011). "Japanese Wind Energy Development Policy: Grand Plan or Group Think?" *Energy Policy*, 39, 6842–6854.

Walker, G., Cass, N., Burningham, K. and Barnett, J. (2010). "Renewable Energy and Sociotechnical Change: Imagined Subjectivities of the Public and their Implications," *Environment and Planning A*, 42(4), 931–947.

Wang, J.Y. (2003). "Present Status and Future Development of Geothermal Energy in China," *Geothermal Resources Council–Transactions*, 27, 65–68.

Wang, Q. (2010). "Effective Policies for Renewable Energy: The Example of China's Wind Power Lessons for China's Photovoltaic Power," *Renewable and Sustainable Energy Reviews*, 14, 702–712.

Wang, R.Z. and Zhai, X.Q. (2012). "Development of Solar Thermal Technologies in China," *Energy*, 35(11), 4407–4416.

Wang, S., Yuan, P., Li, D. and Jiao, Y. (2011). "An Overview of Ocean Renewable Energy in China," *Renewable and Sustainable Energy Reviews*, 15, 91–111.

Wang, T. and Watson, J. (2009). *China's Energy Transition: Pathways for Low Carbon Development*, Brighton: Sussex Energy Group and Tyndall Centre for Climate Change Research.

Wang, Y. (2010). "The Analysis of the Impacts of Energy Consumption on Environment and Public Health in China," *Energy*, 35(11), 4473–4479.

Wang, Z. (2010). "Prospectives for China's Solar Thermal Power Technology Development," *Energy*, 35(11), 4417–4420.

Wang, Z., Qin, H. and Lewis, J. (2012). "China's Wind Power Industry: Policy Support, Technological Achievements, and Emerging Challenges," *Energy Policy*, 51(1), 80–88.

Warner, R. (2010). "Ecological Modernisation Theory: Towards a Critical Ecopolitics of Change?" *Environmental Politics*, 19(4), 538–556.

Watanabe, C. (1972). *Industrial-Ecology: Introduction of Ecology into Industrial Policy*, Tokyo: Ministry of International Trade and Industry.

Watanabe, C. (1993). "Energy and Environmental Technologies in Sustainable Development: A View from Japan," *The Bridge*, 23(2), 8–15.

Watanabe, C. (1995). "Identification of the Role of Renewable Energy: A View from Japan's Challenge, the New Sunshine Programme," *Renewable Energy*, 6(3), 237–274.

Watson, I. (2012). "Contested Meanings of Environmentalism and National Security in Green Korea," *The Pacific Review*, 25(5), 537–560.

Weale, A. (1992). *The New Politics of Pollution*, Manchester: Manchester University Press.

Weber, M. (1947). *The Theory of Social and Economic Organisation*, New York: Oxford University Press.

Weiss, L. (1998). *The Myth of the Powerless State: Governing the Economy in a Global Era*, London: Polity.

Weiss, L. (2010). "The State in the Economy: Neoliberal or Neoactivist?" in J. Campbell, C. Crouch, P. Hull Kristensen, O. K. Pedersen, and R. Whitley (eds). *Oxford Handbook of Comparative Institutional Analysis*, Oxford: Oxford University Press.

Wesley, M. (ed.). (2007). *Energy Security in Asia*, London: Routledge.

West, J., Bailey, I. and Winter, M. (2010). "Renewable Energy Policy and Public Perceptions of Renewable Energy: A Cultural Theory Approach," *Energy Policy*, 38(10), 5739–5748.

White, D.R. (1997). *Postmodern Ecologies*, Albany: State University of New York Press.

Williams, J. and Chang, D.C. (2012). *Taiwan's Environmental Struggle: Green Silicon Island*, London: Routledge.

Windpower.net (2012). *Wind Energy Database*, Buc: Windpower.net.

Wiser, R., and Bolinger, M. (2010). *2009 Wind Technologies Market Report*, Washington, DC: US Department of Energy.

Wishnick, E. (2009). "Competition and Co-operative Practices in Sino-Japanese Energy and Environmental Relations: Towards an Energy Security 'Risk Community?'" *The Pacific Review*, 22(4): 401–428.

Wolff, G., Gallego, B., Tisdale, R. and Hopwood, D. (2008). "CSP concentrates the mind," *Renewable Energy Focus*, 8(1), 42–47.

Wong, R. (2011). "Solar Potential of HDB Blocks in Singapore," *Energy Studies Institute Bulletin on Energy Trends and Development*, 4(3), 6–7.

Woo, H. (2010). "Trends in Ecological River Engineering in Korea," *Journal of Hydro-Environment Research*, 4, 269–278.

Woo-Cumings, M. (ed.). (1999). *The Developmental State*, Ithaca: Cornell University Press.

World Bank. (1993). *The East Asian Miracle: Economic Growth and Public Policy*, Oxford: Oxford University Press.

World Bank. (2001). *China: Air, Land, and Water*, Washington, DC: World Bank.

World Bank/China State Environmental Protection Administration. (2007). *Cost of Pollution in China: Economic Estimates of Physical Damage*, Washington, DC: World Bank.

World Bank. (2010). *Doing a Dam Better: The Lao People's Democratic Republic and the Story of the Nam Theun 2 (NT2)*, Washington, DC: World Bank.

World Bank. (2011). "From Growth to Green Growth: A Framework," *Policy Research Working Paper*, No. 5872, Washington, DC: World Bank.

World Bank. (2012a). *MDBs: Delivering on the Promise of Sustainable Development*, Washington, DC: World Bank.

World Bank. (2012b). *Inclusive Green Growth: The Pathway to Sustainable Development*, Washington, DC: World Bank.

World Coal Association/WCA. (2013). *Coal Statistics 2013*, London: WCA.

World Commission on Dams/WCD. (2000). *Dams and Development: A New Framework for Decision-Making*, London: Earthscan.

World Economic Forum/WEF. (2009). *Green Investing: Towards a Clean Energy Infrastructure*, Geneva: WEF.

World Economic Forum/WEF. (2011). *Green Investing 2011: Reducing the Cost of Financing*, Geneva: WEF.

World Health Organization/WHO. (2012). *Global Burden of Disease Report 2011*, Geneva: WHO.

World Wind Energy Association/WWEA. (2010). *World Wind Energy Report 2009*, Bonn: WWEA Secretariat.

World Wind Energy Association/WWEA. (2011). *World Wind Energy Report 2010*, Bonn: WWEA Secretariat.

World Wind Energy Association/WWEA. (2012a). *Quarterly Bulletin*, No. 1 (March), Bonn: WWEA Secretariat.

World Wind Energy Association/WWEA. (2012b). *Quarterly Bulletin*, No. 2 (June), Bonn: WWEA Secretariat.

World Wind Energy Association/WWEA. (2013a). *Quarterly Bulletin*, No. 1 (March), Bonn: WWEA Secretariat.

World Wind Energy Association/WWEA. (2013b). *Quarterly Bulletin*, No. 2 (June), Bonn: WWEA Secretariat.

Worldwatch Institute. (2006). *Acid Rain Affects One-Third of China*, Washington, DC: Worldwatch Institute

Wu, J.H. and Huang, Y.H. (2006). "Renewable Energy Perspectives and Support Mechanisms in Taiwan," *Renewable Energy*, 31, 1718–1732.

Wu, C.Z., Yin, X.L., Yuan, Z.H., Zhou, Z.Q. and Zhuang, X.S. (2010). "The Development of Bioenergy Technology in China," *Energy*, 35, 4445–4450.

Wu, C.Y. and Mathews, J.A. (2012). "Knowledge Flows in the Solar Photovoltaic Industry: Insights from Patenting by Taiwan, Korea, and China," *Research Policy*, 41, 524–540.

Xia, C. and Song, Z. (2009). "Wind Energy in China: Current Scenario and Future Perspectives," *Renewable and Sustainable Energy Reviews*, 13, 1966–1974.

Xia, D. (2012). "Opportunities and Challenges of the Development of MRE [Marine Renewable Energy] in China," paper presented at the *13th Meeting of United Nations Open-ended Information Consultative Process on Oceans and Law of the Sea*, New York, May 30, 2012.

Xiao, Y. and Hang, Q. (2009). "Wind Power in China: Opportunity Goes with Challenge," *Renewable and Sustainable Energy Reviews*, 14, 2232–2237.

Xie, H., Zhang, C., Hao, B., Liu, S. and Zou, K. (2012). "Review of Solar Obligations in China," *Renewable and Sustainable Energy Reviews*, 16, 113–122.

Xu, Y.C. (2007). "China's Energy Security," *Australian Journal of International Affairs*, 60(2), 265–286.

Yanagisawa, N. (2013). "Geothermal Development Progress in Japan after Earthquake, 2011," paper presented at the *Proceedings Australian Geothermal Energy Conferences 2013*, Brisbane, November 14–15, 2013.

Yang, A. and Cui, Y. (2012). *Global Coal Risk Assessment: Data Analysis and Market Research*, Washington, DC: World Resources Institute.

Yang, C.J. (2010). "Reconsidering Solar Grid Parity," *Energy Policy*, 38, 3270–3273.

Yang, H., Wang, H., Yu, H., Xi, J., Cui, R. and Chen, G. (2003). "Status of Photovoltaic Industry in China," *Energy Policy*, 31, 703–707.

Yang, J., Dai, G., Ma, L., Jia, L., Wu, J. and Wang, X. (2013). "Forest-Based Bioenergy in China: Status, Opportunities, and Challenges," *Renewable and Sustainable Energy Reviews*, 18, 478–485.

Yang, M., Nguyen, F. De T'Serclaes, P. and Buchner, B. (2010). "Wind Farm Investment Risks under Uncertain CDM Benefit in China," *Energy Policy*, 38, 1436–1447.

Yang, M., Patiño-Echeverri, D. and Yang, F. (2012). "Wind Power Generation in China: Understanding the Mismatch between Capacity and Generation," *Renewable Energy*, 41, 145–151.

Yang, X., Zhang W. and Zhu Y. (2011). "Regional Investment Distribution of the Wind Power in China and its Impacts on Wind-Generated Electricity," *Energy Procedia*, 5, 2321–2329.

Yee, W.H., Lo, C.W.H. and Tang, S.Y. (2013). "Assessing Ecological Modernization in China: Stakeholder Demands and Corporate Environmental Management Practices in Guangdong Province," *The China Quarterly*, 213, 101–129.

Yoon, E.S. (2006). "South Korean Environmental Foreign Policy," *Asia Pacific Review*, 13(7), 74–96.

York, R. and Rosa, E. (2003). "Key Challenges to Ecological Modernization Theory," *Organization and Environment*, 16(3), 273–287.

You, C.F. and Xu, X.C. (2010). "Coal Combustion and its Pollution Control in China," *Energy*, 35(11), 4467–4472.

Young, C.Y. and Huang, W.M. (2012). "Review of Taiwan's Climate Policy After Copenhagen," *Renewable and Sustainable Energy Reviews*, 16, 20–28.

Yu, D., Liang, J., Han, X. and Zhao, J. (2011). "Profiling the Regional Wind Power Fluctuation in China," *Energy Policy*, 39, 299–306.

Zhang, D., Chai, Q., Zhang, X., He, J., Yue, L., Dong, X. and Wu, S. (2012). "Economical Assessment of Large-Scale Photovoltaic Power Development in China," *Energy*, 40, 370–375.

Zhang, D., Li, W. and Lin, Y. (2009). "Wave Energy in China: Current Status and Perspectives," *Renewable Energy*, 34, 2089–2092.

Zhang, D., Zhang X., He, J. and Chai, Q. (2011). "Offshore Wind Energy Development in China: Current Status and Future Perspective," *Renewable and Sustainable Energy Reviews*, 15, 4673–4684.

Zhang, D.Q., Tan, S.K. and Gersberg, R.M. (2010). "Municipal Solid Waste Management in China: Status, Problems and Challenges," *Journal of Environmental Management*, 91, 1623–1633.

Zhang, L., Mol, A.P.J. and Sonnenfeld, D.A. (2007). "The Interpretation of Ecological Modernisation in China," *Environmental Politics*, 16(4), 659–668.

Zhang, S. (2012). "International Competitiveness of China's Wind Turbine Manufacturing Industry and Implications for Future Development," *Renewable and Sustainable Energy Reviews*, 16, 3903–3909.

Zhang, S. and Li, X. (2012). "Large Scale Wind Power Integration in China: Analysis from a Policy Perspective," *Renewable and Sustainable Energy Reviews*, 16, 1110–1115.

Zhang, S., Andrews-Speed, P. and Zhao, X. (2013). "Political and Institutional Analysis of the Successes and Failures of China's Wind Power Policy," *Energy Policy*, 56, 331–340.

Zhao, R., Shi, G., Chen, H., Ren, A. and Finlow, D. (2011). "Present Status and Prospects of Photovoltaic Market in China," *Energy Policy*, 39, 2204–2207.

Zhao, X., Feng, T., Liu, L., Liu, P. and Yang, Y. (2011). "International Co-operation Mechanism on Renewable Energy Development in China: A Critical Analysis," *Renewable Energy*, 36, 3229–3237.

Zhao, X., Feng, T. and Wang, M. (2012a). "Large-scale Utilization of Wind Power in China: Obstacles of Conflict between Market and Planning, *Energy Policy*.

Zhao, X., Wang, J., Liu, X., Feng, T. and Liu, P. (2012). "Focus on Situation and Policies for Biomass Power Generation in China," *Renewable and Sustainable Energy Reviews*, 16, 3722–3729.

Zhao, X., Zhang, S., Yang, R. and Wanga, M. (2012c). "Constraints on the Effective Utilization of Wind Power in China: An Illustration from the Northeast China Grid," *Renewable and Sustainable Energy Reviews*, 16, 4508–4514.

Zhao, Y. (2001). "The Present Status and Future of Photovoltaic in China," *Solar Energy Materials and Solar Cells*, 67, 663–671.

Zhao, Y. (2008). *China PV Industry Development Report*, Beijing: China Renewable Energy Development Project Office.

Zhao, Z.Y., Hua J. and Zuo, J. (2009). "Performance of Wind Power Industry Development in China: A Diamond Model Study," *Renewable Energy*, 34, 2883–2891.

Zhao, Z.Y., Zhang, S.Y. and Zuo, J. (2011). "A Critical Analysis of the Photovoltaic Power Industry in China: From Diamond Model to Gear Model," *Renewable and Sustainable Energy Reviews*, 15, 4963–4971.

Zhao, Z.Y., Ling, W.J., Zillante, G. and Zuo, J. (2012b). "Comparative Assessment of Performance of Foreign and Local Wind Turbine Manufacturers in China," *Renewable Energy*, 39, 424–432.

Zheng, K. (2010). "Steady Industrialised Development of Geothermal Energy in China: Country Update Report 2005–2009," paper presented at the *Proceedings World Geothermal Congress*, Bali, Indonesia, April 25–30, 2010.

Zheng, K., Han, Z. and Zhang, Z. (2010). "Steady Industrialized Development of Geothermal Energy in China Country Update Report 2005–2009," paper presented at *Proceedings World Geothermal Congress 2010*, Bali, Indonesia, April 25–29, 2010.

Zheng, N., Song, S. and Huang, Z. (1989). "Small-Scale Hydropower in China," *Biomass*, 20, 77–102.

Zhou, Y., Zhang, B., Zou, J., Bi, J. and Wang, K. (2012). "Joint R&D in Low-Carbon Technology Development in China: A Case Study of the Wind Turbine Manufacturing Industry," *Energy Policy*, 46, 100–108.

Zhuang, J., Gentry, R.W., Yu, G.R., Sayler, G.S., and Bickham, J.W. (2010). "Bioenergy Sustainability in China: Potential and Impacts," *Environmental Management*, 46, 525–530.

Index